Geraghty & Miller's

Groundwater
Bibliography

Fourth Edition

Geraghty & Miller's

Groundwater Bibliography

Compiled and edited by

Frits van der Leeden

Vice President
Geraghty & Miller, Inc.
Plainview, New York

Water Information Center, Inc.
125 East Bethpage Road, Plainview, New York 11803

Water Information Center, Inc.,
a division of
Geraghty & Miller, Inc.,
125 East Bethpage Rd., Plainview, NY 11803
Contact: Fred L. Troise,
Vice President, Marketing.

Library of Congress Catalogue Card Number 87-050095

ISBN 0-912394-20-X

Printed in the United States of America

CONTENTS

Contents

Contents

FOREWORD

This groundwater bibliography had its origin in 1971 when the compiler, a professional hydrogeologist, felt the need for a handy bibliographic reference to publications in the groundwater field. Approximately 1,500 references were selected for the first edition. By the time the second edition was published in 1974 the listing had grown to 1,750 references. This edition contains some 5,000 selected references. The large expansion of this bibliography reflects the tremendous growth of interest in groundwater which has occured in recent years. The importance of groundwater as a water source has finally been realized. More and more public, industrial, and private water supplies depend on groundwater and this resource now accounts for more than 25 percent of all fresh water withdrawn in the nation.

Mathematical modeling of groundwater flow and groundwater quality have also become very popular and this expanding field is reflected in a large increase in publications on the topic. In addition, groundwater quality and the protection of groundwater from contamination have received widespread attention as a result of serious industrial pollution problems and new federal legislation, including the Safe Drinking Water Act and the Resource Conservation and Recovery Act. This has led to the inclusion of a substantial number of references on groundwater contamination. A number of older references have been retained as there is often a need to refer back to certain "classic" contributions in the field, but most papers concerning local hydrogeologic conditions have been omitted. The bibliography is divided into a General Section containing a listing of general bibliographies, periodicals and books followed by a

Subject Section listing 29 topics. Each deals with a particular aspect of the field of hydrogeology. Where necessary, these topics are further broken down into subdivisions. All references are listed only once, alphabetically by author.

The explosion, in general, of books, papers and literature dealing with groundwater topics is so great that it is no longer possible to scan all. No claim to comprehensiveness is made. In preparing this new edition, the author has depended to a great extent on material provided by the editors of the Water Information Center who read and review hundreds of books, reports and papers for their periodicals, The Groundwater Newsletter, Water Newsletter and Research & Development News, and the International Water Report. These reference sources, combined with information obtained from technical publications and colleagues, have enabled the author to "keep up" with most of the literature. It is hoped that this selected bibliography will prove to be helpful to those searching for information on management and protection of the groundwater resource. Comments and suggestions for improvement are welcome.

Westbury, New York   *Frits van der Leeden*
June, 1986

GENERAL SECTION

GENERAL BIBLIOGRAPHIES

AMERICAN GEOPHYSICAL UNION, Washigton, D.C.
1937, *Bibliography of Hydrology, United States of America for the Year 1935 and 1936,* 78 pp.
1938, *Bibliography of Hydrology, United States of America for the Year 1937,* 68 pp.
1939, *Bibliography of Hydrology, United States of America for the Year 1938,* 72 pp.
1940, *Bibliography of Hydrology, United States of America for the Year 1939,* 86 pp.
1941, *Bibliography of Hydrology, United States of America for the Year 1940,* 86 pp.

AMERICAN GEOPHYSICAL UNION, 1952, *Annotated Bibliography on Hydrology, 1941-1950 (United States and Canada),* Fed. Inter-Agency River Basin Comm., Subcomm. on Hydrology, Notes on Hydrologic Activities Bull. 5, 408 pp.

AMERICAN GEOPHYSICAL UNION, 1955, *Annotated Bibliography on Hydrology, 1951-1954, and Sedimentation, 1950-1954 (United States and Canada),* U.S. Inter- Agency Comm., Water Resources Joint Hydrology-Sedementation Bull. 8, 207 pp.

BERGQUIST, W. E., 1976, *Bibliography of Reports Resulting from U.S. Geological Survey Technical Cooperation with other Countries, 1967-74,* Geol. Survey Bull. 1426, 68pp.

BRADBURY, C. E., and others, 1964, *Annotated Bibliography on Hydrology and Sedimentation, 1959-1962 (United States and Canada),* U. S. Inter-Agency Comm. Water Resources Joint Hydrology-Sedimentation Bull. 8, 323 pp.

BUREAU DE RECHERCHES GEOLOGIQUES ET MINIERES, 1971, *Hydrogeologie — Methodologie et Etudes de Synthese, Catalogue Methodique des Rapports Réalisés par le BRGM (1963-1970),* Pub. 71 SGN 101 HYD, 47 pp., B.P. 6009, 45 Orléans (02), France; 1971 supplement Pub. 72 SGN 125 AME, 22 pp.

FOOD & AGRICULTURE ORGANIZATION, 1973, *Water for Agriculture — Annotated Bibliography, 1945-1973,* DC/Sp. 27, FAO, Rome, 400 pp.

GIEFER, G. J., 1976, *Sources of Information in Water Resources,* Water Information Center, Plainview, New York 11803, 290 pp.

GIEFER, G. J., and TODD, D. K. (eds.), 1972, *Water Publications of State Agencies,* Water Information Center, Plainview, New York 11803, 319 pp.

GIEFER, G. J., and TODD, D. K. (eds.), 1976, *Water Publications of State Agencies, First Supplement,* 1971-1974, Water Information Center, Plainview, New York 11803, 189 pp.

HUYAKORN, P. S., and DUDGEON, C. R., 1972, *Groundwater and Well Hydraulics — An Annotated Bibliography,* Australian Water Res. Council, Canberra, Rept. no. 121, 148 pp.

JOHNSON, A. I., 1964, *Selected Bibliography on Laboratory and Field Methods in Ground-Water Hydrology,* U. S. Geol. Survey Water-Supply Paper 1779-Z.

KNAPP, G. L., and GLASBY, J. P., 1973, *Urban Hydrology — A Selected Bibliography with Abstracts,* U. S. Govt. Printing Office, Washington, D. C., 214 pp.

MARGAT, JEAN, 1964, *Guide Bibliographique d'Hydrogéologie*; Ouvrages et Articles en Langue Francaise, Bureau de Recherches Géologiques et Minières, suite Hydrogéologie 113, Paris.

MASSACHUSETTS AUDUBON SOCIETY, 1982, *An Annotated Groundwater Bibliography Covering the Northeast with Emphasis on Massachusetts,* (Kathryn Proko, compiler), Mass. Audubon Soc., Lincoln, Massachusetts 01773, 86 pp.

RALSTON, V. H., 1975, *Water Resources - A Bibliographic Guide to Reference Sources,* Institute of Water Resources The Univ. of Connecticut, Rep. No. 23, 123 pp.

RANDOLPH, J. R., and DEIKE, R. G., 1966, *Bibliography of Hydrology of the United States,* 1963, U. S. Geol. Survey Water-Supply Paper 1863, 166 pp.

RIGGS, H. C., 1962, *Annotated Bibliography on Hydrology and Sedimentation, United States and Canada, 1955-58,* U. S. Geol. Survey Water-Supply Paper 1546, 236 pp.

U. S. ENVIRONMENTAL PROTECTION AGENCY, 1978, *EPA Cumulative Bibliography, 1970-1976: Parts I and II,* A Listing of all Reports Submitted by the EPA and its Predecessor Agencies to the National Technical Information Service, PB -265 920/9WP, NTIS, Springfield, Virginia 22161.

U. S. GEOLOGICAL SURVEY, 1972, *Urban Hydrology: A Selected Bibliography With Abstracts,* Nat. Tech. Info. Service PB 219 105, 216 pp.

U. S. GEOLOGICAL SURVEY, OFFICE OF INTERNATIONAL ACTIVITIES, 1967, *Interim Bibliography of Reports Related to Overseas Activities of the Water Resources Division, 1940-67,* Open file rept., 26 pp.

STOW, D. A. V., and others, 1976, *Preliminary Bibliography on Ground Water in Developing Countries 1970-1976,* Assoc. of Geoscientists for Internat. Development Rept. No. 4, Memorial University, St. John's Newfoundland, A1C 5S7 Canada.

VANEK, VLADIMIR., 1985, *Groundwater and Lakes - A Literature Review,* Inst. of Limnology, Univ. of Lund, Box 65, S - 221 00, Lund, Sweden.

VORHIS, R. C., 1957, *Bibliography of Publications Relating to Ground Water Prepared by the Geological Survey and Cooperating Agencies 1946-55,* U. S. Geol. Survey Water-Supply Paper 1492, 203 pp.

WARD-McLEMORE, ETHEL, 1985, *Selected Bibliography of Ground-Water in the United States,* Publ.10, Geol. Information Library of Dallas, One Energy Square, 4925 Greenville Ave., Dallas, Texas 75206.

WARING, G. A., and MEINZER, O. E., 1947, *Bibliography and Index of Publications Relating to Ground Water Prepared by the Geological Survey and Cooperating Agencies,* U. S. Geol. Survey Water-Supply Paper 992, 412 pp.

WELD, B. A., and others, 1972, *Reports and Maps of the Geological Survey Released only in the Open Files, 1971,* U. S. Geol. Survey Circ. 648, 26 pp.

WELLISCH, H., 1967, *An International Bibliography of Water Resources Development 1950-1965,* Israel Program for Sci. Transl. Jerusalem; also Daniel Davey & Co., Inc., New York.

JOURNALS, BULLETINS AND NEWSLETTERS

AGID NEWS — Monthly Bulletin Assoc. of Geoscientists for International Development, c/o Asian Inst. of Technology, G.P.O. Box 2754, Bangkok 10501, Thailand.

AGRICULTURAL ENGINEERING — Am. Soc. of Agric. Engineers, 2950 Niles Road, St. Joseph, Michigan 49085.

AGUA — Centro de Estudios, Investigacion y Aplicaciones del Agua, Paseo de San Juan 39, Barcelona 9, Spain.

AMERICAN SOCIETY OF CIVIL ENGINEERS, ENVIRONMENTAL ENGINEERING DIVISION JOURNAL, Am. Soc. Civil Engrs., 345 East 47th St., New York, New York 10017.

AMERICAN SOCIETY OF CIVIL ENGINEERS, HYDRAULICS DIVISION JOURNAL — Am. Soc. Civil Engrs., 345 East 47th St., New York, New York 10017.

AMERICAN SOCIETY OF CIVIL ENGINEERS, IRRIGATION AND DRAINAGE DIVISION JOURNAL, Am. Soc. Civil Engrs., 345 East 47th St., New York, New York 10017.

BULLETIN DU BUREAU DE RECHERCHES GEOLOGIQUES ET MINIERES — SECTION HYDROGEOLOGIE, BRGM Service Geologique National, Boite Postale 6009, 45018 Orléans, France.

BULLETIN OF THE AMERICAN ASSOCIATION OF PETROLEUM GEOLOGISTS — Am. Assoc. Petroleum Geol., P. O. Box 979, Tulsa, Oklahoma 74101.

ENVIRONMENTAL RESEARCH BRIEF — U. S. Environmental Protection Agency, Robert S. Kerr Environmental Research Lab., Ada, Oklahoma 74820.

ENVIRONMENTAL SCIENCE & TECHNOLOGY — Am. Chemical Society, 1155 16th St., N.W., Washington, D. C. 20036

EOS — Transactions Am. Geophys. Union, 2000 Florida Ave., N.W., Washington, D. C. 20009.

GEOLOGICAL SOCIETY OF AMERICA BULLETIN — Geol. Soc. Am., 3300 Penrose Place, Boulder, Colorado 80301.

GROUND WATER – Tech. Div. Nat. Water Well Assoc., 6375 Riverside Dr., Dublin, Ohio 43017

GROUND-WATER ABSTRACTS, International Scientific Press (ISP), P.O. Box 14701, Minneapolis, Minnesota 55414.

GROUND WATER AGE – National Trade Publications, Inc., 8 Stanley Circle, Latham, New York 12110

GROUNDWATER DIGEST – Pacific Groundwater Digest, Inc., 2670 Howe Avenue – P. O. Box 255769 Sacramento, California 95825.

GROUND-WATER HEAT PUMP JOURNAL – National Water Well Assoc. Newsletter, 6375 Riverside Dr., Dublin, Ohio 43017

GROUND WATER MICROBIOLOGY NEWSLETTER – Robert S. Kerr. Envir. Research Lab., P. O. Box 1198, Ada, Oklahoma 74820.

GROUND WATER MODELING NEWSLETTER – International Ground Water Modeling Center, Holcomb Research Inst., Butler University, Indianapolis, Indiana 46208.

GROUND-WATER MONITORING REVIEW, Water Well Journal Publ. Co., 6375 Riverside Dr., Dublin, Ohio 43017

GROUND WATER NEWSLETTER – Water Information Center, Inc., 125 East Bethpage Road, Plainview, New York 11803

HIDROLOGIA – Boletin Tecnico Informativo del Instituto de Hidrologia, Po. Bajo Virgen del Puerto, 3, Madrid -5.

HYDROLOGICAL SCIENCE AND TECHNOLOGY – SHORT PAPERS, Am. Institute of Hydrology, 3416 University Ave. S.E., Minneapolis, Minnesota 55414.

HYDROLOGICAL SCIENCES BULLETIN – Internat. Assoc. Hydrological Sciences, Blackwell Scientific Publications Ltd., Osney Mead, Oxford OX20EL, England.

INTERNATIONAL JOURNAL OF MINE WATER, ed. by Dept. of Mining Engineering, Univ. of Nottingham. V. Straskraba, Hydro-Geo Consultants or Prof. Rafael Fernandez-Rubio, Dept. of Hydrogeology, School or Mines, Technical Univ. of Madrid, Rios Rosas 21, Madrid 3, Spain.

INTERNATIONAL WATER REPORT – Water Information Center, Inc., 125 East Bethpage Road, Plainview, New York 11803

JOHNSON DRILLERS JOURNAL – Johnson Div., UOP, Inc., P. O. Box 3118, St. Paul, Minnesota 55165. (no longer published)

JOURNAL AMERICAN WATER WORKS ASSOCIATION — Am. Water Works Assoc., 6666 West Quincy Ave., Denver, Colorado 80235.

JOURNAL OF CONTAMINANT HYDROLOGY, Elsevier Science Publishers, P.O Box 211, 1000 AE Amsterdam, The Netherlands.

JOURNAL OF ENVIRONMENTAL QUALITY — Am. Soc. of Agronomy, Crop Science Soc. of America, and Soil Science Soc. of America, 677 S. Segoe Road, Madison, Wisconsin 53711.

JOURNAL OF GEOPHYSICAL RESEARCH — Am. Geophys. Union, 1909 K Street, N.W., Washington, D. C. 20006.

JOURNAL OF HYDROLOGY — Elsevier Scientific Publishing Co., P. O. Box 211, 1000AE Amsterdam, The Netherlands.

JOURNAL OF RESEARCH OF THE U. S. GEOLOGICAL SURVEY — Supt. of Documents, Government Printing Office, Washington, D. C. 20402.

JOURNAL OF THE NEW ENGLAND WATER WORKS ASSOCIATION — New England Water Works Assoc., 990 Washington Street, Dedham, Massachusetts 02026.

JOURNAL OF WATER SUPPLY AND MANAGEMENT — Pergamon Press, Maxwell House, Fairview Park, Elmsford, New York 10523.

JOURNAL WATER POLLUTION CONTROL FEDERATION — Water Pollution Control Federation, 2626 Pennsylvania Ave., N.W., Washington, D. C. 20037.

OIL AND GAS JOURNAL — Petroleum Publishing Co., P. O. Box 1260, Tulsa, Oklahoma 74101.

SELECTED WATER RESOURCES ABSTRACTS — National Technical Information Service, U. S. Dept. of Commerce, 5285 Port Royal Road, Springfield, Virginia 22151.

SOVIET HYDROLOGY: SELECTED PAPERS — Am. Geophys. Union, 1909 K Street, N.W., Washington, D. C. 20006.

TRANSPORT IN POROUS MEDIA - D. Reidel Publ. Co., c/o Kluwer Academic Publ., 190 Old Derby St., Hingham, Massachusetts 02043.

U.S. WATER NEWS, 230 Main Street, Halstead, KA 67056.

WATER INTERNATIONAL — International Water Resources Assoc., Elsevier Sequoia S.A., P. O. Box 851, 1001 Lausanne 1, Switzerland.

WATER QUALITY BULLETIN - World Health Organization Collaborating Centre on Surface and Ground Water Quality, Canada Centre for Inland Waters, Box 5050, 867 Lakeshore Road, Burlington, Ontario L7R 4A6, Canada

WATER RESOURCES BULLETIN — Am. Water Resources Assoc., St. 5410 Grosvenor Lane, Suite 220, Bethesda, Maryland 20814.

WATER RESOURCES RESEARCH — Am. Geophys. Union, 2000 Florida Ave. N.W., Washington, D. C. 20009.

WATER WELL JOURNAL — Water Well Journal Publ. Co., 6375 Riverside Dr., Dublin, Ohio 43017.

WELL LOG — Newsletter of National Water Well Assoc., 6375 Riverside Dr., Dublin, Ohio 43017.

TEXTS, HANDBOOKS AND DICTIONARIES

AGADJANOV, A. M., 1950, *Hydrogeology and Hydraulics of Ground Water and Oil* (in Russian), Gostoptekhizdat, Moscow, 280 pp.

AMERICAN SOCIETY OF CIVIL ENGINEERS, 1949, *Hydrology Handbook, Manual of Engineering Practice* 28, New York, 184 pp.

ANDERSON, K. E., 1984, *Water Well Handbook,* Missouri Water Well & Pump Contractors Assn., P. O. Box 517, Belle, Missouri, 65013, 281 pp.

ANONYMOUS, 1980, *The International Who's Who in Water Supply,* Marlborough Publ. Ltd., 23 Ramillies Place, London W1A 3BF, England.

ANONYMOUS, 1983, *Glossary of Water Treatment Terms*, TS-69, Mogul, Div. of The Dexter Corp., P.O. Box 200, Chagrin Falls, Ohio 44022, (listing of 165 terms and explanations).

ANONYMOUS, 1984, *The Environmental Glossary*, Government Institutes, Inc., 966 Hungerford Dr., #24, Rockville, Maryland 20850, 295 pp., (3,480 environmental terms and legal definitions).

BALEK, JAROSLAV, *Ground-Water Hydrology and Hydraulics,* Elsevier Publ. Co., New York and Amsterdam.

BEAR, JACOB, 1972, *Dynamics of Fluids in Porous Media,* Am. Elsevier Publ. Co., New York, New York, 764 pp.

BEAR, JACOB, 1980, *Hydraulics of Ground Water,* McGraw-Hill Book Co., 1221 Ave. of the Americas, New York, New York 10020, 567 pp.

BEAR, JACOB, ZASLAVSKY, D., and IRMAY, S., 1968, *Physical Principles of Water Percolation and Seepage.* Arid Zone research XXIX, UNESCO, Paris, 465 pp.

BEAR, JACOB, and CORAPCIOGLU, M. Y., 1985, *Fundamentals of Transport Phenomena in Porous Media*, Kluwer Acad. Publ., 190 Old Derby St., Hingham, Massachusetts 02043, 1,003 pp.

BENNETT, G. D., *Introduction to Ground-Water Hydraulics'*, *Scientific Publ. Co., P.O. Box 23041, Washington, DC, 172 pp., (reprint of Chapter B2, Book 3 of U.S. Geol. Survey Techniques of Water Resources Investigations)*.

BENITEZ, ALBERTO, 1963, *Captación de Aguas Subterráneas*, Editorial Dossat, Madrid, 157 pp.

BOGOMOLOV, G. V., 1957, *Grundlagen der Hydrogeologie*, Deutscher Verl. der Wissenschaften, Berlin, 187 pp.

BOGOMOLOV, G. V., 1965, *Hydrogéologie*, Transl. from Russian by V. Frolov, Libr. du Globe, Paris, 277 pp.

BOGOMOLOV, G. V., and SILIN-BEKCURIN, A. I., 1955, *Special Hydrogeology* (in Russian), Gosgeoltekhizdat, Moscow, 247 pp., transl. into French by E. Jayet and G. Castany, Service d'Information Géol., 1959, Paris, 235 pp.

BOUWER, HERMAN, 1978, *Groundwater Hydrology*, McGraw-Hill Book Co., New York, New York 10020, 480 pp.

BOWEN, ROBERT, 1980, *Ground Water*, John Wiley & Sons, Inc., One Wiley Drive, Somerset, New Jersey 08873, 227 pp.

BUTLER, S. S., 1957, *Engineering Hydrology*, Prentice-Hall, Inc., Englewood Cliffs, New Jersey.

CASTANY, GILBERT, 1963, *Traité Pratique des Eaux Souterraines*, Dunod, Paris, 657 pp.

CASTANY, GILBERT, 1967, *Prospection et Exploitation des Eaux Souterraines*, Dunod, Paris.

CASTANY, GILBERT and MARGAT, JEAN, 1977, *Dictionnaire Français d'Hydrogéologie*, B. R. G. M. Orleans, 249 pp.

CEDERGREN, H. R., 1977, *Seepage, Drainage and Flow Nets*, John Wiley & Sons, Inc., Somerset, New Jersey 08873, 534 pp.

CHAPMAN, R. E., 1981, *Geology and Water — An Introduction to Fluid Mechanics for Geologists*, Martinus Nijhoff Publ. N.V., 2501 CN, The Hague, The Netherlands, 228 pp.

CHILDS, E. C., 1969, *Introduction to the Physical Basis of Soil Water Phenomena*, John Wiley & Sons, Inc., New York.

CHOW, V. T. (ed.), 1964, *Handbook of Applied Hydrology*, McGraw-Hill, New York, 1453 pp.

Texts

COLLINS, R. E., 1961, *Flow of Fluids through Porous Materials,*
Reinhold, New York, 270 pp.

COOLEY, R. L. and others, 1972, *Principles of Ground-Water Hydrology,*
Hydrologic Engineering Methods for Water Resources Development,
v. 10, Corps. of Engrs., U. S. Army, Davis, California, 337 pp.

COREY, T., 1977, *Mechanics of Heterogenous Fluids in Porous Media,*
Water Resources Publications, P.O. Box 303, Fort Collins, Colorado
80521, 259 pp.

CUSTODIO, EMILIO, and LLAMAS, M. R., 1975, *Hidrología
Subterránea,* 2 vols., Ediciones Omega, Barcelona, 2,359 pp.

DACHLER, R., 1936, *Grundwasserströmung,* J. Springer, Vienna 141 pp.

DAVIS, S. N., and DeWIEST, R. J. M., 1966, *Hydrogeology,* John Wiley
& Sons, Inc., New York, 463 pp.

DeWIEST, R. J. M., 1965, *Geohydrology,* John Wiley & Sons, Inc., New
York, 366 pp.

DeWIEST, R. J. M., (ed.), 1969, *Flow Through Porous Media,* Academic
Press, New York, 530 pp.

EKMAN, J. E., 1983, *Groundwater and Distribution Workbook*, EDRS,
P.O. Box 190, Arlington, Virginia 22210, 263 pp. (Student manual
prepared for Wisconsin groundwater training course)

ERIKSSON, E., and others, 1968, *Ground Water Problems,* Pergamon
Press, Inc., Elmsford, New York, Oxford, England.

FETTER, C.W., 1980, *Applied Hydrogeology,* Charles E. Merrill Publ. Co.,
1300 Alum Creek Drive, Columbus, Ohio 43216, 488 pp.

FLEMMING, H. W., 1962, *Die Unterirdische Wasserspeicherung,*
Oldenburg Ver., Munich, 79 pp.

FORCHHEIMER , P., 1930, *Hydraulik,* B. G. Teubner
Verlagsgesellschaft, Berlin.

FOURMARIER, P., 1958, *Hydrogéologie,* 2nd ed., Paris, Masson & Cie.,
294 pp.

FREEZE, R. A. and BACK, WILLIAM, 1983, *Physical Hydrogeology*,
Hutchinson Ross Publ. Co., 135 W. 50th St., New York, New York
10020, 448 pp., (collection of classic papers).

FREEZE, R. A., and CHERRY, J. A. 1979, *Ground Water,* Prentice-Hall,
Inc., Englewood Cliffs, New Jersey.

GERAGHTY, J.J., and others, 1973, *Water Atlas of the United States — Basic Facts About the Nation's Water Resources,* Water Information Center, Plainview, New York 11803, 122 plates.

GLOVER, R. E., 1974, *Transient Ground Water Hydraulics,* Water Resources Publications, P. O. Box 303, Fort Collins, Colorado 80522, 413 pp.

GRAY, D. M., 1973, *Handbook on the Principles of Hydrology,* Water Information Center, Inc., Plainview, New York 11803, 720 pp.

GREENKORN, R. A., 1985, *Flow Phenomena in Porous Media,* Marcel Dekker, Inc., 270 Madison Ave., New York, New York 10016, 560 pp.

HALEK, VACLAV and SVEC, JAN, 1979, *Ground-Water Hydraulics,* Developments in Water Science, v. 7. Elsevier North Holland, Inc., New York, New York 10017, 620 pp.

HAMILTON, C.E. (ed.), 1983, *Manual on Water,* Special Tech. Publ. 442A, Am. Soc. for Testing and Materials, Sales Service Dept., 1916 Race St., Philadelphia, Pennsylvania 19103, 471 pp.

HARR, M. E., 1977, *Ground Water and Seepage,* McGraw-Hill Book Co. Inc., New York, 315 pp.

HEATH, R. C., and TRAINER, F. W., 1968, *Introduction to Ground-Water Hydrology,* John Wiley & Sons, Inc., New York, 284 pp; Revised ed. 1981, Water Well Journal, Dublin, Ohio 43017

HEATH, R.C., 1983, *Basic Ground-Water Hydrology,* U.S. Geol. Survey, Water-Supply Paper 2220.

HEIJ, C.J., and MEINARDI, C.R., 1984, *A Groundwater Primer,* Internat. Reference Center, P.O. Box 5500, 2280 HM Rijswijk, The Netherlands, 119 pp.

HILLIEL, D., 1971, *Soil and Water, Physical Principles and Process,* Academic Press, New York, New York, 288 pp.

HUISMAN, L., 1972, *Groundwater Recovery,* Winchester Press Inc., New York, New York, 365 pp.

IMBEAUX, E., 1930, *Essai d'Hydrogéologie,* Dunod, Paris, 704 pp.

INGERSOLL, L. R., and others, 1948, *Heat Conduction with Engineering and Geological Applications,* McGraw-Hill Book Co. Inc., New York.

Texts

JOHNSON DIVISION, 1986, *Ground Water and Wells,* Signal Environmental Systems, Inc., P.O. Box 64118, St. Paul, Minnesota 55164, 1100 pp.

KARR, W. V., 1969, *Ground Water; Methods of Extraction and Construction,* Internat. Underground Water Inst., Bexley, Ohio.

KASHEF, A. I, 1985, *Groundwater Engineering,* McGraw-Hill Book Co., New York, New York 10020, 512 pp.

KAZMANN, R. G., 1972, *Modern Hydrology,* Harper & Row, Inc., New York, 635 pp.

KEILHACK, K., 1935, *Lehrbuch der Grundwasser- und Quellenkunde,* 3rd ed., Gebr. Borntraeger, Berlin, 575 pp.

KLIMENTOV, P.P., 1983, *General Hydrogeology,* Transl. from Russian by K.G. Gurevich, Mir Publishers, Moscow, U.S. distributor: Imported Publications, Chicago, 240 pp.

KOEHNE, W., 1928, *Grundwasserkunde,* E. Nagele, Stuttgart, 291 pp.

KOVACS, G. and others, 1981, *Subterranean Hydrology,* Water Resources Publications, P. O. Box 2841, Littleton, Colorado 80161, 978 pp.

KUENEN, P. H., 1963, *Realms of Water,* John Wiley & Sons, Inc., New York, 327 pp.

LI, W. H., 1973, *Differential Equations of Hydraulic Transients, Dispersion and Ground-Water Flow,* Prentice-Hall, Englewood Cliffs, New York, 317 pp.

LINSLEY, R. K., and FRANZINI, J. B., 1972, *Water Resources Engineering,* 2nd ed., McGraw-Hill Book Co., New York, New York, 690 pp.

LOHMAN, S. W., 1972, *Definitions of Selected Ground-Water Terms — Revisions and Conceptual Refinements,* U. S. Geol. Survey Water-Supply Paper 1988, 21 pp.

LUTHIN, J. N. (ed.), 1957, *Drainage of Agricultural Lands,* American Society of Agronomy, Madison, Wisconsin. Water Information Center, Plainview, New York 11803.

MANDEL, SAMUEL, and SHIFTAN, Z. L., 1981, *Groundwater Resources — Investigation and Development,* Academic Press, New York, New York 10003, 288 pp.

MARINO, M. A., and LUTHIN, J. N., 1981, *Seepage and Groundwater,* Elsevier Science Publ. Co., Inc., New York, New York 10017, 492 pp.

MATTHESS, G., 1973, *Die Beschaffenheit des Grundwassers,* in Richter, W. (ed.), Lehrbuch der Hydrogeologie, v. 2, Gebrüder Borntraeger, Berlin, 324 pp.

McWORTHER, D. B. and SUNADA, D.K., 1981, *Ground-Water Hydrology and Hydraulics,* Water Resources Publications, P.O. Box 2841, Littleton, Colorado 80151.

MEINZER, O. E. (ed.), 1942, *Hydrology,* McGraw-Hill Book Co., Inc., New York, 712 pp.

MUSKAT, M., 1937, *The Flow of Homogeneous Fluids through Porous Media,* McGraw-Hill Book Co., Inc., New York, 737 pp. (2nd printing, Edwards Brothers, Ann Arbor, 1946.)

MUSKAT, M., 1949, *Physical Principles of Oil Production,* McGraw-Hill Book Co., Inc., New York 922 pp.

NATIONAL WATER WELL ASSOCIATION, 1985, *Ground Water ... Defined,* NWWA, Dublin, Ohio 43017, (reference for 111 terms and 204 definitions used in the ground water industry).

NATIONAL WATER WELL ASSOCIATION, 1986, *Ground Water Hydrology for Water Well Contractors,* NWWA, Dublin, Ohio 43017, 288 pp.

OGIL'VI, N. A., and FEDOROVICH, D. I., 1966, *Groundwater Seepage Rates,* Plenum Publ. Corp., New York.

OLIVER, HENRY, 1972, *Irrigation and Water Resources Engineering,* Crane, Russak and Co., New York, New York, 190 pp.

PEREIRA, H. C., 1973, *Land Use and Water Resources,* Cambridge Univ. Press., New York, New York, 246 pp.

PFANNKUCH, H. O., 1969, *Dictionary of Hydrogeology,* (Eng., Fr., Ger.), Am. Elsevier Publ. Co.

PLOTNIKOV, N. A., 1962, *Ressources en Eaux Souterraines: Classification et Méthodes d'Evaluation,* transl. from Russian by M. Laronde, Gauthier-Villars, Paris, 194 pp.

POLUBARINOVA-KOCHINA, P. YA., 1962, *Theory of Ground-Water Movement,* Princeton Univ. Press, Princeton, New Jersey, 613 pp.

POWERS, J. P., 1984, *Construction Dewatering: A Guide to Theory and Practice,* John Wiley & Sons, Somerset, New Jersey, 494 pp.

PRICE, MICHAEL, 1985, *Introducing Groundwater,* Allen & Unwin, Inc., 8 Winchester Place, Winchester, Massachusetts 01830, 195 pp.

PRINZ, E., 1923, *Handbuch der Hydrologie,* 2nd ed., J. Springer, Berlin, 422 pp.

RAGHUNATH, H. M., 1982, *Groundwater,* John Wiley & Sons, Inc., Somerset, New Jersey 08873, 456 pp.

RANDKIVI, A. J., and CALLANDER, R. A., 1976, *Analysis of Groundwater Flow,* John Wiley & Sons, New York, 214 pp.

RAU, J.L., 1984, *Ground Water Hydrology for Water Well Contractors,* Natl. Water Well Assoc., Dublin, Ohio, 288 pp.

RETHATI, L., 1984, *Groundwater in Civil Engineering,* Elsevier Science Publishers, 52 Vanderbilt Ave., New York, New York 10017, 474 pp.

RICHTER, W., and WAGER, R., 1969, *Hydrogeologie,* in Bentz, A. and H. J. Martini (eds.), Lehrbuch der Angewandten Geologie, v. II, pt. 2, pp. 1357-1546, Ferdinand Enke, Stuttgart.

ROSENSHEIN, J.S., and BENNETT G.D., (ed.), 1984, *Groundwater Hydraulics,* American Geophys. Union, 2000 Florida Ave., N.W., Washington, DC 20009, 407 pp. (papers on state-of-the art in hydraulics, transport and modeling).

SALEEM, Z. A. (ed.), 1976, *Advances in Groundwater Hydrology,* Am. Water Resources Assoc., Minneapolis, 333 pp.

SCHEIDEGGER, A. E., 1974, *The Physics of Flow through Porous Media,* 3rd ed., Univ. of Toronto, Toronto, 353 pp.

SCHNEEBELZ, T., 1966, *Hydraulique Souterraine,* Eyrolles, Paris.

SCHOELLER, R. H., 1962, *Les Eaux Souterraines,* Masson et Cie., Paris, 642 pp.

SILIN-BEKCURIN, A. I., 1961, *Hydrogeology of Irrigated Lands,* Foreign Languages Publ. House, Moscow, 110 pp.

SILIN-BEKCURIN, A. I., 1965, *Dynamics of Underground Water,* (with Elements of Hydraulics), 2nd ed., Moscow University, Moscow, 380 pp.

TEBBUTT, T. H. Y., 1973, *Water Science and Technology,* Barnes and Noble Books, New York, New York, p. 240.

THOMAS, H. E., 1951, *The Conservation of Ground Water,* McGraw-Hill Book Co., Inc., New York, 327 pp.

THURNER, A., 1967, *Hydrogeologie,* Springer Verlag, Vienna, 350 pp.

TODD, D. K., 1970, *The Water Encyclopedia,* Water Information Center, Plainview, New York 11803, 559 pp.

TODD, D. K., 1980, *Groundwater Hydrology,* John Wiley & Sons, Inc., New York, 535 pp.

TOLMAN, C. F., 1937, *Ground Water,* McGraw-Hill Book Co., Inc., New York, 593 pp.

UNESCO, 1978, *International Glossary of Hydrogeology*, Quadrilingual: English/French/Spanish/Russian, Publ. DC.77.WS.71.

U.S. DEPARTMENT OF THE INTERIOR, 1981, *Ground Water Manual; A Guide for the Investigation, Development, and Management of Ground-Water Resources,* U.S. Govt. Printing Office, Washington, D.C., 480 pp.

U. S. CONGRESS, 1955, *Water (The Yearbook of Agriculture),* 84th Cong., 1st sess., 751 pp.

VAN DER KAMP, G. S. J. P., 1973, *Periodic Flow of Groundwater,* Rodopi, Amsterdam, 121 pp.

VAN DER LEEDEN, FRITS, 1975, *Water Resources of the World,* Water Information Center, Inc., Plainview, New York 11803, 568 pp.

VAN SCHILFGAARDE, J. (ed.), 1974, *Drainage for Agriculture,* Agronomy Monograph no. 17, Amer. Soc. Agronomy, Madison, Wisconsin, 700 pp.

VEN TE CHOW, (ed.), 1964, *Handbook of Applied Hydrology,* McGraw-Hill Book Co. Inc., New York.

VERRUIJT, A., 1970, *Theory of Ground Water Flow,* Gordon & Breach, Inc., New York.

WALTON, W. C., 1970, *Groundwater Resource Evaluation,* McGraw-Hill Book Co., Inc., 664 pp.

WIENER, A., 1972, *The Role of Water in Development,* McGraw-Hill Book Co., New York, New York, 483 pp.

WILSON, JAMES, 1982, *Ground Water: A Non-Technical Guide,* Acad. of Natural Sciences, Philadelphia PA, 105 pp.

Texts

WISLER, C. O., and BRATER, E. F., 1959, *Hydrology,* John Wiley & Sons, Inc., New York.

SUBJECT SECTION

HISTORY AND DEVELOPMENT OF GROUNDWATER HYDROLOGY

ADAMS, F. D., 1928, *Origin of Springs and Rivers — A Historical Review,* Fennia, v. 50, no. 1, 16 pp.

ARISTOTLE, *Meteorologica,* translated by E. W. Webster, in W. D. Ross (ed.), The Oxford Translation of the Complete Works of Aristotle, 1923-1955, v. 3, Clarendon Press, Oxford.

BACK, WILLIAM and STEPHENSON, D. A. (eds.), 1979, *Contemporary Hydrogeology: The George Burke Maxey Memorial Volume,* J. Hydrology, v. 43, no. 1/4, Oct. (Special Issue). Also publ. by Elsevier North-Holland, Inc., 52 Vanderbilt Ave., New York, New York 11017, 570 pp.

BAKER, M. N., and HORTON, R. E., 1936, *Historical Development of Ideas Regarding the Origin of Springs and Ground Water,* Trans. Am. Geophys. Union, v. 17, pp. 395-400.

BISWAS, A. K., 1970, *History of Hydrology,* American Elsevier, 336 pp.

CARLSTON, C. W., 1943, *Notes on the Early History of Water-Well Drilling in the United States,* Econ. Geol., v. 38, pp. 119-136.

FANCHER, G., 1956, *Henry Darcy — Engineer and Benefactor of Mankind,* Jour. Petr. Tech., v. 8, pp. 12-14.

FERRIS, J. G., and SAYRE, A. N., 1955, *The Quantitative Approach to Groundwater Investigations,* Economic Geology, 50th anniversary volume.

HALL, H. P., 1954, *A Historical Review of Investigations of Seepage Toward Wells,* J. Boston Soc. Civil Engrs., July, pp. 251-311.

JONES, P. B., and others, 1963, *The Development of the Science of Hydrology,* Texas Water Comm. Circ. 63-03, 35 pp.

MAXEY, G. B., 1979, *The Meinzer Era of U. S. Hydrogeology, 1910-1940,* J. Hydrology, v. 43, no. 1/4, Oct.

MEINZER, O. E., 1934, *The History and Development of Ground-Water Hydrology,* J. Washington Acad. Sci., v. 24, pp. 6-32.

MEYBOOM, P., 1966, *Current Trends in Hydrogeology,* Earth-Science Reviews, v. 2, pp. 354-364.

NACE, R. L., 1978, *Development of Hydrology in North America,* Water International, Sept., pp. 20-26.

NARASIMHAN, T.N., (ed.), 1983, *Recent Trends in Hydrogeology*, Spec. Paper 189, Publ. Sales, Geol. Society of America, P.O. Box 9140, Boulder, Colorado 80301, 450 pp.

PALISSY, B. *Discours Admirables,* translated by A. La Rocque, The Admirable Discourses of Bernard Palissy, Univ. of Illinois Press, Urbana, Illinois.

PARIZEK, R. R., 1963, *The Hydrologic Cycle Concept,* Mineral Ind. Penn. State Univ., v. 32, no. 7.

PARKER, G. G., and JOHNSON, A. I., 1974, *Research and Advances in Ground-Water Resources Studies, 1964-1974,* in *A Decade of Progress in Water Resources,* Proc. no. 19, Am. Water Resources Assoc., Urbana, Illinois, pp. 42-75.

PERRAULT, PIERRE, 1967, *On the Origin of Springs*, Transl. by Aurele LaRocque, Hafner Publ. Co., New York, 209 pp.

GROUNDWATER ENVIRONMENTS/ GROUNDWATER SYSTEMS

BIBLIOGRAPHY

ANONYMOUS, 1982, *Ground Water Aquifers, 1977-June, 1982 (173 Citations from the Selected Water Resources Abstracts Data Base),* PB82-867755, NTIS, Springfield, Virginia 22161, 240 pp.

LAMOREAUX, P. E., and others, 1970, *Hydrology of Limestone Terranes — Annotated Bibliography of Carbonate Rocks,* Geol. Survey Alabama, Bull. 94, Pt. A, 242 pp.

WILLIAMS, J. R., 1965, *Ground Water in Permafrost Regions: An Annotated Bibliography,* U. S. Geol. Survey Water-Supply Paper 1792, 294 pp.

GENERAL WORKS

ANDREWS, C. B., and ANDERSON, M. P., 1978, *Impact of a Power Plant on the Ground-Water System of a Wetland,* Ground Water, v. 16, no. 2, pp. 105-111.

ANONYMOUS, 1984, *Permafrost;* Proceedings, Fourth International Conference, National Academy Press, 2101 Constitution Ave. NW, Washington, DC 20419, 1,524 pp.

ANONYMOUS, 1985, *Proceedings of the Ogallala Aquifer Symposium II,* Water Resources Center, Texas Tech Univ., P.O. Box 4630, Lubbock, Texas 79409, 596 pp.

ANONYMOUS, 1985, *The Hydrogeology of Rocks of Low Permeability,* IAH-U.S. Committee, c/o E.S. Simpson, 200 Douglass Bldg., Dept. of Hydrology and Water Resources, U. of Arizona, Tucson AZ, 85721.

AYERS, J.F., and VACHER, H.L., 1986, *Hydrogeology of an Atoll Island: A Conceptual Model from Detailed Study of a Micronesian Example,* Ground Water, v. 24, no. 2, pp. 185-198.

BALDWIN, G. V., and McGUINNESS, C. L., 1963, *A Primer on Ground Water,* U. S. Geol. Survey, 26 pp.

BENTALL, Ray, and others, 1972, *Water Supplies and the Land — the Elkhorn River Basin of Nebraska,* Nebraska Conserv. and Survey Div. Resource Atlas No. 1.

BLANK, N. R., and SCHROEDER, M. C., 1973, *Geologic Classification of Aquifers,* Ground Water, v. 11, no. 2, pp. 3-5.

BOSWELL, E. H., 1976, *The Lower Wilcox Aquifer in Mississippi,* U. S. Geol. Survey Water Resources Inv. 60-75.

BOSWELL, E. H., 1976, *The Meridian-Upper Wilcox Aquifer in Mississippi,* U. S. Geol. Survey Water-Resources Inv. 76-79.

BROWN, R. M., MILLER, J. A., and SWAIN, F. M., 1972, *Structural and Stratigraphic Framework and Spatial Distribution of Permeability of the Atlantic Coastal Plain, North Carolina to New York,* U. S. Geol. Survey Prof. Paper 796, 79 pp.

BUCHAN, S., 1963, *Geology in Relation to Ground Water,* J. Inst. Water Engineers, v. 17, pp. 153-164.

CARTWRIGHT, KEROS and HUNT, C. S., 1978, *Hydrogeology of Underground Coal Mines in Illinois,* Illinois State Geol. Survey Reprint 1978 N, Urbana, Illinois 61801.

CEDERSTROM, D. J., and others, 1953, *Occurrence and Development of Ground Water in Perma-Frost Regions,* U. S. Geol. Survey Circ. 275, 30 pp.

CEDERSTROM, D. J., 1972, *Evaluation of Fields of Wells in Consolidated Rocks, Virginia to Maine,* U. S. Geol. Survey Water-Supply Paper 2021, 38 pp.

CUSHMAN, R. M., and others, 1980, *Sourcebook of Hydrologic and Ecological Features — Water Resource Regions of the Conterminous United States,* Ann Arbor Science Publ., Inc., P. O. Box 1425, Ann Arbor, Michigan, 126 pp.

CUSTODIO, EMILIO, 1976, *La Geohidrologia de Formaciones Volcanicas: Estado Actual de Conocimientos,* Asociacion de Geológos Españoles I Simposio Nacional de Hidrogeologia, Valencia (Spain) Oct. 1976, v. II, pp. 848-870.

DARTON, N. H., 1897, *Preliminary Report on Artesian Waters of a Portion of the Dakotas,* U. S. Geol. Survey 17th Ann. Rept. Part 2, pp. 1-92.

DESHPANDE, B. G., and SEN GUPTA, S. N., 1956, *Geology of Ground Water in the Deccan Traps and the Application of Geophysical Methods,* India Geol. Survey Bull., ser. B, no. 8, 22 pp.

DIJON, ROBERT, 1971, *Ground Water in Africa,* U. N. Dept. of Economic and Social Affairs, Nat. Resources/Water Series, Sales No. E.71.II.A.16, U. N. New York, New York.

DOYLE, F. L., 1973, *Environmental Geology and Hydrology, Madison Area, Madison County Alabama,* Ala. Geol. Survey Atlas Series 5.

DUGAN, J.T., and PECKENPAUGH, J.M., 1985, *Effects of Climate, Vegetation, and Soils on Consumptive Water Use and Ground-Water Recharge to the Central Midwest Regional Aquifer System, Mid-Continent United States,* U.S. Geol. Survey, WRI 85-4236, Lincoln, NE, 78 pp.

EMERY, P. A., and others, 1971, *Hydrology of San Luis Valley, South-Central Colorado,* U. S. Geol. Survey Hydrol. Invest. Atlas HA-381.

ENSLIN, J. F., 1943, *Basins of Decomposition in Igneous Rocks, their Importance as Underground Water Reservoirs and their Location by the Electrical Resistivity Method,* Trans. Geol. Soc. South Africa, v. 46, pp. 1-12.

FAULKNER, G. L., 1975, *Geohydrology of the Cross-Florida Barge Canal Area with Special Reference to the Ocalla Vicinity,* U. S. Geol. Survey Water-Resour. Invest. 1-73, 257 pp.

FERNANDEZ-RUBIO, R., (ed.), 1979, *Water in Mining and Underground Work.* 3-Volume Proc. First World Congress on Water in Mines and Underground Work, Granada, Spain Sept. 18-22, 1978, Work Group of Hydrogeology, Sciences Faculty, Apartado de Correos 556, Granada, Spain.

FUJIMURA, F. N., and CHANG, W. B. C., (eds.), 1981, *Groundwater in Hawaii — A Century of Progress,* Proc. Artesian Water Centennial Symp., Univ. Press of Hawaii, Honolulu, Hawaii 96822, 260 pp.

GOVERNMENT OF CANADA, 1974, *Hydrogeological Considerations in Northern Pipeline Development,* (SSC50), UNIPUB, 345 Park Ave. South, New York, New York 10010, 35 pp.

HELGESEN, J. O., and others, 1982, *Regional Study of the Dakota Aquifer (Darton's Dakota Revisited),* Ground Water, v. 20, no. 4, pp. 410-414.

HOPKINS, D. M., and others, 1955, *Perma-frost and Ground Water in Alaska,* U. S. Geol. Survey Prof. Paper 264-F, pp. 113-146.

INTERNATIONAL ASSOC. OF SCIENTIFIC HYDROLOGY, 1967, *Hydrology of Fractured Rocks,* Proc. Dubrovnik Symp., Oct. 1965.

INTERNATIONAL ASSOC. OF HYDROLOGICAL SCIENCES and HUNGARIAN GEOL. INSTITUTE, 1978, *Hydrogeology of Great Sedimentary Basins,* Proc. of the Budapest Symposium May-June 1976, IAHS Publ. No. 120., 200 Florida Ave. NW, Washington, D. C. 20009.

INTERNATIONAL GEOLOGICAL CONGRESS, 1972, *Proceedings 24th IGC Montreal 1972, Hydrogeology, Section II,* Organizing Committee, 601 Booth St., Ottawa, Canada, 330 pp.

JONES, P. H., and WALLACE, JR., R. H., 1974, *Hydrogeologic Aspects of Structural Deformation in the Northern Gulf of Mexico Basin,* U. S. Geol. Survey Jour. Res., v. 2, no. 5.

KANE, D.L., and STEIN, JEAN., 1983, *Water Movement into Seasonally Frozen Soils,* Water Resources Res., v. 19, no. 6.

KANIWETSKY, ROMAN, 1979, *Regional Approach to Estimating the Ground-Water Resources of Minnesota,* Minnesota Geol. Survey Rep. of Investigations 22 St. Paul 55108.

KEECH, C. F., and BENTALL, Ray, 1971, *Dunes on the Plains — The Sand Hills Region of Nebraska,* Nebr. Conserv. and Survey Div. Resource Report 4.

KOHOUT, F. A., and others, 1976, *Fresh Ground Water Found Deep Beneath Nantucket Island, Massachusetts,* J. of Research U. S. Geol. Survey, v. 4, no. 5, Sep. — Oct.

KOHOUT, F. A., and others, 1977, *Fresh Groundwater Stored in Aquifers Under the Continental Shelf: Implications From a Deep Test, Nantucket Island, Massachusetts,* Water Resour. Bull., v. 13, no. 2, pp. 373-386.

LARSSON, INGEMAR, 1984, *Ground Water in Hard Rocks,* Div. of Water Sciences, UNESCO, 7 Place de Fontenoy, 75700 Paris, France, 228 pp.

LeGRAND, H. E., 1949, *Sheet Structure, a Major Factor in the Occurrence of Ground Water in the Granites of Georgia,* Econ. Geol., v. 44, pp. 110-118.

LaMOREAUX, P. E., 1971, *Environmental Geology and Hydrology, Madison Country, Alabama,* Geol. Survey of Alabama Atlas Series 1.

LLOYD, J. W., and others, 1980, *A Ground-Water Resources Study of a Pacific Ocean Atoll — Tarawa, Gilbert Islands,* Water Resources Bull. v. 16, no. 4, pp. 646-653.

LLOYD, J. W., (ed.), 1982, *Case Studies in Groundwater Resources Evaluation,* Oxford Univ. Press, New York, New York 10016. MARINOV, N.A, 1982, *Distribution of Fresh Ground-Water in the Earth's Interior',* H *Water Resources, v. 9, no. 4, Jul. - Aug.*

MAXEY, G. B., and HACKETT, J. E., 1963, *Applications of Geohydrologic Concepts in Geology,* J. Hydrology, v. 1, pp. 35-46.

McGUINNESS, C. L., 1963, *The Role of Ground Water in the National Water Situation,* U. S. Geol. Survey Water Supply Paper 1800, 1121 pp.

MOTTS, W. S., and HEELEY, R. W., 1973, *Wetlands and Ground Water,* in Larson, (ed.), *A Guide to Important Characteristics and Value of Freshwater Wetlands in the Northeast,* Water Resources Research Center Publ. 31, Univ. of Massachusetts, Amherst, Massachusetts.

NEWCOMB, R. C., 1959, *Some Preliminary Notes on the Ground Water of the Columbia River Basalt,* Northwest Sci., v. 33, pp. 1-18.

OBERDORFER, J.A., and BUDDEMEIER, R.W., 1983, *Hydrogeology of a Great Barrier Reef: Implications for Groundwater in Reef and Atoll Islands,* EOS, v. 64, p. 703.

O'BRIEN, A. L. and MOTTS, W. S., 1980, *Hydrogeologic Evaluation of Wetland Basins for Land Use Planning,* Water Resources Bull., v. 16, no. 5, October.

PETERSON, F. L., 1972, *Water Development on Tropic Volcanic Islands — Type Example: Hawaii,* Ground Water, v. 10, no. 5, pp. 18-23, Sep.-Oct.

RAHN, P. H., and PAUL, H. A., 1975, *Hydrogeology of a Portion of the Sand Hills and Ogallala Aquifer, South Dakota and Nebraska,* Ground Water, v. 13, no. 5, pp. 428-437.

STEARNS, H. T., 1946, *Geology and Ground-Water Resources of the Island of Hawaii,* Terr. Hawaii Div. Hydrography Bull. v. 9, 368 pp.

TAKASAKI, K.J., and MINK, J.F., 1985, *Evaluation of Major Dike-Impounded Ground-Water Reservoirs, Island of Oahu,* U.S. Geol. Survey Water-Supply Paper 2217, 77 pp.

TEMPERLEY, B. N., 1960, *A Study of the Movement of Groundwater in Lava-Covered Country,* Overseas Geol. and Mineral Resources, v. 8, pp. 37-52.

THEIS, C. V., 1965, *Ground Water in Southwestern Region,* in *Fluids in Subsurface Environments,* Am. Assoc. Petroleum Geol. Mem. 4, pp. 327-342.

THOMAS, H. E., and LEOPOLD, L. B., 1964, *Groundwater in North America,* Science, v. 143, pp. 1001-1006.

TISDEL, F. W., 1964, *Water Supply from Ground Water Sources in Permafrost Areas of Alaska,* in *Science in Alaska 1963,* 14th Alaskan Science Conference, Am. Assoc. Advancement of Sci., Alaska Div., pp. 113-124.

TSYTOVICH, N.A., 1984, *The Mechanics of Frozen Ground,* Hemisphere Publ. Corp., 19 West 44th St., New York, New York 10036, 426 pp.

UNESCO, 1967, *Hydrology of Fractured Rocks,* 2 vol., 689 pp. UNIPUB, 345 Park Ave. South, New York, New York 10010.

UNESCO, 1969, *Water in the Unsaturated Zone,* 2 vol., 995 pp., UNIPUB, 345 Park Ave. South, New York, New York 10010.

UNESCO, 1984, *Ground Water in Hard Rocks,* Studies and Reports in Hydrology, No. 33, Paris, 228 pp.

UNITED NATIONS, 1976, *Ground Water in the Western Hemisphere,* U. N. Dept. of Economic and Social Affairs, Nat. Resources/Water Series No. 4 ST/ESA/35, New York, New York, 337 pp.; also UNIPUB, 345 Park Ave. South, New York, New York, 10010.

UNITED NATIONS, 1979, *A Review of the United Nations Ground-Water Exploration and Development Program in the Developing Countries 1962-1977,* (UN79/2/2A4), UNIPUB, 345 Park Ave. South, New York, New York 10010, 84 pp.

UNITED NATIONS ECONOMIC COMMISSION FOR EUROPE, 1978, *Selected Water Problems in Islands and Coastal Areas,* Committee on Water Problems of the UNECE, 1978 Seminar, Malta, Pergamon Press, Maxwell House, Fairview Park, Elmsford, New York 10523, 522 pp.

U.S. GENERAL ACCOUNTING OFFICE, 1977, *Ground Water: An Overview (Consequences of Mining of Ground Water, Land Subsidence and Salt Water Intrusion),* GAO Rep. CED-77-69, Washington, D. C.

VACHER, H. L., 1978, *Hydrogeology of Bermuda — Significance of an Across-the-Island Variation in Permeability*, J. of Hydrology, v. 39, no. 3/4, Dec.

WALTON, W. C., and SCUDDER, G. D., 1960, *Ground-Water Resources of the Valley-Train Deposits in the Fairborn Area, Ohio*, Ohio Div. Water, Tech. Rept. no. 3.

WATSON, L. J., 1964, *Development of Ground Water in Hawaii*, J. Hydraulics Div., Amer. Soc. Civil Engrs. v. 90, no. HY6, pp. 185-202.

WHEATCRAFT, S.W. and BUDDEMEIER, R.W., *Atoll Island Hydrology*, Ground Water, v. 19, pp. 311-320.

WHITEHEAD, B. R., 1980, *Hydrogeology of Sedimentary Rocks — Clastics*, Canadian Water Well, v. 6, no. 2, May.

WILLIAMS, J. R., 1970, *Ground Water in the Permafrost Regions of Alaska*, U. S. Geol. Survey Prof. Paper 696, 83 pp.

WINOGRAD, I. J., 1971, *Hydrogeology of Ash Flow Tuff: A Preliminary Statement*, Water Resources Res., v. 7, no. 4, pp. 994-1006.

WINQVIST, G., 1953, *Ground Water in Swedish Eskers*, Lindstahl, Stockholm, 91 pp.

GROUNDWATER SYSTEMS IN THE UNITED STATES

HEATH, R. C., 1982, *Classification of Ground-Water Systems of the United States*, Ground Water, v. 20, no. 4, pp. 393-401.

HEATH, R.C., 1984, *Ground-Water Regions of the United States*, Water-Supply Paper 2242, Supt. of Doc., U.S. Govt. Printing Office, Washington, DC 20402, 78 pp.

MEINZER, O. E., 1939, *Ground Water in the United States, A Summary*, U. S. Geol. Survey Water-Supply Paper 836-D, pp. 157-232.

MEINZER, O. E., 1959, *The Occurrence of Ground Water in the United States*, U. S. Geol. Survey Water Supply Paper 489, 321 pp.

THOMAS, H. E., 1952, *Ground-water Regions of the United States — their Storage Facilities*, in *The Physical and Economic Foundation of Natural Resources*, v. 3, House Interior and Insular Affairs Comm., U. S. Congress, Washington, D. C.

TODD, D.K., 1983, *Ground-Water Resources of the United States,* Premier Press, P.O. Box 4428 Berkeley, California 94704, 749 pp. (compiled from U.S. Geol. Survey Prof. Paper 813).

U.S. GEOLOGICAL SURVEY, *Summary Appraisals of the Nation's Ground-Water Resources,* Prof. Paper 813 (by Water Resources Region; see index map; letter indicates chapter designation)

A-Ohio Region, by Bloyd, R. M., Jr., 1974.
B-Upper Mississippi Region, by Bloyd, R. M., Jr., 1975.
C-Upper Colorado Region, by Price, D. and Arnow, T., 1974.
D-Rio Grande Region, by West, S.W., and Broadhurst, W. L., 1975.
E-California Region, by Thomas, H. E., and Phoenix, D. A., 1976.
F-Texas-Gulf Region, by Baker, E. T., Jr., and Wall, J. R., 1977.
G-Great Basin Region, by Eakin, T. E., Price, D. and Harrill, J. R., 1976.
H-Arkansas-White-Red Region, by Bedinger, M. S., and Sniegocki, R. T., 1976.
I-Mid-Atlantic Region, by Sinnott, Allen, and Cushing, E. M., 1978.
J-Great Lakes Region, by Weist, W. G., Jr., 1978.
K-Souris-Red-Rainy Region, by Reeder, H. O., 1978.
L-Tennessee Region, by Zurawski, Ann, 1978.
M-Hawaii Region, by Takasaki, K. J., 1978.
N-Lower Mississippi Region, by Terry, J. E., Hosman, R. L., and Bryant, C. T., 1979.
O-South Atlantic-Gulf Region, by Cederstrom, D. J., Boswell, E. H., and Tarver, G. R., 1979.
P-Alaska Region, by Zenone, Chester, and Anderson, G. S., 1978.
Q-Missouri Basin Region, by Taylor, O. J., 1978.
R-Lower Colorado Region, by Davidson, E. S., 1979.
S-Pacific Northwest Region, by Foxworthy, B. L., 1979.
T-New England Region, by Sinnott, Allen, 1980.
U-Caribbean Region, by Gomez-Gomez, Fernando and Heisel, J. E., 1980.

UNITED STATES GEOLOGICAL SURVEY, 1985, *National Water Summary 1984: Hydrologic Events, Selected Water-Quality Trends, and Ground-Water Resources,* Water Supply Paper 2275, USGS, 604 S. Pickett St., Alexandria, Virginia 22304, (contains state-by-state summaries).

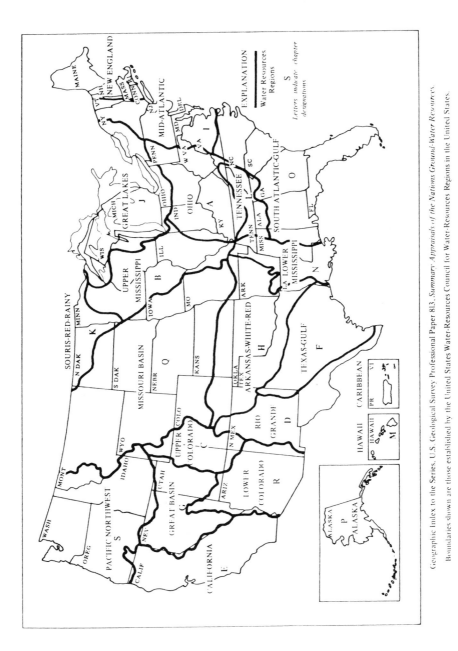

Geographic Index to the Series. U.S. Geological Survey Professional Paper 813. *Summary Appraisals of the Nation's Ground-Water Resources.*

Boundaries shown are those established by the United States Water-Resources Council for Water-Resources Regions in the United States.

33

ARID ZONE HYDROGEOLOGY/QANATS

AMBROGGI, R. P., 1966, *Water Under the Sahara,* Sci. Am., v. 214, no. 5, pp. 21-29.

BEAUMONT, P., 1971, *Qanat Systems in Iran,* Bull. Internat. Assoc. Sci. Hydrology, v. 16, pp. 39-50.

BOGOMOLOV, G. V., 1961, *Conditions of Formation of Fresh Waters under Pressure in Certain Desert Zones of North Africa, the U.S.S.R., and South-West Asia,* in *Salinity Problems in the Arid Zones;* pp. 37-41, UNESCO, Paris.

BYORDI, MOHAMMED, 1974, *Ghanats of Iran: Drainage of Sloping Aquifer,* J. Irrigation and Drainage Div., Am. Soc. Civil Engrs., v. 100, no. IR3, Sep.

CRESSEY, G. B., 1958, *Qanats, Karez and Foggaras,* Geogr. Review, v. 48, pp. 27-44.

DAVIDSON, E. S., 1973, *Geohydrology and Water Resources of the Tucson Basin, Arizona,* U. S. Geol. Survey Water-Supply Paper 1939-E.

GISCHLER, CHRISTIAAN, 1979, *Water Resources in the Arab Middle East and North Africa,* MENAS Ltd., Gallipoli House, Wisbech, Cambridge PE 148TN, England, 132 pp.

KUNIN, V. V., 1957, *Conditions of the Formation of Underground Waters in Deserts,* Internat. Assoc. Sci. Hydrology, General Assembly Toronto, Pub. 44, v. 2, pp. 502-516.

LOELTZ, O. J., and others, 1975, *Geohydrologic reconnaissance of the Imperial Valley, California,* U. S. Geol. Surv. Prof. Paper 486-K, K1-K54.

MAXEY, G. B. 1968, *Hydrology of Desert Basins,* Ground Water, v. 6, no. 5, pp. 10-22.

MITHAL, R. S., and SINGHAL, B. B. S., (eds.), 1969, Proc. Symposium on *Ground Water Studies in Arid and Semiarid Regions,* Dept. of Geology and Geophysics, Univ. of Roorkee, Roorkee, India.

NEAL, J. T., 1969, *Playa Variation,* in McGinnies, W. G. and Goldman, B. J., *Arid Lands in Perspective,* Univ. of Arizona Press, pp. 15-44.

ORTIZ, N. V., and others, 1980, *Ground-Water Flow and Uranium in Colorado Plateau,* Ground Water, v. 18, no. 6, pp. 596-606.

PICARD, L., 1953, *Outline of Ground-Water Geology in Arid Regions,* Proc. Ankara Symp. on Arid Zone Hydrology, UNESCO, Paris, pp. 165-176.

ROBAUX, A., 1953, *Physical and Chemical Properties of Ground Water in the Arid Countries,* Proc. Ankara Symp. on Arid Zone Hydrology, UNESCO, Paris, pp. 17-28.

UNESCO, 1962, *The Problems of the Arid Zone,* Proc. Paris Symp., Arid Zone Research 18, UNESCO, Paris, 481 pp.

WINOGRAD, I.J., and THORDARSON, W., 1975, *Hydrogeologic and Hydrochemical Framework, South-Central Great Basin, Nevada-California, with Special Reference to Nevada Test Site,* U. S. Geol. Survey Prof. Paper 712-C, pp. C1-C126.

WULFF, H. E., 1968, *The Qanats of Iran,* Sci. Am., v. 218, pp. 94-100, 105.

KARST HYDROGEOLOGY

ABD-EL-AL, IBRAHIM, 1953, *Statics and Dynamics of Water in the Syro-Lebanese Limestone Massifs,* Proc. Ankara Symp. on Arid Zone Hydrology, UNESCO, Paris, pp. 60-76.

BENNETT, G. D., and GIUSTI, E. V., 1976, *Water Resources of the North Coast Limestone Area, Puerto Rico,* U. S. Geol. Survey Water Resources Inv. 75-42, 47 pp.

BRUCKER, R. W., HESS, J. W., and WHITE, W. B., 1972, *Role of Vertical Shafts in the Movement of Ground Water in Carbonate Aquifers,* Ground Water, v. 10, no. 6, pp. 5-13.

BURDON, D. J., 1967, *Hydrogeology of some Karstic Areas of Greece,* in *Hydrology of Fractured Rocks,* Proc. Dubrovnik Symp. Oct. 1965, Internat. Assoc. Sci. Hydrology, Pub. 73, pp. 308-317.

BURDON, D. J., and SAFADI, C., 1964, *The Karst Groundwaters of Syria,* J. Hydrology, v. 2, pp. 324-347.

BURDON, D. J., and PAPAKIS, N., 1963, *Handbook of Karst Hydrogeology with Special Reference to the Carbonate Aquifers of the Mediterranean Region,* United Nations Food and Agricultural Organization, Inst. for Sub-Surface Res., Athens, 276 pp.

BURGER, A., and DUBERTRET, L., 1984, *Hydrogeology of Karstic Terrains: Case Histories,* UNESCO/IHP Project, Verlag Heinz Heise GmbH, P.O. Box 2746, 3000 Hannover, 1, F.R. Germany, 264 pp.

CHINESE INSTITUTE OF HYDROGEOLOGY AND ENGINEERING GEOLOGY, CHINESE ACADEMY OF GEOLOGICAL SCIENCES, 1976, *Karst in China,* Shanghai People's Publishing House.

DAVIES, W. E., and LeGRAND, H. E., 1971, -Karst of the United States,' Chap. 15, in -Karst of the World,' ed. by M. Herak and V. T. Stringfield: Amsterdam, Elsevier, pp. 467-505.

DAVIS, W. M., 1930, *Origin of Limestone Caverns,* Geol. Soc. Am. Bull., v. 41, pp. 475-628.

FOX, I. A., and RUSHTON, K. R., 1976, *Rapid Recharge in a Limestone Aquifer,* Ground Water, v. 14, no. 1, pp. 21-27.

GEZÉ, B., 1965, *Les Conditions Hydrogéologiques des Roches Calcaires,* Chronique d'Hydrogéologie no. 7, B.R.G.M., pp. 9-39.

GIUSTI, E. V., 1978, *Hydrogeology of the Karst of Puerto Rico,* U. S. Geol. Survey Prof. Paper 1012, 68 pp.

HARVEY, E. J., 1980, *Ground Water in the Springfield-Salem Plateaus of Southern Missouri and Northern Arkansas,* U. S. Geol. Survey Water Resources Inv. 80-101, NTIS, Springfield, Virginia 22161, 66 pp.

HOWARD, A. D., 1963, *The Development of Karst Features,* Natl. Speleological Soc. Bull., v. 25, pp. 45-65.

HOWARD, A. D., 1964, *A Model for Cavern Development under Artesian Ground-Water Flow, with Special Reference to the Black Hills,* Bull. Natl. Speleological Soc. v. 26, pp. 7-16.

KIERSCH, G. A., and HUGHES, P. W., 1952, *Structural Localization of Ground Water in Limestones — Big Bend District, Texas-Mexico,* Econ. Geol. v. 47, pp. 794-806.

LAMOREAUX, P. E., and POWELL, W. J., 1960, *Stratigraphic and Structural Guides to the Development of Water Wells and Well Fields in a Limestone Terrane,* Internat. Assoc. Sci. Hydrology Pub. no. 52, pp. 363-375.

LAMOREAUX, P.E., WILSON, B.M., and MEMON, B.A., 1985, *Guide to the Hydrology of Carbonate Rocks', Div. of Water Sciences, UNESCO, 7 Place de Fontenoy, 75700 Paris, France, 345 pp.*

LATTMAN, L. H., and PARIZEK, R. R., 1964, *Relationship between Fracture Traces and the Occurrence of Ground Water in Carbonate Rocks,* J. Hydrology, v. 2, pp. 73-91.

LeGRAND, H. E., 1973, *Concepts of Karst Development in Relation to Interpretation of Surfce Runoff,* U. S. Geol. Survey Jour. Research, v. 1, no. 3, pp. 351-360.

LeGRAND, H. E., 1973, *Hydrological and Ecological Problems of Karst Regions,* Science, 179, pp. 859-864.

LeGRAND, H. E., and STRINGFIELD, V. T., 1973, *Karst Hydrology — A Review:* J. Hydrology, 19, pp. 1-23.

MEISLER, HAROLD, 1963, *Hydrogeology of the Carbonate Rocks of the Lebanon Valley, Pennsylvania,* Pa. Geol. Survey, 4th Ser., Bull. W18, Harrisburg, Pennsylvania, 81 pp.

MILANOVIC, P., 1981, *Karst Hydrogeology,* Water Resources Publications, Distrib. Div., P.O. Box 2841, Littleton, Colorado 80161, 444 pp.

MILLER, W. R., 1976, *Water in Carbonate Rocks of the Madison Group in Southeastern Montana — A Preliminary Evaluation,* U. S. Geol. Survey Water-Supply Paper W2043.

MONROE, W. H., 1970, *A Glossary of Karst Terminology,* U. S. Geol. Survey Water-Supply Paper 1899-K, 26 pp.

NORRIS, S. E., and FIDLER, R. E., *Availability of Ground Water from Limestone and Dolomite Aquifers in Northwest Ohio and its Relation to Geologic Structure,* in *Geological Survey Research 1971,* U. S. Geol. Survey Prof. Paper 750-B, pp. B229-B235.

PARIZEK, R. R., 1979, *Carbonate Hydrogeologic Environments: Their Relationship to Land Use, Water Resources Development and Management,* PB-292 337, NTIS, Springfield, Virginia 22161.

RADZITZKY d'OSTROWICK, I., 1953, *L'Hydrogéologie des Roches Calcareuses,* Dinant, L. Bordeaux-Capelle, 199 pp.

SIDDIQUI, S. H., and PARIZEK, R. R., 1971, *Hydrogeologic Factors Influencing Well Yields in Folded and Faulted Carbonate Rocks in Central Pennsylvania,* Water Resources Res., v. 7, no. 5, pp. 1295-1312.

STRINGFIELD, V. T., 1966, *Artesian Water in Tertiary Limestone in the Southeastern States,* U. S. Geol. Survey Prof. Paper 517, 226 pp.

STRINGFIELD, V. T., and LeGRAND, H. E., 1966, *Hydrology of Limestone Terranes in the Coastal Plain of the Southeastern United States,* Geol. Soc. Am. Spec. Paper 93, 46 pp.

STRINGFIELD, V. T., and LeGRAND, H. E., 1969, *Hydrology of Carbonate Rock Terranes — A Review with Special Reference to the United States,* J. Hydrology, v. 8, no. 3,4, pp. 349-417.

WHITE, W. B., 1969, *Conceptual Models for Carbonate Aquifers,* Ground Water, v. 7, no. 3, pp. 15-22.

YEVJEVICH, VUJICA, (ed.), 1976, *Karst Hydrology and Water Resources,* Proc. U. S.-Yugoslavian Symp., Dubrovnik, June 2-7, 1975. 2 Vols. Water Resources Publications, Fort Collins, Colorado 80522.

EXPLORATION TECHNIQUES/MAPPING AND DATA COLLECTION

BIBLIOGRAPHY

ANONYMOUS, 1977, *Index of Well Sample and Core Repositories of the United States and Canada,* U. S. Geol. Survey Open-File Rep. 77-567, U. S. Geol. Survey, Denver, Colorado 80225, 195 pp.

JONES, N. E., 1966, *Bibliography of Remote Sensing of Resources,* U. S. Army Corps of Engrs., Ft. Belvoir, Va., for NASA, Earth Resources Program Office, N68-1 870, 40 pp.

LLAVERIAS, R. K., 1970, *Remote Sensing Bibliography for Earth Resources,* 1966-67, U. S. Geol. Survey, Clearinghouse Fed. Sci. and Tech. Inf., PB 192863, Springfield, Virginia.

OFFICE OF WATER DATA COORDINATION, 1972, Catalog of Information on Water Data (21 volumes by region), U. S. Geol. Survey.

PAUSZEK, F., 1973, *Digest of the Catalog of Information on Water Data,* Office of Water Data Coordination, U. S. Geol. Survey Water Resources Inv. 63-73, 83 pp.

U.S. GEOLOGICAL SURVEY, 1968, *Bibliography of Remote Sensing of Earth Resources for Hydrological Applications 1960-1967,* U. S. Geol. Survey open-file Interagency rept. NASA-134.

U. S. GEOLOGICAL SURVEY, 1981, *Index to Water-Data Acquisition in the Coal Provinces of the United States,* Third Volume USGS, 417 National Center, Reston, Virginia 22092, 2133 pp. (Availability of water resources data in Northern Great Plains and Rocky Mountains).

U. S. GEOLOGICAL SURVEY, 1981, *A Guide to Obtaining Information from the USGS, 1981,* Circ. 777, USGS., Alexandria, Virginia 22304, 42 pp.

GENERAL WORKS

AGRICULTURAL RESEARCH SERVICE, 1962, *Field Manual for Research in Agricultural Hydrology,* Agric. Handbk. no. 224, U. S. Dept. of Agric., 215 pp.

AMERICAN WATER WORKS ASSOCIATION, 1976, *Ground Water,* AWWA Manual M21, 6666 West Quincy Avenue, Denver, Colorado 80235.

BIRMAN, J. H., 1969, *Geothermal Exploration for Ground Water,* Geol. Soc. Am. Bull., v. 80, pp. 617-630, Apr.

BRASSINGTON, RICK, *Finding Water,* Rookery Books, 12 Culcheth Hall Dr., Culcheth, Warrington, Cheshire, WA3 4PS, UK.

BROWN, R. H. (ed.), and others, 1977, *Ground-Water Studies — An International Guide for Research and Practice,* UNIPUB, New York, New York 10016.

BUREAU DE RECHERCHES GEOLOGIQUES ET MINIERES, 1962, *Methods d'Etudes et de Recherches des Nappes Aquifères,* Paris, 158 pp.

CANADA INLAND WATERS BRANCH, 1969, *Instrumentation and Observation Techniques,* (Remote Sensing) Proc. Hydrology Symposium no. 7, Dept. of Energy, Mines and Resources, Ottawa, 343 pp.

CARTWRIGHT, KEROS, 1968, *Thermal Prospecting for Ground Water,* Water Resources Res. v. 4, no. 2, pp. 395-403.

CARTWRIGHT, KEROS, 1970, *Groundwater Discharge in the Illinois Basin as Suggested by Temperature Anomalies,* Water Resources Res., v. 6, no. 3, pp. 912-918.

CARTWRIGHT, KEROS, 1974, *Tracing Shallow Groundwater Systems by Soil Temperatures,* Water Resources Res., v. 10, pp. 847-855.

CHIKISHEV, A. G. (Ed.), 1965, *Plant Indicators of Soils, Rocks, and Subsurface Waters.* Consultants Bureau, New York, 210 pp.

DaCOSTA, J. A., and FALCON MORENO, E., 1963, *Manual de Métodos Cuantitativos en el Estudio de Agua Subterránea, Parte 1 — Principios Fundamentales, Mecánica de Acuiferos, Parte 2 — Organizacion y Realizacion de Pruebas de Acuiferos, Métodos, Ejemplos,* U. S. Geol. Survey, Phoenix, Arizona, 122 pp.

DAVIS, G. H., HEATH, R. C., and SEABER, P. R., 1977, *Planning and Design of Ground-Water Level Networks,* in Ground-Water Studies — An International Guide for Research and Practice. Supplement No. 3; UNIPUB, Box 433, Murray Hill Station, New York, New York 10016, 11 pp.

FENT, O. S., 1949, *Use of Geologic Methods in Ground-Water Prospecting,* J. Am. Water Works Assoc., v. 41, pp. 590-598.

HAUN, J. D. and LeROY, L. W., (eds.), 1958, *Subsurface Geology in Petroleum Exploration,* Colorado School of Mines, Golden, Colorado.

HEALY, H. G., 1978, *Appraisal of Uncontrolled Flowing Artesian Wells in Florida,* PB80-102239, NTIS Springfield, Virginia 22151.

HEATH, R. C., 1976, *Design of Ground-Water Level Observation-Well Programs,* Ground Water, v. 14, no. 2, pp. 71-77.

LEGGETTE, R. M., 1950, *Prospecting for Ground Water-Geologic Methods,* J. Am. Water Works Assoc. v. 42, pp. 945-946.

LeROY, L. W., 1951, *Subsurface Geologic Methods,* 2nd. ed., Colorado School of Mines, Golden, 1156 pp.

LUCKEY, R. R., 1972, *Analyses of Selected Statistical Methods for Estimating Groundwater Withdrawal,* Water Resources Res., v. 8, no. 1, pp. 05-210, Feb.

MAZLOUM, S., 1953, *Boring and Prospecting for Ground-Water in Arid Zones,* Proc. Ankara Symp. on Arid Zone Hydrology, pp. 184-187, UNESCO, Paris.

MEINZER, O. E., 1927, *Plants as Indicators of Ground Water,* U. S. Geol. Survey Water Supply Paper 577, 95 pp.

NATIONAL RESEARCH COUNCIL, 1977, *Environmental Monitoring, Volume IV.* National Academy of Sciences, Washington, D. C.

PARSONS, M. L., 1971, *Groundwater Thermal Regime in a Glacial Complex,* Water Resources Res., v. 6, no. 6, pp. 1701-1720.

PISKIN, K., 1971, *Methods of Evaluating Thickness and Texture of Glacial Till in Studies of Ground-Water Recharge,* Illinois Acad. Sci. Trans., v. 64, no. 4, pp. 337-346.

ROBERTSON, C. E., 1963, *Well Data for Water Well Yield Map,* Missouri Geol. Survey and Water Resources, 23 pp.

SCHAAP, W., (ed.), 1984, *Methods and Instrumentation for the Investigation of Groundwater Systems,* Proc. Symp. Noordwijkerhout, Netherlands, May 1983, TNO, P.O. Box 297, 2501 BD, The Hague, The Netherlands, 690 pp.

SCHNEIDER, R., 1962, *An Application of Thermometry to the Study of Ground Water,* U. S. Geol. Survey Water-Supply Paper 1544-B. 16 pp.

SNIPES, D.S., and others, 1984, *Indicators of Ground Water Quality and Yield for a Public Water Supply in Rock Fracture Zones of the Piedmont,* Clemson Univ., PB85-233765/WEP, NTIS, Springfield, Virginia 22161, 100 pp.

STRAUB, C. P., GOPPERS, V. M., and DuCHENE, ALAIN, 1977, *Water Quality Status and Trends in Minnesota — Indices for Water Supply and Ground Water Pollution,* Water Resources Res. Center, University of Minnesota, Minneapolis, Minnesota, 144 pp.

THORPE, T. W., 1950, *Prospecting for Ground Water-Test Drilling,* J. Am. Water Works Assoc., v. 42, pp. 957-960.

UNITED NATIONS, 1967, *Method and Techniques of Ground-Water Investigation and Development,* ECAFE/UNESCO Water Resources Series No. 33, Sales no. E.68.II.F.6, 206 pp.

UNITED NATIONS WATER RESOURCES CENTER, 1960, *Large-Scale Ground-Water Development,* Pub. no. 60, II.B. 3, UNIPUB, New York, 84 pp.

VIKTOROV, D. B. and others, 1964, *Short Guide to Geo-Botanical Surveying,* Pergamon, Oxford, 158 pp.

WALLIN, T. R., 1977, *Illinois Redesigns its Ambient Water Quality Monitoring Network,* Illinois Environmental Protection Agency, 116 pp. plus Appendix.

WALTON, W. C., and NEILL, J. C., 1963, *Statistical Analysis of Specific-Capacity Data for a Dolomite Aquifer,* J. Geophys. Res. v. 68, pp. 2251-2262.

AERIAL PHOTOGRAPHY/REMOTE SENSING

AVERY, T. E., 1977, *Interpretation of Aerial Photographs,* Burgess Publishing, Minneapolis, 392 pp.

BOETTCHER, A. J., and others, 1976, *Use of Thermal-Infrared Imagery in Ground-Water Investigations, Northwestern Montana,* U. S. Geol. Survey J. Res., v. 4, no. 6, pp. 727-732.

BOWDEN, L. W., and PRUIT, E. L., (eds.), 1975, *Manual of Remote Sensing, v. 2, Interpretation and Applications,* Amer. Soc. Photogrammetry, Falls Church, Virginia, pp. 869-2144.

DRACUP, J. A., and SHAPIRO, K. A., 1967, *Remote Sensing of Hydrologic Data by Satellites,* Proc. Third. Ann. Conf. Am. Water Resources Assoc., San Francisco, California, pp. 472-484.

ESTES, J. E. and SIMONETT, D. S., 1978, *Remote Sensing Detection of Perched Water Tables,* Calif. Water Resources Center Contrib. No. 175, Univ. of Calif., 475 Kerr Hall, Davis, 95616.

FOSTER, K. E., and others, 1980, *The Use of Landsat Imagery in Ground-Water Exploration,* Water Resources Bull. v. 16, no. 5, October.

HOWE, R. H. L., 1960, *The Application of Aerial Photographic Interpretation to the Investigation of Hydrologic Problems,* Photogramm. Eng., v. 26, no. 1, pp. 85-95.

HOWE, R. H. L., and others, 1956, *Application of Air Photo Interpretation in the Location of Ground Water,* J. Am. Water Works Assoc., v. 48, no. 11, pp. 1380-1390.

HUNTLEY, DAVID, 1978, *On the Detection of Shallow Aquifers Using Thermal Infrared Imagery,* Water Resources Res., v. 14, no. 6, Dec.

IDSO, S. B., and others, 1975, *Detection of Soil Moisture by Remote Surveillance,* Amer. Sci., v. 63, pp. 549-557.

LATTMAN, L. H., 1958, *Techniques of Mapping Geologic Fracture Traces and Lineaments on Aerial Photographs,* Photogram, Engineering, v. 24, pp. 568-576.

LATTMAN, L. H., and PARIZEK, R. R., 1964, *Relationship Between Fracture Traces and the Occurrence of Ground Water in Carbonate Rocks,* J. Hydrology, v. 2, pp. 73-91.

LAYTON, J. P. (ed.), 1970, *Proceedings of the Princeton University Conference on Aerospace Methods for Revealing and Evaluating Earth's Resources,* Sept. 25-26, 1969, Princeton, New Jersey.

LEE, KENNAN, 1969, *Infrared Exploration for Shoreline Springs at Mono Lake, California Test Site,* Stanford Univ. Remote Sensing Lab., Tech. Rept. 69-7.

LEPLEY, L. K., and PALMER, L. A., 1967, *Remote Sensing of Hawaiian Coastal Springs Using Multispectral and Infrared Techniques,* Univ. of Hawaii Water Resources Res. Center Tech. Report 18.

LOHMAN, S. W., and ROBINOVE, C. J., 1964, *Photographic Description and Appraisal of Water Resources,* Photogrammetria, v. 19, no. 3, 1962-1964.

LOWE, R. H., and others, 1956, *Application of Air Photo Interpretation in the Location of Ground Water,* J. Am. Water Works Assoc., v. 48, p. 559, May.

MANN, J. F., Jr., 1958, *Estimating Quantity and Quality of Ground Water in Dry Regions Using Airphotographs,* Internat. Assoc. Sci. Hydrology, General Assembly of Toronto, v. 2, pp. 125-134.

MEER, G. E., 1969, *Aerial Photographic Method for Studying Ground Water,* Army Foreign Sci. and Technology Center, Clearinghouse for Fed. Sci. and Tech. Info. Accession no. AD-690613, Springfield, Virginia.

MOLLARD, J. D., 1968, *The Role of Photo-Interpretation in Finding Groundwater Sources in Western Canada,* The Queen's Printer, Ottawa, pp. 57-75.

MOORE, G. K., and DEUTSCH, MORRIS, 1975, *ERTS Imagery for Ground Water Investigations,* Ground Water, v. 13, no. 2, pp. 214-226.

NEFEDOV, K. Ye., 1964, *Hydrogeologic Mapping by Means of Aerial-Photographic Survey Data,* Doklady Acad. Sciences U.S.S.R., Earth Sciences Section, Am. Geol. Institute, v. 148, pp. 90-92.

NIGHTINGALE, H.I., 1975, *Ground-Water Recharge Rates from Thermometry,* Ground Water, v. 13, no. 4, pp. 340-344.

PLUHOWSHI, E. J., 1972, *Hydrologic Interpretations Based on Infrared Imagery of Long Island, New York,* U. S. Geol. Survey Water-Supply Paper 2009-B, 20 pp.

POWELL, W. J., and others, 1970, *Delineation of Linear Features and Application to Reservoir Engineering Using Apollo 9 Multispectral Photography,* Geol. Survey of Alabama, Inf. ser. 41.

RAY, R. G., 1960, *Aerial Photographs in Geologic Interpretation and Mapping,* U. S. Geol. Survey Prof. Paper 373, 227 pp.

ROBINOVE, C. J., 1965, *Infrared Photography and Imagery in Water Resources Research,* J. Am. Water Works Assoc., v. 57, no. 7, pp. 834-840.

SETZER, J., 1966, *Hydrologic Significance of Tectonic Fractures Detectable on Airphotos,* Ground Water, v. 4, no. 4, pp. 23-27.

STRANDBERG, C. H., 1967, *Color Aerial Photography for Water Supply and Pollution Control Reconnaissance,* Proc. Third Ann. Conf. on Remote Sensing of Air and Water Pollution, North Am. Aviation, Inc., Autonetics Div., Anaheim, California, pp. 13-1 to 13-10.

TELFORD, W. M., KING, W. F., and BECKER, A., 1977, *VLF Mapping of Geological Structure,* Paper 76-25, Geol. Survey of Canada, Ottawa.

WARREN, W. M., and WIELCHOWSKY, C. W., 1973, *Aerial Remote Sensing of Carbonate Terranes in Shelby County,* Geol. Survey of Ala. Reprint Ser. 26.

MAPPING TECHNIQUES

BOGOMOLOV, G. V., and PLOTNIKOV, N. A., 1957, *Classification of Underground Water Resources and their Plotting on Maps,* Internat. Assoc. Sci. Hydrology, General Assembly Toronto, Pub. 44, v. 2, pp. 86-97.

CASTANY, GILBERT, and MARGAT, JEAN, 1965, *Les Cartes Hydrologiques,* Internat. Assoc. Sci. Hydrology Bull., v. 10, no. 1, pp. 74-81.

CHURINOV, M. V., 1957, *Hydrogeological Maps and their Role in Estimating the Water-Bearing Capacity of Rocks and Subsoil Water Resources,* Internat. Assoc. Sci. Hydrology, Pub. 44, v. 2, pp. 62-67 (1958).

DaCOSTA, J. A., 1960, *Presentation of Hydrologic Data on Maps in the United States of America,* Internat. Assoc. Sci. Hydrology, Pub. 52, pp. 143-186.

DAVIS, P. R., and MATLOCK, W. G., 1973, *The Effect of Data Density on Ground-Water Contouring Accuracy,* Trans. Am. Soc. Agric. Engrs., v. 16, no. 6, Nov.-Dec.

HANSON, H. C., 1972, *The Accuracy of Groundwater Contour Maps,* Water Resources Res., v. 8, no. 1, pp. 201-204.

INTERNATIONAL ASSOCIATION OF SCIENTIFIC HYDROLOGY/
UNESCO/INSTITUTE OF GEOLOGICAL SCIENCES, 1970,
International Legend for Hydrogeological Maps, (English/French/
Spanish/Russian), 101 pp.

MARGAT, JEAN, 1966, *La Cartographie Hydrogéologique,* Bur. Rech.
Geol Min., Publ. no. DS 66A 130.

McGUINNESS, C. L., 1964, *Generalized Map Showing Annual Runoff
and Productive Aquifers in the Conterminous United States,* U. S.
Geol. Survey Hydrol. Investigations Atlas HA-194.

MEYBOOM, P., 1961, *A Semantic Review of the Terminology of
Groundwater Maps,* Bull. Internat. Assoc. Sci. Hydrology, v. 6, no. 1,
pp. 29-36.

PETTYJOHN, W. A., and RANDICH, P. G., 1966, *Geohydrologic Use of
Lithofacies Maps in Glaciated Areas,* Water Resources Res., Fourth
Quarter, v. 2, no. 4, pp. 679-689.

UNESCO, 1970, *International Legend for Hydrogeological Maps,*
UNIPUB, 345 Park Ave. South, New York, New York 10010, 101
pp.

UNESCO, 1977, *Hydrological Maps: A Contribution to the International
Hydrological Decade,* UNIPUB, 345 Park Ave. South, New York,
New York 10010, 204 pp.

U.S. GEOLOGICAL SURVEY, 1967, *Productive Aquifers and
Withdrawals from Wells* (Map), U. S. Geol. Survey National Atlas
Sheet 126.

U.S. GEOLOGICAL SURVEY, various dates, *Geologic Map Indices of the
States* (includes maps issued by U. S. Geol. Survey, state and
commercial organizations, universities and professional societies at a
scale of 1:250,000 or larger), U. S. Geol. Survey, 907 National
Center, Reston, Virginia 22092.

DATA BASE DESIGN AND MANAGEMENT

BENTALL, RAY, 1963, *Methods of Collecting and Interpreting Ground-
Water Data,* U. S. Geol. Survey Water Supply Paper 1544-H, 97 pp.

DOYEL, W. W., 1974, *Basic Data — A New Awareness,* Water Resources
Bull., v. 10, no. 4, pp. 710-718.

DOYEL, W. W., JOHNSON, A. I., and LANG, S. M., 1972, *Federal Information and Data Centers for Engineers and Scientists,* Materials Research and Standards, v. 12, no. 1, pp. 17-20.

DOYEL, W. W., and LANG, S. M., 1972, *NAWDEX — A System for Improving Accessibility to Water Data,* Proc. Symp. on Watersheds in Transition, Am. Water Resources Assoc., pp. 91-97.

DUCKSTEIN, L., and KISIEL, C. C., 1971, *Efficiency of Hydrologic Data Collection Systems, Role of Type I and Type II Errors,* Water Resources Bull., v. 7, no. 3, pp. 592-604.

DUTCHER, L. C., 1972, *Proposed Criteria for Design of a Data Collection System for Ground-Water Hydrology in California 1970-2000,* Water Resources Res., v. 8, no. 1, pp. 188-193.

EDWARDS, M. D., 1974, *The Processing and Storage of Water Quality in the National Water Data Storage and Retrieval System,* U. S. Geol. Survey, Water Resources Division, Reston, Virginia, 85 pp.

FERRARA, R. and NOLAN, R. L., 1973, *New Look at Computer Data Entry,* Journal of Systems Management, Association for Systems Management, pp. 24-33, Feb.

FOSTER, K. E. and DeCOOK, J., 1974, *Implementation of Arizona Water Information System (AWIS) Remote Terminal Accessible Hydrologic Data Sets on DEC-10 Computer,* University of Arizona, Tucson, Arizona, 21 pp.

FRIEDRICHS, D. R., 1972, *Information Storage and Retrieval System for Well Hydrograph Data-User's Manual,* Battelle Pacific Northwest Laboratories, Richland, Washington, 23 pp.

GILLILAND, J. A. and TREICHEL, A., 1968, *GOWN — A Computer Storage System for Groundwater Data,* Canadian Journal of Earth Sciences, v. 5, pp. 1518-1524, Sept.

GILLILAND, J. A., 1972, *Principles of Groundwater Data Acquisition,* Water Resorces Res., v. 8, no. 1, pp. 182-187, Feb.

GUENTHER, G., MINCAVAGE, D. and MORLEY, F., 1973, *Michigan Water Resources Enforcement and Information System,* U. S. Environmental Protection Agency, Office of Research and Monitoring, Socioeconomic Environmental Studies Series, EPA-R5-73-020, Washington, D. C., 161 pp.

HICKEY, J. J., 1972, *Important Considerations in the Process of Designing a Groundwater Data Collection Program,* Water Resources Res, v. 8, no. 1, pp. 178-181, Feb.

47

HOUSE, W. C., (ed.), 1974, *Data Base Management,* Mason and Lipscomb Publishers, Inc., New York, New York, 470 pp.

HOWEN, C. R., and BAKER, C., 1977, *Computer Processing of Ground-Water Data,* in *Ground-Water Studies — An International Guide for Research and Practice,* Supplement No. 3, UNIPUB, Box 433, Murray Hill Station, New York, New York 10016, 5 pp.

JOHNSON, A. I., 1972, *Symposium on Planning and Design of Groundwater Data Programs — Introduction,* Water Resources Res., v. 8, no. 1, pp. 177-241.

JOHNSON, S.B., 1986, *Development of an Inexpensive Data Acquisition System,* Ground Water Monitoring Rev., v. 6, no. 2, pp. 106-107.

KENT, D. C., and others, 1973, *An Approach to Hydrogeologic Investigations of River Alluvium by the Use of Computerized Data Processing Techniques,* Ground Water, v. 11, no. 4,. pp. 30-41, Jul-Aug.

McNELLIS, J. M., and others, 1968, *Digital Computer Applications that Facilitate Collection and Interpretation of Ground-Water Data,* Reprint from Tucson Symp., Dec. 1968, U. S. Geol. Survey Lawrence, Kansas.

MERCER, M. W., and MORGAN, C. O., 1981, *Storage and Retrieval of Ground-Water Data at the U. S. Geological Survey,* Ground Water, v. 13, no. 5, pp. 543-551.

PETERS, H. J., 1972, *Criteria for Groundwater Level Data Networks for Hydrologic and Modeling Purposes,* Water Resources Res., v. 8, no. 1, pp. 194-200.

PETTYJOHN, W. A., 1967, *Evaluation of Basic Data and a Variety of Techniques Needed in Hydrologic Systems Analysis,* in *System Approach to Water Quality in the Great Lakes,* Proc. Am. Symp. Water Resources Res., Ohio State Univ. Water Resorce Center, 3d.

PFANNKUCH, H. O., 1976, *Study of Criteria and Models Establishing Optimum Levels of Hydrogeologic Information for Ground-Water Basin Management,* NTIS Accession No. PB-244 719.

PRICE, W. E. and BAKER, C. H., 1974, *Catalog of Aquifer Names and Geologic Unit Codes Used by the Water Resources Division,* U. S. Geol. Survey, Water Resources Division, Reston, Virginia, 306 pp.

SETMIRE, J.G., 1985, *A Conceptual Ground-Water Quality Monitoring Network for San Fernando Valley, California,* U.S. Geol. Survey Water Resources Inv. Rept. 84-4128, USGS Sacramento or San Diego, California, 49 pp.

STALLMAN, R. W., 1972, *Data Needs for Predicting Problems Caused by the Use of Subsurface Reservoirs,* Water Resources Res., v. 8, no. 1, pp. 238-241.

TAKASAKI, K. J., 1977, *Elements Needed in Design of a Ground-Water Quality Monitoring Network in the Hawaiian Islands,* U. S. Geol. Survey, Water-Supply Paper 2041, Washington, D. C., 23 pp.

TEMPLIN, W.E., 1985, *Regional Ground-Water-Quality Network Design,* Proc. Symp., Groundwater Contamination and Reclamation, A.W.R.A. Tucson, Arizona, Aug., 1985, pp. 37-44.

U. S. ENVIRONMENTAL PROTECTION AGENCY, 1971, *Storage and Retrieval of Water Quality Data Training Manual,* PB-214, 580, Washington, D. C. 302 pp.

U.S. GEOLOGICAL SURVEY, 1973, *Recommended Methods for Water-Data Acquisition,* Office of Water Data Coordination, U. S. G. S., 2100 M St., N.W., Washington, D. C. 20244.

U. S. SUBCOMMITTEE ON HYDROLOGY, INTER-AGENCY COMMITTEE ON WATER RESOURCES, 1966, *Inventory of Federal Sources of Ground-Water Data,* Bull. no. 12, Washington, D. C., 294 pp.

WINNER, M.D., Jr., 1981, *An Observation Well Network as Applied to North Carolina,* NTIS, PB-82 106089, Springfield, Virginia 22161.

WINTER, T. C., 1972, *An Approach to the Design of Statewide or Regional Ground-Water Information Systems,* Water Resources Res., v. 8, no. 1, pp. 222-230, Feb.

GEOPHYSICAL INVESTIGATIONS

GENERAL WORKS

ADAMS, W. M., 1969, *Geophysical Studies for Volcanological Geohydrology,* Hawaii Water Resources Res. Center Contrib. no. 18.

ANONYMOUS, 1971, *Geophysics and Ground Water,* Water Well Jour., v. 25, no. 7, pp. 43-60, no. 8, pp. 35-50.

ASTIER, J. L., 1971, *Géophysique Appliquée à l'Hydrogéologie,* Masson, Paris, 270 pp.

BAYS, C. A. and FOLK, S. H., 1944, *Developments in the Application of Geophysics to Ground-Water Problems,* Illinois Geol. Survey Circ. 108, Urbana, 25 pp.

BAYS, C. A., 1950, *Prospecting for Ground Water-Geophysical Methods,* J. Am. Water Works Assoc., v. 42, pp. 947-956.

BREUSSE, J. J., 1963, *Modern Geophysical Methods for Subsurface Water Exploration,* Geophysics, v. 28, pp. 633-657.

COSTELLO, R.L., 1985, *Field Testing of Geophysical Techniques*, NTIS, AD-A123-996/1, 5285 Port Royal Rd., Springfield, Virginia 22161, 63 pp.

COSTELLO, R.L., 1985, *Identification and Description of Geophysical Techniques*, AD-A123-939/1, NTIS, 5285H Port Royal Rd., Springfield, Virginia 22161, 231 pp.

DOBRIN, M. B., 1965, *Introduction to Geophysical Prospecting,* 3rd ed. McGraw-Hill, New York 583 pp.

FOSTER, J. W., and BUHLE, M. B., 1951, *An Integrated Geophysical and Geological Investigation of Aquifers in Glacial Drift near Champaign-Urbana, Illinois,* Econ. Geol., v. 46, pp. 368-397.

FRISCHKNECHT, F.C., and RAAB, P.V., 1984, *Location of Abandoned Wells with Geophysical Methods,* Env. Monitoring Systems Lab., Las Vegas, NV, PB85-122638/WEP, NTIS, Springfield, Virginia 22161, 59 pp.

GRANT, F. S. and WEST, G. F., 1965, *Interpretation Theory in Applied Geophysics,* McGraw-Hill, New York, 583 pp.

GRIFFITHS, D. H. and KING, R. F., 1966, *Applied Geophysics for Engineers and Geologists,* Pergamon Press, New York, 223 pp.

HALL, D. H., and HAJUAL, Z., 1962, *The Gravimeter in Studies of Buried Valleys,* Geophysics, v. 27, no. 6, pt. 2, pp. 939-951.

HATHERTON, T., and others, 1966, *Geophysical Methods in Geothermal Prospecting in New Zealand,* Bul. Volcanol, v. 29, pp. 485-498.

HEILAND, C. A., 1937, *Prospecting for Water with Geophysical Methods,* Trans. Am. Geophys. Union, v. 18, pp. 574-588.

HEILAND, C. A., 1946, *Geophysical Exploration,* Prentice-Hall Inc., New York, 1013 pp.

IBRAHIM, A., and HINZE, W. J., 1972, *Mapping Buried Bedrock Topography with Gravity,* Ground Water, v. 10, no. 3, pp. 10-17, May-June.

JAKOSKY, J. J., 1950, *Exploration Geophysics,* Trija Publ. Co., Los Angeles, 1195 pp.

JOINER, T. J., and SCARBROUGH, W. L., 1969, *Hydrology of Limestone Terranes — Geophysical Investigations,* Univ. of Alabama, Bull. 94, part D.

JONES, P. H., and SKIBITZKE, H. E., 1956, *Subsurface Geophysical Methods in Ground-Water Hydrology,* Advances in Geophys., v. 3, pp. 241-300.

LEE, F. W., 1936, *Geophysical Prospecting for Underground Waters in Desert Areas,* Inf. Circ. 6899, U. S. Bureau Mines, Washington, D. C., 27 pp.

McDONALD, H. R., and WANTLAND, D., 1961, *Geophysical Procedures in Ground Water Study,* Trans. Am. Soc. Civil Engrs., v. 126, pt. III, pp. 122-135.

McGINNIS, L. D., and KEMPTON, J. P., 1961, *Integrated Seismic, Resistivity, and Geologic Studies of Glacial Deposits,* Illinois Geol. Survey Circ. 323, 23 pp.

McGINNIS, L. D., and others, 1963, *Relationship of Gravity Anomalies to a Drift-Filled Bedrock Valley System in Northern Illinois,* Ill. State Geol. Survey Circ. 354, 23 pp.

Geophysics

MEINZER, O. E., 1937, *The Value of Geophysical Methods in Ground-Water Studies,* Trans. Am. Geophys. Union, v. 18, pp. 385-387.

MORLEY, L. W. (ed.), 1969, *Mining and Groundwater Geophysics/1967,* Econ. Geol. Rept. 26, Dept. of Energy, Mines and Resources, Ottawa, Canada, 722 pp.

NATIONAL WATER WELL ASSOCIATION, 1971, *Geophysics and Ground Water: A Primer, Part 1.* Water Well J., Jul., pp. 43-60, *Part 2,* Water Well J., Aug., pp. 35-50.

NATIONAL WATER WELL ASSOCIATION, 1985, *Proceedings of the February 1984 Conference on Surface and Borehole Geophysical Methods in Ground-Water Investigations,* NWWA, Dublin, Ohio 43017.

NATIONAL WATER WELL ASSOCIATION, 1985, *Surface and Borehole Geophysical Methods in Ground Water Investigations,* Proc. 2nd Nat. Conf., February, 1985, Fort Worth, Texas, NWWA, Dublin, Ohio 43017, 415 pp.

PIN, F.G., and KETELLE, R.H., 1984, *Mapping Subsurface Pathways for Contaminant Migration at a Proposed Low Level Waste Disposal Site Using Electromagnetic Methods,* CONF-840245-5, Oak Ridge National Lab., TN; DE84007538, NTIS, Springfield, Virginia 22161, 9 pp.

ROBERTSHAW, J., and BROWN, P. D., 1955, *Geophysical Methods of Exploration and their Application to Civil Engineering Problems,* Proc. Inst. Civil Engrs., pt. 1, v. 4, pp. 644-690.

SHAW, S. H., 1963, *Some Aspects of Geophysical Surveying for Ground Water,* J. Inst. Water Engrs., London, v. 17, pp. 175-188.

SMITH, W. O., and NICHOLS, H. B., 1953, *Mapping Water — Saturated Sediments by Sonic Methods,* Sci. Monthly, v. 77, no. 1, July.

SPANGLER, D. P., and LIBBY, F. J., 1968, *Application of the Gravity Survey Method to Watershed Hydrology,* Ground Water, v. 6, no. 6, pp. 21-27.

STEWART, M. T., 1982, *Evaluation of Electromagnetic Methods for Rapid Mapping of Salt-Water Interfaces in Coastal Aquifers,* Ground Water, v. 20, no. 5, pp. 538-545.

STICKEL, J. F., Jr., and others, 1952, *Geophysics and Water,* J. Am. Water Works Assoc., v. 44, pp. 23-35.

TODD, D. K., 1955, *Investigating Ground Water by Applied Geophysics,* Proc. Am. Soc. Civil Engrs., v. 81, 625, 14 pp.

WEST, R. E., and SUMNER, J. S., 1972, *Ground-Water Volumes from Anomalous Mass Determinations for Alluvial Basins,* Ground Water, v. 10, no. 3, pp. 18-23, May-June.

WILSON, G. V., and others, 1970, *Evaluation, by Test Drilling, of Geophysical Methods Used for Ground-Water Development in the Piedmont Area, Alabama,* Geol. Survey of Alabama Circ. 65, 15 pp.

WOOLLARD, G. P., and HANSON, G. F., 1954, *Geophysical Methods Applied to Geologic Problems in Wisconsin,* Wisc. Geol. Survey Bull. 78, Sci. ser. 15, 255 pp.

YAZICIGIL, HASAN, and SENDLEIN, L. V. A., 1982, *Surface Geophysical Techniques in Ground-Water Monitoring: Part II* Ground Water Monitoring Review, v. 2, no. 1, pp. 56-62.

ZOHDY, A. A. R., EATON, G. P., and MABEY, D. R., 1974, *Application of Surface Geophysics to Ground-Water Investigations,* U. S. Geol. Survey Techniques of Water Resources Inv. Chap. D1, Bk.2, 116 pp.

RESISTIVITY METHODS

BARNES, H. E., 1952, *Soil Investigations Employing a New Method of Layer-Value Determination for Earth Resistivity Interpretation,* Highway Res. Bd. Bull, 65, pp. 26-36.

BHATTACHARYA, P. K., and PATRA, H. P., 1969, *Direct Current Geoelectric Sounding,* American Elsevier Publ. Co., Inc., New York, 135 pp.

BUGG, S. F. and LLOYD, J. W., 1976, *A Study of Fresh Water Lens Configuration on the Cayman Islands Using Resistivity Methods,* Quarterly Jour. Engrng. Geol., v. 9, pp. 291-302.

BUHLE, M. B., 1953, *Earth Resistivity in Ground-Water Studies in Illinois,* Trans. Am. Inst. Mining Engrs., Tech. Paper 3496L, Mining Eng., pp. 395-399.

CARRINGTON, T. J., and WATSON, D. A., 1981, *Preliminary Evaluation of an Alternate Electrode Array for Use in Shallow-Subsurface Electrical Resistivity Studies,* Ground Water, v. 19, no. 1, Jan.-Feb., pp. 48-57.

CARTWRIGHT, KEROS, and McCOMAS, M. R., 1968, *Geophysical Surveys in the Vicinity of Sanitary Landfills in Northeastern Illinois,* Ground Water, v. 6, no. 5, Sept.-Oct., pp. 23-31.

CARTWRIGHT, KEROS, and SHERMAN, F. B., 1972, *Electrical Earth Resistivity Surveying in Landfill Investigations,* Illinois Geol. Survey Reprint Series 1972-U.

COMPAGNIE GENERALE DE GEOPHYSIQUE, 1963, *Master Curves for Electrical Sounding,* European Assoc. of Exploration Geophysicists, The Hague, The Netherlands, 49 pp.

CONWELL, C. N., 1951, *Application of the Electrical Resistivity Method to Delineation of Areas of Seepage Along a Canal — Wyoming Canal — Riverton Project,* Geol. Rep. G-114, U. S. Bureau Reclamation, Denver, Colorado, 10 pp.

DIZIOGLU, M. Y., 1953, *Underground-Water Investigations by Means of Geophysical Methods (Particularly Electrical) in Central Anatolia,* Proc. Ankara Symp. on Arid Zone Hydrology, pp. 199-215, UNESCO, Paris.

FLATHE, H., 1955, *A Practical Method of Calculating Geoelectrical Model Graphs for Horizontally Stratified Media,* Geophys. Prospecting, v. 3, pp. 268-294.

FLATHE, H., 1963, *Five-Layer Master Curves for the Hydrogeological Interpretation of Geoelectrical Resistivity Measurements above a Two-Story Aquifer,* Geophys. Prospecting, v. 11, pp. 471-508.

FRETWELL, J. D., and STEWART, M. T., 1981, *Resistivity Study of a Coastal Karst Terrain, Florida,* (study of salt-water interface), Ground Water, v. 19, no. 2, March-April, pp. 156-162.

FROHLICH, R. K., 1973, *Detection of Fresh-Water Aquifers in the Glacial Deposits of North-Western Missouri by Geoelectrical Methods,* Water Resources Bull., v. 9, no. 4, p. 783.

GISH, O. H., and ROONEY, W. J., 1925, *Measurement of Resistivity of Large Masses of Undisturbed Earth,* Terrestrial Magnetism, v. 30, no. 4, pp. 161-188.

HACKBARTH, D. A., 1971, *Field Study of Subsurface Spent Sulfite Liquor Movement Using Earth Resistivity Measurements,* Ground Water, v. 9, no. 3, pp. 11-16.

HACKETT, J. E., 1956, *Relation between Earth Resistivity and Glacial Deposits near Shelbyville, Illinois,* Ill. State Geol. Survey Circ. 223.

HALLENBECK, F., 1953, *Geo-Electrical Problems of the Hydrology of West German Area,* Geophys. Prospecting, v. 1, pp. 241-249.

JONES, P. H., and BUFORD, T. B., 1951, *Electric Logging Applied to Groundwater Exploration,* Geophysics, v. 6., no. 1.

KELLER, G. V., and FRISCHKNECHT, F. C., 1966, *Electrical Methods in Geophysical Prospecting,* Permagon, Oxford, 517 pp.

KELLY, S. F., 1962, *Geophysical Exploration for Water by Electrical Resistivity,* J. New England Water Works Assoc., v. 76, pp. 118-189.

KELLY, W. E., 1976, *Geoelectric Sounding for Delineating Ground-Water Contamination,* Ground Water, v. 14, no. 1, pp. 6-10.

KELLY, W. E., 1977, *Geoelectric Sounding for Estimating Aquifer Hydraulic Conductivity,* Ground Water, v. 15, no. 6, pp. 420-425.

KLEFSTAD, G., SENDLEIN, L. V. A., and PALMQUIST, R. C., 1975, *Limitations of the Electrical Resistivity Method in Landfill Investigations,* Ground Water, v. 13, no. 5, pp. 418-427.

KOSINSKI, W. K., and KELLY, W. E., 1981, *Geoelectric Soundings for Predicting Aquifer Properties,* Ground Water, v. 19, no. 2, March-April, pp. 163-171.

LANDES, K. K., and WILSON, J. T., 1943, *Ground-Water Exploration by Earth-Resistivity Methods,* Papers Mich. Acad. Arts, Sci., Let., v. 29, pp. 345-354.

MERKEL, R. H., 1972, *The Use of Resistivity Techniques to Delineate Acid Mine Drainage in Ground Water,* Ground Water, v. 10, no. 5, pp. 38-42, Sep.-Oct.

MERKEL, R. H., and KAMINSKI, J. T., 1972, *Mapping Ground Water by Using Electrical Resistivity with a Buried Concrete Source,* Ground Water, v. 10, no. 2, pp. 18-25.

MOONEY, H. M., and WETZEL, W. W., 1956, *The Potentials About a Point Electrode and Apparent Resistivity Curves for a Two-, Three-, and Four- Layered Earth,* Univ. Minn. Press, Minneapolis, 146 pp.

MOORE, R. W., 1957, *Applications of Electrical Resistivity Measurements to Subsurface Investigations,* Public Roads, v. 29, no. 7.

MORRIS, D. B., 1964, *The Application of Resistivity Methods to Groundwater Exploration of Alluvial Basins in Semi-Arid Areas,* Jour. Instn. Water Engrs., v. 18, pp. 59-65.

ORELLANA, ERNESTO, 1961, *Criterios Eróneos en la Interpretación de Sondeos Eléctricos:* Revista de Geofísica, v. 20, pp. 207-227.

ORELLANA, ERNESTO and MOONEY, H. M., 1966, *Master Tables and Curves for Vertical Elecrical Sounding over Layered Structures,* INTERCIENCIA, Madrid, 150 pp., 66 tables.

ORELLANA, ERNESTO, and MOONEY, H. M., 1972, *Two-Layer and Three-Layer Master Curves and Auxiliary Point Diagrams for Vertical Electrical Sounding Using Wenner Electrode Arrangement,* INTERCIENCA, Madrid -13.

PAGE, L. M., 1968, *Use of the Electrical Resistivity Method for Investigating Geologic and Hydrologic Conditions in Santa Clara County, California,* Ground Water, v. 6, no. 5, pp. 31-40.

PAVER, G. L., 1945, *On the Application of the Electrical Resistivity Method of Geophysical Surveying to the Location of Underground Water, with Examples from the Middle East,* Proc. Geol. Soc. London, pp. 46-51, Apr.

PAVER, G. L., 1948, *Iso-Resistivity Mapping for the Investigation of Underground Water Supplies,* General Assembly Oslo, Internat. Assoc. Sci. Hydrology, v. 3, pp. 290-295.

PAVER, G. L., 1950, *The Geophysical Interpretation of Underground Water Supplies, a Geological Analysis of Observed Resistivity Data,* J. Inst. Water Engrs., v. 4, pp. 237-266.

PRIDDY, R. R., 1955, *Fresh-Water Strata of Mississippi as Revealed by Electric Studies,* Miss. Geol. Survey Bull. 83.

ROGERS, R. B., and KEAN, W. F., 1980, *Monitoring Ground-Water Contamination at a Fly Ash Disposal Site Using Surface Electrical Resistivity Methods,* Ground Water, v. 18, no. 5, Sept.-Oct., pp. 472-478.

ROMAN, IRWIN, 1952, *Resistivity Reconnaissance,* Am. Soc. Testing Mater. Spec. Tech. Pub. 122.

SAYRE, A. N., and STEPHENSON, E. L., 1937, *The Use of Resistivity-Methods in the Location of Salt-Water Bodies in the El Paso, Texas Area,* Trans. Am. Geophys. Union, v. 18, pp. 393-398.

SPICER, H. C., 1952, *Electrical Resistivity Studies of Subsurface Conditions near Antigo, Wisconsin,* U. S. Geol. Survey Circ. 181, 19 pp.

STALLMAN, R. W., 1963, *Type Curves for the Solution of Single-Boundary Problems,* U. S. Geol. Survey Water Supply Paper 1545-C, pp. 45-47.

STOLLAR, R. L., and ROUX, PAUL, 1975, *Earth Resistivity Surveys — A Method for Defining Ground-Water Contamination,* Ground Water, v. 13, no. 2, pp. 145-150.

SWARTZ, J. H., 1937, *Resistivity Studies of some Salt-Water Boundaries in the Hawaiian Islands,* Trans. Am. Geophys. Union, v. 18, pp. 387-393.

SWARTZ, J. H., 1939, *Geophysical Investigations in the Hawaiian Islands,* Trans. Am. Geophys. Union, v. 20, pp. 292-298.

SWEENEY, J.T., 1985, *Comparison of Electrical Resistivity Methods for Investigation of Ground Water Conditions at a Landfill Site,* Ground Water Monitoring Rev., v. 4, no. 1, pp. 52-59.

TAGG, G. F., 1934, *Interpretation of Resistivity Measurements,* Trans. Am. Inst. Mining Met. Eng., Geophys. Prospecting, pp. 135-145.

TATTAM, C. M., 1937, *The Application of Electrical Resistivity Prospecting to Ground Water Problems,* Colorado School of Mines Quart., v. 32, pp. 117-138.

URISH, D.W., 1983, *The Practical Application of Surface Electricity Resistivity to Detection of Ground-Water Pollution,* Ground Water, v. 21, no. 2, pp. 144-152.

VACQUIER, V., and others, 1956, *Prospecting for Ground Water by Induced Electrical Polarization,* New Mexico Inst. of Mining and Technology, Res. and Devel. Div., 41 pp.

VAN DAM, J. C., 1955, *Geo-Electrical Investigations in the Delta Area of the Netherlands,* Serv. for Water Resources Dev. (Rijkswaterstaat), The Netherlands.

VAN NOSTRAND, R. G., and COOK, K. L., 1966, *Interpretation of Resistivity Data,* U. S. Geol. Survey Prof. Paper 499, 310 pp.

VOLKER, ADRIAN, and DIJKSTRA, J., 1955, *Détermination des Salinités des Eaux dans le Sous-Sol du Zuiderzee par Prospection Géophysique,* Geophys. Prospecting, v. 3, pp. 111-125.

VOLKER, ADRIAN, and VAN DAM, J. C., 1954, *Geo-Elektrisch Onderzoek bij Uitvoering van Waterbouwkundige Werken,* Serv. for Water Resources Dev. (Rijkswaterstaat), The Netherlands.

WARNER, D., 1969, *Preliminary Field Studies Using Earth Resistivity Measurements for Delineating Zones of Contaminated Ground Water,* Ground Water, v. 7, no. 1, pp. 9-16.

WAY, H. J. R., 1942, *An Analysis of the Results of Prospecting for Water in Uganda by the Resistivity Method,* Trans. Inst. Min. and Met. Engrs., v. 51, pp. 285-310.

WENNER, F., 1916, *A Method of Measuring Earth-Resistivity,* Bull. Bureau Standards, v. 12, Washington, D. C., pp. 469-478.

WETZEL, W. W., and McMURRY, H. V., 1937, *A Set of Curves to Assist in the Interpretation of the Three-Layer Resistivity Problem,* Geophysics, v. 2, pp. 329-341.

WORKMAN, L. E., and LEIGHTON, M. M., 1937, *Search for Ground Water by the Electrical Resistivity Method,* Trans. Am. Geophys. Union, v. 18, pp. 403-409.

SEISMIC METHODS

BERSON, I. S., and others, 1959, *Wave Refraction by Aquiferous Sands,* U.S.S.R. Acad. Sci. Bull., Geophysics Ser., Jan. and Feb. (English transl. by the Am. Geophys. Union), pp. 17-29 and pp. 115-118.

BONINI, W. E., 1959, *Seismic-Refraction Method in Ground-Water Exploration,* Trans. Am. Inst. Mining Metall. Engrs., v. 211, pp. 485-488.

BURWELL, E. B., Jr., 1940, *Determination of Ground-Water Levels by the Seismic Method,* Trans. Am. Geophys. Union, v. 21, pp. 439-440.

DUGUID, J. O., 1968, *Refraction Determination of Water Table Depth and Alluvium Thickness,* Geophysics, v. 33, no. 3, pp. 481-488.

HAENI, F.P., 1986, *Application of Continuous Seismic Reflection Methods to Hydrologic Studies,* Ground Water, v. 24, no. 1, pp. 23-31.

JOHNSON, R. B., 1954, *Use of the Refraction Seismic Method for Differentiating Pleistocene Deposits in the Arcola and Tuscola Quadrangles, Illinois,* Ill. State Geol. Survey Rept. Invest. 176.

LINEHAN, D., 1951, *Seismology Applied to Shallow Zone Research,* Am. Soc. Test. Materials, Spec. Tech. Pub. 122, pp. 156-170.

LINEHAN, D., and KEITH, S., 1949, *Seismic Reconnaissance for Ground Water Development,* J. New England Water Works Assoc., v. 63, no. 1, pp. 76-95.

McQUILLIN, R., BACON, M., and BARCLAY, W., 1985, *An Introduction to Seismic Interpretation,* Gulf Publ. Co., P.O. Box 2608, Houston, Texas 77001, 256 pp.

MEIDAV, T., 1968, *A Multilayer Seismic Refraction Nomogram,* Geophysics, v. 33, pp. 524-526.

MISSIMER, T. M., and GARDNER, R. A., 1976, *High Resolution Seismic Reflection Profiling for Mapping Shallow Aquifers in Lee County, Florida,* U. S. Geol. Survey Water-Resour. Invest., 74-65, 35 pp.

MUSGRAVE, A. W. (ed.), 1967, *Seismic Refraction Prospecting,* Soc. Explor. Geophysicists, Tulsa, 604 pp.

SANDER, J. E., 1978, *The Blind Zone in Seismic Ground-Water Exploration,* Ground Water, v. 16, no. 6, Nov.-Dec., pp. 394-398.

SHEPARD, E. R., and WOOD, A. E., 1940, *Application of the Seismic Refraction Method of Subsurface Exploration to Flood-Control Projects,* Trans. Am. Inst. Min. and Met. Engrs., v. 138, pp. 312-325.

WARRICK, R. E., and WINSLOW, J. D., 1960, *Application of Seismic Methods to a Ground-Water Problem in Northeastern Ohio,* Geophysics, v. 25, no. 2, pp. 505-519.

WELL LOGGING AND SOIL SAMPLING

BIBLIOGRAPHY

TAYLOR, T. A., and DEY, J. A., 1985, *Bibliography of Borehole Geophysics as Applied to Ground-Water Hydrology,* U.S. Geol. Survey Circ. 0926, 62 pp.

GENERAL WORKS

ACKER, W. L., 1974, *Basic Procedures for Soil Sampling and Core Drilling,* Acker Drill Co., Inc., Scranton, Pennsylvania.

ALGER, R.P., 1966, *Interpretation of Electric Logs in Freshwater Wells in Unconsolidated Formations,* Soc. Prof. Well Log Analysts, 7th Ann. Logging Symp., Houston, Texas, Trans. pp. CC1-CC25.

ANONYMOUS, 1967, *Aquifer Evaluation with Radioisotope Well Logs,* Proc. Am. Water Works Assoc., Ser. no. 4, pp. 319-328.

ARCHIE, G. E., 1942, *The Electrical Resistivity Log as an Aid in Determining some Reservoir Characteristics,* Trans. Am. Inst. Min. and Met. Engrs., v. 146, pp. 54-62.

ASQUITH, GEORGE, and GIBSON, CHARLES, 1982, *Basic Well Log Analysis for Geologists,* Cat. no. 617, Am. Assoc. Petroleum Geol., P.O. Box 979, Tulsa, Oklahoma 74101, 216 pp. (geophysical logging).

BAFFA, J. J., 1948, *The Utilization of Electrical and Radioactivity Methods of Well Logging for Ground-Water Supply Development,* J. New England Water Works Assoc., v. 62, pp. 207-219.

BARLITT, H. R., 1976, *How to Get Good Samples by Rotary Drilling,* The Johnson Drillers Journal, Jan-Feb.

BARNES, B. A., and LIVINGSTON, P., 1947, *Value of the Electrical Log for Estimating Ground-Water Supplies and the Quality of Ground Water,* Trans. Am. Geophys. Union, v. 28, pp. 903-911.

BARTH, D.S., and STARKS, T.H., 1985, *Sediment Sampling Quality Assurance User's Guide*, Nevada Univ., PB85-233542/WEP, NTIS, Springfield, Virginia 22161, 129 pp.

BARTON, C. M., 1974, *Borehole Sampling of Saturated Uncemented Sands and Gravels*, Ground Water, v. 12, no. 3, pp. 170-181.

BATEMAN, R.M., 1984, *Log Quality Control*, IHRDC Press, 137 Newbury St., Boston, Massachusetts 02116, 386 pp.

BENNETT, G. D., and PATTEN, E. P. Jr., 1960, *Borehole Geophysical Methods for Analyzing Specific Capacity of Multiaquifer Wells*, U. S. Geol. Survey Water-Supply Paper 1536-A, 25 pp.

BIRDWELL, 1973, *Geophysical Well Log Interpretation, Birdwell Div.*, P. O. Box 1590, Tulsa, Oklahoma 74102.

BLACK, W.H., and others, 1983, *The Piezometric Permeability Profiler*, Ground Water Monitoring Rev., v. 3, no. 4, pp. 17-25.

BLANKENNAGEL, R. K., 1968, *Geophysical Logging and Hydraulic Testing*, Pahute Mesa, Nevada Test Site, Ground Water, v. 6, no. 4, pp. 24-32.

BREDEHOEFT, J. D., and PAPADOPULOS, I. S., 1965, *Rates of Vertical Ground Water Movement Estimated from Earth's Thermal Profile*, Water Resources Res., v. 1, pp. 325-328, (temperature logging).

BROWN, D. L., 1971, *Techniques for Quality-of-Water Interpretations from Calibrated Geophysical Logs, Atlantic Coast Area*, Ground Water, v. 9, no. 4, pp. 25-38.

BRYAN, F. L., 1950, *Application of Electric Logging to Water Well Problems*, Water Well J., v. 4, no. 1, pp. 3-7.

CALLAHAN, J. T., and others, 1963, *Television — A New Tool for the Ground-Water Geologist*, Ground Water, v. 1, no. 4, pp. 4-6.

CROFT, M. G., 1971, *A Method for Calculating Permeability from Electric Logs*, U. S. Geol. Prof. Paper 750-B, pp. B265-B269.

CROSBY, J. W. III, and ANDERSON, J. V., 1971, *Some Applications of Geophysical Well Logging to Basalt Hydrogeology*, Ground Water, v. 9, no. 5, pp. 12-20, Sep.-Oct.

DESBRANDES, ROBERT , 1985, *Enclyclopedia of Well Logging*, Gulf Pub. Co., Book Div., P.O. Box 2608, Houston, Texas 77001, 584 pp.

DOLL, H. G., 1949, *The S. P. Log: Theoretical Analysis and Principles of Interpretation,* Trans. Am. Inst. Min. and Met. Engrs., v. 179, pp. 146-185.

ERICKSON, C. R., 1946, *Vertical Water Velocity in Deep Wells,* J. Am. Water Works Assoc., v. 38, pp. 1263-1272.

FERTL, W. H., and others, 1974, *A Look at Cement Bond Logs,* Journal Petroleum Technology, v. 26, pp. 607-617.

FIEDLER, A. G., 1928, *The Au Deep-Well Current Meter and its Use in the Roswell Artesian Basin,* New Mexico, U. S. Geol. Survey Water Supply Paper 596, pp. 24-32.

FRIMPTER, M. H., 1969, *Casing Detector and Self-Potential Logger,* Ground Water, v. 7, no. 5, pp. 24-27, Sep.-Oct.

GORDER, Z. A., 1963, *Television Inspection of a Gravel Pack Well,* J. Am. Water Works Assoc., v. 55, pp. 31-34.

GROSMANGIN, M., and others, 1961, *A Sonic Method for Analyzing the Quality of Cementation of Borehole Casings,* J. Petroleum Technology, pp. 165-171, Feb.

GUYOD, HUBERT, 1952, *Electrical Well Logging Fundamentals,* Houston, 164 pp.

GUYOD, HUBERT, and PRANGLIN, J. A., 1959, *Analysis Charts for the Determination of True Resistivity from Electric Logs,* Houston, 202 pp.

GUYOD, HUBERT, 1965, *Interpretation of Electric and Gamma Ray Logs in Water Wells,* Am. Geophys. Union, Tech. Paper, Mandrel Industries Inc., Houston, Texas.

HESS, A.E., 1984, *Use of a Low Velocity Borehold Flowmeter in the Study of Hydraulic Conductivity in Fractured Rock, in Surface and Borehole Geophysical Methods in Ground Water Investigations,* (D.M. Nielsen, ed.), pp. 812-832. National Water Well Assoc. Dublin, Ohio.

HILCHIE, D.W., 1982, *Advanced Well Log Interpretation,* Douglas W. Hilchie, Inc., Golden Colorado.

IHD WORKING GROUP ON NUCLEAR TECHNIQUES IN HYDROLOGY, 1971, *Nuclear Well Logging in Hydrology,* Tech. Rep. Series No. 126, STI/DOC/10/126, Internat. Atomic Energy Agency, Vienna.

INESON, J., and GRAY, D. A., 1963, *Electrical Investigations of Borehole Fluids,* J. Hydrology, v. 1, pp. 204-218.

INTERNATIONAL ATOMIC ENERGY AGENCY, 1971, *Nuclear Well Logging in Hydrology,* Tech. Repts. Ser. 126, 90 pp., UNIPUB, Inc., P. O. Box 433, New York, New York 10016.

JOHNSON, A. I., 1968, *An Outline of Geophysical Logging Methods and their Uses in Hydrological Studies,* U. S. Geol. Survey Water-Supply Paper 1892, pp. 158-164.

JONES, P. H., and BUFORD, T. D., 1951, *Electrical Logging Applied to Ground-Water Exploration,* Geophysics, v. 16, no. 1.

JONES, P.H., and SKIBITZKE, H. E., 1956, *Subsurface Geophysical Methods in Ground-Water Hydrology,* in *Advances in Geophysics* (H. E. Landsberg, ed.), v. 3, Academic Press, New York, pp. 241-300.

KELLY, D. R., 1969, *A Summary of Major Geophysical Logging Methods,* Bull. M61, Pennsylvania Geol. Survey, Harrisburg, 82 pp.

KELLY, W. E., and FROHLICH, R.K., 1985, *Relations Between Aquifer Electrical and Hydraulic Properties,* Ground Water, v. 23, no. 2, pp. 182-189.

KELLY, W. E., and REITER, P.F., 1984, *Influence of Anisotropy on Relations Between Aquifers Hydraulic and Electrical Properties,* J. Hydrology, v. 3/4.

KEYS, W. S., 1968, *Well Logging in Ground-Water Hydrology,* Ground Water, v. 6, no. 1, pp. 10-19.

KEYS, W. S., and BROWN, R. F., 1971, *The Use of Well Logging in Recharge Studies of the Ogallala Formation in West Texas,* U. S. Geol. Survey Prof. Paper 750-B, pp. B270-277.

KEYS, W. S., and BROWN, R. F., 1974, *Role of Borehole Geophysics in Underground Waste Storage and Artificial Recharge,* Underground Waste Management and Artificial Recharge, Am. Assoc. Petrol. Geol., pp. 147-191.

KEYS, W. S., and BROWN, R. F., 1978, *The Use of Temperature Logs to Trace the Movement of Injected Water,* Ground Water, v. 16, no. 1, pp. 32-48.

KEYS, W. S., and MacCARY, L. M., 1971, *Application of Borehole Geophysics to Water-Resources Investigations,* Ch. E 1-*Techniques of Water-Resources Investigations,* Book 2, Collection of Environmental Data, U. S. Geol. Survey, 124 pp.

KEYS, W. S., and MacCARY, L. M. 1973, *Location and Characteristics of the Interface Between Brine and Fresh Water from Geophysical Logs of Boreholes in the Upper Brazos River Basin, Texas:* U. S. Geol. Survey Prof. Paper 809-B, pp. B1-B23.

KEYS, W. S., 1984, *Synthesis of Borehole Geophysical Data at the Underground Research Laboratory, Manitoba, Canada,* BMI/OCRD-15, U.S. Geol. Survey, Denver, CO; DE84015470, NTIS, Springfield, Virginia 22161, 47 pp.

KIERSTEIN, R.A., 1984, *True Location and Orientation of Fractures Logged with Acoustic Televiewer,* WRI 83-4275, U.S. Geol. Survey, 73 pp.

KIRBY, M. E., 1954, *Improve Your Work with Drilling-Time Logs,* Johnson Drillers' J., v. 26, no. 6, pp. 6-7, 14.

KWADER, THOMAS, 1982, *Interpretation of Borehole Geophysical Logs and Their Application to Water Resources Investigations,* Florida State U., Unpubl. Dissertation.

KWADER, THOMAS, 1986, *The Use of Geophysical Logs for Determining Formation Water Quality,* Ground Water, v. 24, no. 1, pp. 11-15.

LAO, C. and others, 1969, *Application of Well Logging and Other Well Logging Methods in Hawaii,* Tech. Rept. 21, Water Resources Research Center, Univ. of Hawaii, Honolulu, 108 pp.

LIVINGSTON, P. P., and LYNCH, W., 1937, *Methods of Locating Salt-Water Leaks in Water Wells,* U. S. Geol. Survey Water-Supply Paper 796-A, 20 pp.

LOW, J. W., 1958, *Examination of Well Cuttings,* in *Subsurface Geology in Petroleum Exploration,* by Haun, J. D., and LeRoy, L. W., Colorado School of Mines, pp. 17-58.

LYNCH, E. J., 1962, *Formation Evaluation,* Harper & Row, New York, New York, 422 pp.

MacCARY, L. M., 1971, *Resistivity and Neutron Logging in Silurian Dolomite of Northwest Ohio,* U. S. Geol. Survey Prof. Paper 750-D, pp. D190-D197.

MacCARY, L. M., 1979, *Interpretation of Well Logs in a Carbonate Aquifer,* PB-288 508 NTIS, 5285 Port Royal Rd, Springfield, Virginia 22151.

MacCARY, L. M., 1980, *Use of Geophysical Logs to Estimate Water Quality Trends in Carbonate Aquifers,* NTIS PB80-224124, Springfield, Virginia 22161.

MacCARY, L. M., 1983, *Geophysical Logging in Carbonate Aquifers,* Ground Water, v. 21, no. 3, pp. 334-342.

MAHER, J. C., 1959, *The Composite Interpretive Method of Logging Drill Cuttings,* Oklahoma Geol. Survey Guide Book 8, 48 pp.

MARSH, C. R., and PARIZEK, R. R., 1963, *Induction-Tuned Method to Determine Casing Lengths in Hydrogeologic Investigations,* Ground Water, v. 6, no. 6, pp. 11-17, Nov.-Dec.

MASON, B. J., 1983, *Preparation of Soil Sampling Protocol: Techniques and Strategies,* Nevada Univ., Las Vegas, EPA-600/4-83-020, NTIS, PB83-206979, 112 pp.

MATLOCK, G. C. A., and others, 1976, *Well Cuttings Analysis in Ground-Water Resources Evaluation,* Ground Water, v. 14, no. 5, Sept.-Oct., pp. 272-277.

McCARDELL, W. M., and others, 1953, *Origin of the Electric Potential Observed in Wells,* Trans. Am. Inst. Min. and Met. Engrs., v. 198, pp. 41-50.

MEYER, W. R., 1963, *Use of a Neutron Moisture Probe to Determine the Storage Coefficient of an Unconfined Aquifer,* U. S. Geol. Survey Prof. Paper 450-E, pp. 174-176.

MORAHAN, T. J., and DORRIER, R. C., 1984, *The Application of Television Borehole Logging to Ground Water Monitoring Programs,* Ground Water Monitoring Rev., v. 4, no. 4., pp. 183-187.

MOUNCE, W. D., and RUST, W. M., Jr., 1945, *Natural Potentials in Well Logging,* Trans. Am. Inst. Min. and Met. Engrs., v. 164, pp. 288-294.

MUNCH, JOSEPH, and KILLEY, R. W. D., 1985, *Equipment and Methodology for Sampling and Testing Cohesionless Sediments,* Ground Water Monitoring Rev., v. 5, no. 1, pp. 38-42.

MURPHY, W. C., undated, *The Interpretation and Calculation of Formation Characteristics from Formation Test Data,* Halliburton Services, Duncan, Oklahoma.

MYLANDER, H. A., 1953, *Oil-Field Techniques Used for Water-Well Drilling,* J. Am. Water Works Assoc., v. 45, pp. 764-772.

NORRIS, S. E., and SPIEKER, A. M., 1966, *Ground-Water Resources of the Dayton Area, Ohio,* U. S. Geol. Survey Water-Supply Paper 1808, 167 pp. (geophysical logging).

NORRIS, S. E., 1972, *The Use of Gamma Logs in Determining the Character of Unconsolidated Sediments and Well Construction Features,* Ground Water, v. 10., no. 6, pp. 14-21, Nov.-Dec.

PAILLET, F. L., and KEYS, W. S., 1984, *Applications of Borehole Geophysics in Characterizing the Hydrology of Fractured Rocks,* in *Surface and Borehole Geophysical Methods in Ground Water Investigations,* (D. M. Nielson, ed.), pp. 743-761, National Water Well Assoc, Dublin, Ohio.

PARIZEK, R. P., and SIDDIQUI, S. H., 1970, *Determining the Sustained Yield of Wells in Carbonate and Fractured Aquifers,* Ground Water, v. 8, no. 5, pp. 12-20, (caliper logging).

PATTEN, E. P., Jr., and BENNETT, G. D., 1962, *Methods of Flow Measurement in Well Bores,* U. S. Geol. Survey Water-Supply Paper 1544-C, 28 pp.

PATTEN, E. P., Jr., and BENNETT, G. D., 1963, *Application of Electrical and Radioactive Well Logging to Ground-Water Hydrology,* U. S. Geol. Survey Water-Supply Paper 1544-D, 60 pp.

PETERSON, F. L., 1974, *Neutron Well Logging in Hawaii,* Hawaii Water Resources Res. Center Tech. Rept. No. 75.

PETERSON, F. L., and LAO, CHESTER, 1970, *Electric Well Logging of Hawaiian Basaltic Aquifers,* Ground Water, v. 8, no. 2, pp. 11-19.

PETERSOON, J. K., and others, 1985, *Well Logging in Permafrost,* Spec. Rept. Corps. Engrs, U. S. Army, Cold Regions Res. and Engrng. Lab., v. 85-5, pp. 68-70.

PICKELL, J. J., and HEACOCK, J. G., 1960, *Density Logging,* Geophysics, v. 25, pp. 891-904.

PICKETT, G. R., 1960, *The Use of Acoustic Logs in the Evaluation of Sandstone Reservoirs,* Geophysics, v. 25, pp. 250-274.

PIRSON, S. J., 1973, *Handbook of Well Log Analysis,* Prentice-Hall, Inc., Englewood Cliffs, New Jersey.

POLAND, J. F., and MORRISON, R. B., 1940, *An Electrical Resistivity Apparatus for Testing Well Waters,* Trans. Am. Geophys. Union, v. 21, pp. 35-46.

PRYOR, W. A., 1956, *Quality of Groundwater Estimated from Electric Resistivity Logs*, Circ. 215, Illinois State Geol. Survey, 15 pp.

RALSTON. D. R.. 1967. *Influences of Water Well Design on Neutron Logging*, unpublished M.S. Thesis, University of Arizona, Tucson, Arizona.

RAMACHANDAR RAO, M. B., 1953, *Self-Potential Anomalies due to Subsurface Water Flow at Garimenapenta, Madras State, India*, Min. Eng., v. 5, pp. 400-403.

RUSSELL, W. L., 1941, *Well Logging by Radioactivity*, Am. Assoc. Petroleum Geologists Bull., v. 25, pp. 1768-1788.

SAMMEL, E. A., 1968, *Convective Flow and its Effect on Temperature Logging in Small-Diameter Wells*, Geophysics, v. 33, no. 6, pp. 1004-1012.

SCHLUMBERGER, 1972-78, *Log Interpretation*, Volume 1 — Principles, 113 pp.; Volume 2 — Applications, 116 pp; Volume 3 — Log Interpretation Charts, 83 pp., Schlumberger Ltd., P. O. Box 2175, Houston, Texas 77001.

SHAMEY, L., and ADAMS, W. 1971, *Interpretation of Electrical Resistivity Logs in a Two-Zone Cylindrically Symmetric Geometry*, Hawaii Water Resources Res. Center Tech. Rep. 46.

SHAMEY, L., and ADAMS, W., 1973, *Three Dimensional Zone Model Log Interpretation*, Hawaii Water Resources Res. Center Tech. Rep. No. 69.

SHEPHERD, G. F., 1958, *Drilling Time Logging*; in *Subsurface Geology in Petroleum Exploration*, by Haun, J. D., and LeRoy, L. W., Colorado School of Mines, pp. 367-388.

SOREY, M. L., 1971, *Measurement of Vertical Ground-Water Velocity from Temperature Profiles in Wells*, Water Resources Res., v. 7, no. 4, pp. 963-970.

STRATTON, E. F., and FORD, R. D., 1951, *Electric Logging*, in *Subsurface Geologic Methods* (L. W. LeRoy, ed.) 2nd ed., Colorado School of Mines, Golden, pp. 364-392.

SWANSON, ROGER, 1983, *Sample Examination Manual.* Cat. no. 603, Am. Assoc. Petroleum Geol., P.O. Box 979, Tulsa, Oklahoma 74101, 100 pp.

SYMS, M. C., 1982, *Downhole Flowmeter Analysis Using an Associated Caliper Log*, Ground Water, v. 20, no. 5, pp. 606-610.

TAPPER, WILFRED, 1958, *Caliper and Temperature Logging*; in *Subsurface Geology in Petroleum Exploration,* by Haun, J. D., and LeRoy, L. W., Colorado School of Mines, pp. 345-356.

TEXAS AGRIC. and MECH. COLLEGE, 1946, *Well Logging Methods Conference,* Texas Eng. Exp. Sta. Bull. 93, College Station, 171 pp.

TOPP, G. C., and others, 1984, *The Measurement of Soil Water Content using a Portable TDR Hand Probe,* Canadian J. Soil Sci., v. 64, no. 3, pp. 313-321.

TRAINER, F. W., 1968, *Temperature Profiles in Water Wells as Indicators of Bedrock Fractures,* U. S. Geol. Survey Prof. Paper 600-B, pp. B210-B214.

TSELENTIS, G. A., 1985, *The Processing of Geophysical Well Logs by Microcomputers as Applied to the Solution of Hydrogeological Problems,* J. Hydrology, v. 80, no. 3/4, Oct. 15.

TURCAN, A. N., Jr., 1962, *Estimating Water Quality from Electrical Logs,* U. S. Geol. Survey Prof. Paper 450-C, pp. 135-136.

TURCAN, A. N., Jr., 1966, *Calculation of Water Quality from Electrical Logs — Theory and Practice,* Louisiana Geol. Survey, Water Resources Pamphlet 19, 23 pp.

TURCAN, A. N., Jr., and WINSLOW, A. G., 1970, *Quantitative Mapping of Salinity, Volume and Yield of Saline Aquifers Using Borehole Geophysical Logs,* J. Water Resources Res., v. 6, no. 5, pp. 1478-1481, Oct.

U. S. WATER and POWER RESOURCES SERVICE, 1974, *Procedures for Sampling, Classification, and Testing of Soils and Installation of Instruments,* In Appendix of *Earth Manual,* 2nd ed., U. S. Govt Printing Office, Washington, D. C. 20402, pp. 327-386.

VONHOF, J. A., 1966, *Water Quality Determination from Spontaneous-Potential Electric Log Curves,* J. Hydrology, v. 4, pp. 341-347.

WALSTROM, J. E., 1952, *The Quantitative Aspects of Electric Log Interpretation,* Trans. Am. Inst. Min. and Met. Engrs., v. 195, pp. 47-58.

WILLIAMS, J. H., and others, 1984, *Characterization of Ground Water Circulation in Selected Fractured Rock Aquifers Using Borehole Temperature and Flow Logs,* in *Surface and Borehole Geophysical Methods in Ground Water Investigations,* (D.M. Nielson, ed.), National Water Well Assoc., Dublin, Ohio, pp. 842-852.

WOODWARD, D. G., 1984, *Lithologic Changes in Aquifers in Southeastern Minnesota as Determined from Natural Gamma Borehole Logs,* in *Surface and Borehole Geophysical Methods in Ground Water Investigations* (D. M. Nielsen, ed.), National Water Well Assoc. Dublin, Ohio, pp. 788-800.

WYLLIE, M. R. J., 1949, *Statistical Study of Accuracy of Some Connate-Water Resistivity Determinations Made from Self-Potential Log Data,* Am. Assoc. Petroleum Geologists Bull., v. 33, 1892-1900.

WYLLIE, M. R. J., 1963, *The Fundamentals of Well Log Interpretation,* Academic Press, New York, 238 pp.

ZEMANEK, J. and others, 1970, *Formation Evaluation by Inspection with the Borehole Televiewer,* Geophysics, v. 35, no. 2, p. 254.

HYDRAULIC CHARACTERISTICS OF SOILS AND AQUIFERS

BIBLIOGRAPHY

FISHEL, V. C., 1942, *Bibliography on Permeability and Laminar Flow,* U. S. Geol. Survey Water-Supply Paper 887, pp. 20-50.

SOILS/UNSATURATED ZONE

AMERICAN SOCIETY FOR TESTING MATERIALS, 1967, *Permeability and Capillarity of Soils,* ASTM Spl. Tech. Publ. no. 417, Philadephia, Pennsylvania, 210 pp.

ANONYMOUS, 1986, *Proceedings of the Conference on Characterization of the Vadose Zone,* National Water Well Assoc., Dublin, Ohio 43017 (42 papers).

BARBER, E. S., 1955, *Symposium on Permeability of Soils,* Am. Soc. Testing Materials, Spec. Pub. no. 163, 136 pp.

BAVER, L. D., and others, 1972, *Soil Physics,* 4th ed., John Wiley & Sons, New York, 498 pp.

BLACK, C. A., (ed.), 1965, *Methods of Soil Analysis,* Part 1, Agronomy Monograph no. 9, Am. Soc. Agronomy, Madison, Wisconsin, 770 pp.

BOERSMA, L., 1965, *Field Measurement of Hydraulic Conductivity Above a Water Table,* Methods of Soil Analysis, Pt. 1. Agronomy, no. 9, pp. 234-252, Am. Soc. Agronomy, Madison, WI.

BOUWER, H., and JACKSON, R. D., 1974, *Determining Soil Properties, Drainage for Agriculture,* Agronomy, no. 17, pp. 611-672, Am. Soc. Agronomy, Madison, WI.

BUCKMAN, H. O., and BRADY, N. C., 1969, *The Nature and Properties of Soils,* Macmillan Publ. Co., New York, 653 pp.

CHILDS, E. C., 1969, *An Introduction to the Physical Basis of Soil Water Phenomena,* John Wiley & Sons, London, 493 pp.

CHRISTIANSEN, J. E., 1944, *Effect of Entrapped Air upon the Permeability of Soils,* Soil Sci., v. 58, pp. 355-365.

ELLIS, B. G., 1973, *The Soil as a Chemical Filter,* in *Recyling Treated Municipal Wastewater and Sludge through Forest and Cropland,* W. E. Sopper and L. T. Kardos, (eds.), Pennsylvania State University Press.

EVERETT, L. G., and McMILLION, L. G., 1985, *Operational Ranges for Suction Lysimeters,* Ground Water Monitoring Rev., v. 5, no. 3, pp. 51-61.

FULLER, W. H., 1977, *Movement of Selected Metals, Asbestos and Cyanide in Soil: Application to Waste Disposal Problems,* U. S. Environmental Protection Agency, EPA-600/2-77-020, U. S. Government Printing Office.

GILLHAM, R. W., 1984, *The Capillary Fringe and Its Effect on Water-Table Response,* J. Hydrology, v. 67, no. 1/4.

HILLEL, D., 1971, *Soil and Water, Physical Principles and Processes,* Academic Press, New York.

IRMAY, S., 1954, *On the Hydraulic Conductivity of Unsaturated Soils,* Trans. Amer. Geophysical Union, v. 35, pp. 463-467.

JACKSON, R. D., and VAN BAVEL, C. H. M., 1965, *Solar Distillation of Water from Soil and Plant Materials: a Simple Desert Survival Technique,* Science, v. 149, pp. 1377-1379.

JOHNSON, A. I., 1962, *Methods of Measuring Soil Moisture in the Field,* U. S. Geol. Survey Water-Supply Paper 1619-U, 25 pp.

JORGENSEN, D. G., 1980, *Relationships Between Basic Soil-Engineering Equations and Basic Ground-Water Flow Equations,* U. S. Geol. Survey Water-Supply Paper W 2064.

KHOSLA, B. K., 1980, *Comparison of Calculated and In Situ Measured Unsaturated Hydraulic Conductivity,* J. of Hydrology, v. 47, no. 3/4, July.

KIRKHAM, Don, and POWERS, W. L., 1972, *Advanced Soil Physics,* John Wiley and Sons, New York, New York, 534 pp.

KLUTE, A., 1965, *Laboratory Measurement of Hydraulic Conductivity of Unsaturated Soil,* in *Methods of Soil Analyses,* Black, C. A., (ed.), Agronomy, no. 9, pp. 253-261, American Society of Agronomy, Madison, Wisconsin.

LEHMAN, G. S., 1968, *Soil and Grass Filtration of Domestic Sewage Effluent for the Removal of Trace Elements,* Unpublished Ph.D. Dissertation, University of Arizona.

McNEIL, B. L., 1974, *Soil Salts and their Effect on Water Movement* in *"Drainage for Agriculture,"* J. van Schilfgaarde (ed.) Agronomy, no. 17, Am. Society of Agronomy, Madison, Wisconsin, pp. 409-433.

MURRMANN, R. P. and KOUTZ, F. R., 1972, *Role of Soil Chemical Processes in Reclamation of Wastewater Applied to Land, Wastewater Management by Disposal on the Land,* U. S. Army Corps of Engineers, Cold Regions Research and Engineering Lab Specialty Rep. 171.

MYHRE, D. L., SANFORD, J. O., and JONES, W. F., 1972, *Apparatus and Techniques for Installing Access Tubes in Soil Profiles to Measure Soil Water,* Soil Science, v. 108, no. 4, pp. 296-999.

NIELSEN, D. R., BIGGAR, J. W., and ERH, K. T., 1973, *Spatial Variability of Field-Measured Soil-Water Properties,* Hilgardia, v. 42, no. 7, pp. 215-260.

REYNOLDS, W. D., ELRICK, D. E., and CLOTHIER, B. E., 1985, *The Constant Head Well Permeameter: Effect of Unsaturated Flow,* Soil Sci. v. 139, no. 2, pp. 172-180.

REYNOLDS, W. D., and ELRICK, D. E., 1986, *A Method for Simultaneous In Situ Measurement in the Vadose Zone of Field-Saturated Hydraulic Conductivity, Sorptivity and the Conductivity-Pressure Head Relationship,* Ground Water Monitoring Rev., pp. 84-95.

RUNNELLS, D. D., 1976, *Wastewaters in the Vadose Zone of Arid Regions: Geochemical Interactions,* Ground Water, v. 14, no. 6, pp. 374-385.

SOIL SURVEY STAFF, 1951, *Soil Survey Manual,* U. S. Dept. Agriculture Handbook no. 18, 503 pp.

STEPHENS, D. B., LAMBERT, K., and WATSON, D., 1984, *Influence of Entrapped Air on Field Determinations of Hydraulic Properties in the Vadose Zone,* Proc. Conf. Characterization 2nd Monitoring of Vadose Zone, NWWA, 6375 Riverside Dr., Dublin, Ohio 43017, pp. 57-76.

VACHAUD, G., 1967, *Determination of the Hydraulic Conductivity of Unsaturated Soils from an Analysis of Transient Flow Data,* Water Resources Research, v. 3, pp. 697-705.

WILSON, L. G., 1980, *Monitoring in the Vadose Zone — A Review of Technical Elements and Methods,* General Electric Company — TEMPO, GE79TMP-55, Santa Barbara, California.

YARON, B., GOLDSHMID, J., and DAGAN, G., (eds.), 1984, *Pollutants in Porous Media: The Unsaturated Zone Between Soil Surface and Groundwater,* Springer-Verlag New York, Inc., P.O. Box 2485, Secaucus, New Jersey 07094, 300 pp.

AQUIFERS/SATURATED ZONE

ATHY, L. F., 1930, *Density, Porosity, and Compaction of Sedimentary Rocks,* Am. Assoc. Petroleum Geologists Bull., v. 14, pp. 1-24.

BAIR, E. S., and PARIZEK, R. R., 1978, *Detection of Permeability Variations by a Shallow Geothermal Technique,* Ground Water, v. 16, no. 4, pp. 254-263.

BEDINGER, M. S., 1961, *Relation Between Median Grain Size and Permeability in the Arkansas River Valley,* Arkansas, U. S. Geol. Survey Prof. Paper 424-C, pp. 31-32.

BLES, J. L., and FEUGA, B., 1984, *Fracturing of Rocks,* North Oxford Academic Publ. Co., 242-245 Banbury Rd, Oxford OX3 7DW, U. K., 150 pp.

BOSAZZA, V. L., 1952, *On Storage of Water in Rocks in Situ,* Trans. Am. Geophys. Union, v. 33, pp. 42-48.

BRACE, W. F., 1980, *Permeability of Crystalline and Argillaceous Rocks,* Internat. J. Rock Mech. Min. Sci. and Geomech., Abstract 17, pp. 241-251.

BRADBURY, K. R., and ROTHCHILD, E. R., 1985, *Computer Notes—A Computerized Technique for Estimating the Hydraulic Conductivity of Aquifers from Specific Capacity Data,* Ground Water, v. 23, no. 2, pp. 240-246.

BROOKS, R. H. and COREY, A. T., 1966, *Properties of Porous Media Affecting Fluid Flow,* J. Irrig. Drain. Div., Amer. Soc. Civil Engrs., v. 92, no. IR2, pp. 61-88.

BRUCKER, R. W., and others, 1972, *Role of Vertical Shafts in the Movement of Ground Water in Carbonate Aquifers,* Ground Water, v. 10, no. 6, pp. 5-13, Nov.-Dec.

CASTILLO, E., and others, 1972, *Unconfined Flow through Jointed Rock,* Water Resources Bull., v. 8, pp. 266-281.

COHEN, P., 1963, *Specific-Yield and Particle-Size Relations of Quaternary Alluvium Humboldt River Valley, Nevada*, U. S. Geol. Survey Water-Supply Paper 1669-M, 24 pp.

DAGAN, G., 1967, *A Method of Determining the Permeability and Effective Porosity of Unconfined Anisotropic Aquifers*, Water Resources Res., v. 3, pp. 1059-1071.

DAVIS, S. N., 1969, *Porosity and Permeability of Natural Materials*, in Flow through Porous Media, Academic Press (New York), de Wiest, R. J. M. (ed.), pp. 53-89.

DENSON, K. H., and WU, P. K., 1971, *High Head Permeability of Sand with Dispersed Clay Particles*, Water Resources Res., v. 7, no. 6, pp. 1661-1662.

DE RIDDER, N. A., and WITT, K. E., 1965, *A Comparative Study on the Hydraulic Conductivity of Unconsolidated Sediments*, J. of Hydrology, v. 3, pp. 180-206.

DOS SANTOS, A. G., Jr., and YOUNGS, E. G., 1969, *A Study of the Specific Yield in Land-Drainage Situations*, J. of Hydrology, v. 8, no. 1, pp. 59-81.

DREISS, S. J., 1984, *Effects of Lithology on Solution Development in Carbonate Aquifers*, J. Hydrology, v. 70, no. 1/4.

ELZEFTAWY, ATEF, and CARTWRIGHT, KEROS, 1981, *Evaluating the Saturated and Unsaturated Hydraulic Conductivity of Soils*, Illinois State Geol. Survey Reprint, Champaign, Illinois 61820.

FISHEL, V. C., 1935, *Further Tests of Permeability with Low Hydraulic Gradients*, Trans. Am. Geophys. Union, v. 16, pp. 499-503.

FRANZINI, J. B., 1951, *Porosity Factor for Case of Laminar Flow through Granular Media*, Trans. Am. Geophys. Union, v. 32, pp. 443-446.

FRASER, H. J., 1935, *Experimental Study of Porosity and Permeability of Clastic Sediments*, J. Geology, v. 43, pp. 910-1010.

GAITHER, ALFRED, 1953, *A Study of Porosity and Grain Relationships in Experimental Sands*, J. Sed. Petrology, v. 23, pp. 180-195.

GRATON, L. C., and FRASER, H. J., 1935, *Systematic Packing of Spheres with Particular Relation to Porosity and Permeability, and Experimental Study of the Porosity and Permeability of Clastic Sediments*, J. Geol., v. 43.

HARLEMAN, D. R. F., and others, 1963, *Dispersion-Permeability Correlation in Porous Media*, J. Hydraulics Div., Amer. Soc. Civil Engrs., v. 89, no. HY2, pp. 67-85.

HAZEN, A., 1893, *Some Physical Properties of Sands and Gravels with Special Reference to Their Use in Filtration*, 24th Ann. Rept., Mass. State Bd. Health, Boston.

HEIGOLD, P. C., and others, 1979, *Aquifer Transmissivity from Surficial Electrical Methods*, Ground Water, v. 17, no. 4, Jul.-Aug., pp. 388-345.

INTERA ENVIRONMENTAL CONSULTANTS, INC., 1983, *Porosity, Permeability, and Their Relationship in Granite, Basalt, and Tuff*, Accession DE83-011519, NTIS, Springfield, Virginia 22161.

JOHNSON, A. I., 1967, *Specific Yield — Compilation of Specific Yields for Various Materials*, U. S. Geol. Survey Water-Supply Paper 1662-D, 74 pp.

JONES, O. R., and SCHNEIDER, A. D., 1969, *Determining Specific Yield of the Ogallala Aquifer by the Neutron Method*, Water Resources Res., v. 5, no. 6, p. 1267-1272.

KELLY, W. E., 1980, *Porosity Permeability Relationships in Stratified Glacial Deposits (ABS)*, EOS, v. 61, no. 17, p. 232.

KLINKENBERG, L. J., 1941, *The Permeability of Porous Media to Liquids and Gases, Drilling and Production Practice*, Am. Petroleum Inst., New York, pp. 200-214.

KRUMBEIN, W. C., and MONK, C. D., 1943, *Permeability as a Function of the Size Parameters of Unconsolidated Sand*, Am. Inst. Min. and Met. Engrs., Trans. Petroleum Div., v. 151, pp. 153-163.

KWADER, THOMAS, 1985, *Estimating Aquifer Permeability from Formation Resistivity Factors*, Ground Water, v. 23, no. 6, pp. 762-766.

LANGBEIN, W. B., 1959, *Water Yield and Reservoir Storage in the United States*, U. S. Geol. Survey Circ. 409.

LEGRAND, H. E., and STRINGFIELD, V. T., 1971, *Development and Distribution of Permeability in Carbonate Aquifers*, Water Resources Res., v. 7, no. 5, pp. 1284-1294.

LEWIS, D. C., and BURGY, R. H., 1964, *Hydraulic Characteristics of Fractured and Jointed Rock*, Ground Water, v. 2, no. 3, pp. 4-9.

LONG, J. C. S., and others, 1982, *Porous Media Equivalents for Networks of Discontinuous Fractures,* Water Resources Res., v. 18, no. 3, Jun.

LOVELL, R. E., DUCKSTEIN, L., KISIEL, C. C., 1972, *Use of Subjective Information in Estimation of Aquifer Parameters,* Water Resources Res., v. 8, no. 3, pp. 680-690.

MANGER, G. E., 1963, *Porosity and Bulk Density of Sedimentary Rocks,* U. S. Geol. Survey Bull., 1144-E, 55 pp.

MASCH, F. D., and DENNY K. J., 1966, *Grain Size Distribution and its Effect on the Permeability of Unconsolidated Sands,* Water Resources Research, v. 2, pp. 665-677.

McQUEEN, I. S., 1974, *Evaluating the Reliability of Specific-Yield Determinations,* J. Research U. S. Geol. Survey, v. 1, pp. 371-376.

MEINZER, O. E., 1928, *Compressibility and Elasticity of Artesian Aquifers,* Econ. Geol. v. 23, pp. 263-291.

MERCADO, A., and HALEVY, E., 1966, *Determining the Average Porosity and Permeability of a Stratified Aquifer with the Aid of Radioactive Tracers,* Water Resources Res., v. 2, no. 3, pp. 525-531.

MOORE, G. K., 1973, *Hydraulics of Sheetlike Solution Cavities,* Ground Water, v. 11, no. 4, pp. 4-11, Jul.-Aug.

MORRIS, D. A., and JOHNSON, A. I., 1967, *Summary of Hydrologic and Physical Properties of Rock and Soil Materials, as Analyzed by the Hydrologic Laboratory of the U. S. Geological Survey, 1948-60,* U. S. Geol. Survey Water-Supply Paper 1839-D, 42 pp.

MURRAY, R. C., 1960, *Origin of Porosity in Carbonate Rocks,* J. Sed. Petrology, v. 30, pp. 59-84.

NORRIS, S. E., 1963, *Permeability of Glacial Till,* U. S. Geol. Survey Prof. Paper 450-E, pp. 150-151.

NWANKWOR, G. I., CHERRY, J. A. and GILLHAM, R. W., 1984, *A Comparative Study of Specific Yield Determinations for a Shallow Sand Aquifer,* Ground Water, v. 22, no. 6, pp. 764-772.

PARKES, M. E., and WATERS, P. A., 1980, *Comparison of Measured and Estimated Unsaturated Hydraulic Conductivity,* Water Resources Res., v. 16, no. 4, Aug.

PRICE, MICHAEL, 1977, *Specific Yield Determinations from a Consolidated Sandstone Aquifer,* J. Hydrology, v. 3, pp. 147-156.

PRILL, R. C., 1965, *Specific Yield-Laboratory Experiments Showing the Effect of Time on Column Drainage,* U. S. Geol. Survey Water-Supply Paper 1662-B, 55 pp.

RAMSAHOYE, L. E., and LANG, S. M., 1961, *A Simple Method for Determining Specific Yield from Pumping Tests,* U. S. Geol. Survey Water Supply Paper 1536-C.

RASMUSSEN, W. C., 1963, *Permeability and Storage of Heterogeneous Aquifers in the United States,* Internat. Assoc. Sci. Hydrology, Pub. 64, pp. 317-325.

REHM, B. W., GROENEWOLD, G. H., and MORIN, K. A., 1980, *Hydraulic Properties of Coal and Related Materials, Northern Great Plains,* Ground Water, v. 18, no. 6, Nov.-Dec., pp. 551-561.

RODE, A. A., 1955, *Hydraulic Properties of Soils and Rocks,* (in Russian), Acad. Sci. U.S.S.R., Moscow, 131 p.

SAGAR, BUDHI and RUNCHAL, AKSHAI, 1982, *Permeability of Fractured Rock: Effect of Fracture Size and Data Uncertainties,* Water Resources Res., v. 18, no. 2, Apr.

SAHNI, B. M., and HARBHAJAN, S. S., 1979, *Storage Coefficients from Ground-Water Maps,* J. Irrigation and Drainage Div., Am. Soc. Civil Engrs., v. 105, no. IR2, June.

SAYRE, A. N., and SMITH, W. O., 1962, *Retention of Water in Silts and Sands,* U. S. Geol. Survey Prof. Paper 450-C.

SKAGGS, R. W., 1976, *Determination of the Hydraulic Conductivity — Drainable Porosity Ratio from Water Table Measurements,* Trans. Am. Soc. Agric. Engrs., v. 19, no. 1, pp. 73-80.

SNOW, D. T., 1967, *Rock Fracture Spacings, Openings and Porosities,* Am. Soc. Civil Engrs., Seattle Meeting, Conf. Preprint 515, 42 pp.

STEARNS, N. D., 1928, *Laboratory Tests on Physical Properties of Water-Bearing Materials,* U. S. Geol. Survey Water-Supply Paper 596, pp. 121-176.

STERNBERG, Y. M., 1971, *Parameter Estimation for Aquifer Evaluation,* Water Resources Bull., v. 7, no. 3, pp. 447-456.

STEVENS, P. R., 1963, *Examination of Drill Cuttings and Application of Resulting Information to Solving of Field Problems on the Navajo Indian Reservation, New Mexico and Arizona,* U. S. Geol. Survey Water-Supply Paper 1544-H.

STEWART, J. W., 1962, *Relation of Permeability and Jointing in Crystalline Metamorphic Rocks near Jonesboro, Georgia,* U. S. Geol. Survey Prof. Paper 450-D, pp. 168-170.

STONER, J. D., 1981, *Horizontal Anisotropy Determined by Pumping in Two Powder River Basin Coal Aquifers, Montana,* Ground Water, v. 19, no. 1, Jan.-Feb., pp. 34-40. (hydraulic conductivity of coal aquifers)

STRINGFIELD, V. T., and LeGRAND, H. E., 1971, *Development and Distribution of Permeability in Carbonate Aquifers,* Water Resources Res., v. 7, no. 5, pp. 1284-1294.

STRINGFIELD, V. T., and LeGRAND, H. E., 1971, *Effects of Karst Features on Circulation of Water in Carbonate Rocks in Coastal Areas,* J. Hydrology, 14, pp. 139-157.

STRINGFIELD, V. T., Le GRAND, H. E., and LA MOREAUX, P. E., 1977, *Development of Karst and its Effect on the Permeability and Circulation of Water in Carbonate Rocks with Special Reference to the Southeastern States,* Alabama Geol. Surv. Bull. 94, Pt. G, 68 pp.

SUMMERS, W. K., 1973, *Specific Capacities of Wells in Crystalline Rocks,* Ground Water, v. 10, no. 6, pp. 37-47, Nov.-Dec.

TICKELL, F. G., and HIATT, W. N., 1938, *Effect of Angularity of Grain on Porosity and Permeabiliy of Unconsolidated Sands,* Am. Assoc. Petroleum Geologists Bull., v. 22, pp. 1272-1279.

URISH, D. W., 1981, *Electrical Resistivity-Hydraulic Conductivity Relationships in Glacial Outwash Aquifers,* Water Resources Res., v. 17, no. 5, Oct.

WAY, SHAO-CHIH, and McKEE, C. R., 1982, *In-Situ Determination of Three-Dimensional Aquifer Permeabilities,* Ground Water, v. 20, no. 5, pp. 594-603.

WEEKS, E. P., 1969, *Determining the Ratio of Horizontal to Vertical Permeability by Aquifer-Test Analysis,* Water Resources Res., v. 5, pp. 196-214.

WORTHINGTON, P. F., 1976, *Hydrogeophysical Equivalence of Water Salinity, Porosity and Matrix Conduction in Arenaceous Aquifers,* Ground Water, v. 14, no. 4, pp. 224-232.

WORTHINGTON, P. F., 1977, *Influence of Matrix Conduction upon Hydrogeophysical Relationship in Arenaceous Aquifers,* Water Resources Res., v. 13, no. 1, pp. 87-92.

YEH, W. W.-G., 1981, *Aquifer Parameter Identification*, J. Hydraul. Div. Am. Soc. Civ. Eng., v. 101 (HY9), pp. 1197-1209.

WATER LEVELS AND FLUCTUATION

BLANCHARD, F. B., and BYERLY, P., 1935, *A Study of a Well Gauge as a Seismograph,* Seismological Soc. Am. Bull., v. 25, pp. 313-321.

BLOEMEN, G. W., 1968, *Determination of Constant Rate Deep Recharge or Discharge from Ground-Water Level Data,* J. Hydrology, v. 6, no. 1, Jan.

BLOOMSBURG, G. L., 1971, *Groundwater Pressure Wave in Confined Porous Media,* Univ. of Idaho, Water Resources Res. Inst., National Tech. Information Center PB 208814., 5 pp.

BREDEHOEFT, J. D., 1967, *Response of Well-Aquifer Systems to Earth Tides,* J. Geophys. Res., v. 72, no. 12, pp. 3075-3087.

CABRERA, G., and MARINO, M. A., 1976, *Dynamic Response of Aquifer Systems to Localized Recharge,* Water Resour. Bull., v. 12, no. 1, pp. 49-63.

CARR, P. A., 1971, *Use of Harmonic Analysis to Study Tidal Fluctuations in Aquifers near the Sea,* Water Resources Res., v. 7, no. 3, pp. 632-643.

CARR, P. A., and VAN DER KAMP, G. S., 1969, *Determining Aquifer Characteristics by the Tidal Method,* Water Resources Res., v. 5, no. 5, p. 1023.

CLARK, W. E., 1967, *Computing the Barometric Efficiency of a Well,* J. Hydraulics Div., Amer. Soc. Engrs., v. 93, no. HY4, pp. 93-98.

COOPER, H. H., Jr., and others, 1965, *The Response of Well-Aquifer Systems to Seismic Waves,* J. Geophys. Res., v. 70, pp. 3915-3926.

DaCOSTA, J. A., 1964, *Effect of Hebgen Lake Earthquake on Water Levels in Wells in the United States,* U. S. Geol. Survey Prof. Paper 435-o, pp. 167-178.

DAVIES, W. E., and LeGRAND, H. E., 1971, *Water Levels in Carbonate Rock Terranes,* Ground Water, v. 9, no. 3, pp. 4-10.

DOMINICK, T. F., and ROBERTS, H., 1971, *Mathematical Model for Beach Groundwater Fluctuations,* Water Resources Res., v. 7, no. 6, pp. 1626-1634.

FERRIS, J. G., 1951, *Cyclic Fluctuations of Water Level as a Basis for Determining Aquifer Transmissibility,* General Assembly Brussels, Internat. Assoc. Sci. Hydrology, Pub. 33, v. 2, pp. 148-155.

FETTER, C. W., 1977, *Statistical Analysis of the Impact of Groundwater Pumpage on Low-flow Hydrology,* Water Resour. Bull., v. 13, no. 2, pp. 309-323.

FISHEL, V. C., 1956, *Long-Term Trends of Ground-Water Levels in the United States,* Trans. Am. Geophys. Union, v. 37, pp. 429-435.

FRANKE, O. L., 1968 *Double-Mass Curve Analysis of the Effects of Sewering on Ground-Water Levels on Long Island, New York,* U. S. Geol. Survey Research 1968, Prof. Paper 600-B, pp. B205-209.

GARBER, M. S., and WOLLITZ, L. E., 1969, *Measuring Underground-Explosion Effects on Water Levels in Surrounding Aquifers,* Ground Water, v. 7, no. 4.

GEORGE, W. O., and ROMBERG, F. E., 1951, *Tide-Producing Forces and Artesian Pressures,* Trans. Am. Geophys. Union, v. 32, pp. 369-371.

GILLILAND, J. A., 1969, *A Rigid Plate Model of the Barometric Effect,* J. Hydrology, v. 7, pp. 233-245.

GRANTZ, A., and others, 1964, *Alaska's Good Friday Earthquake, March 27, 1964, a Preliminary Geologic Evaluation,* U. S. Geol. Survey Circ. 491, 35 pp.

GREGG, D. O., 1966, *An Analysis of Ground-Water Fluctuations Caused by Ocean Tides in Glynn County, Georgia,* Ground Water, v. 4, no. 3, pp. 24-32.

GUYTON, W. F., 1958, *Fluctuations of Water Levels and Artesian Pressure in Wells in the United States: their Measurement and Interpretation,* Internat. Assoc. Sci. Hydrology, General Assembly Oslo, Pub. 31, v. 3, pp. 85-92.

HAGERTY, D. J., and LIPPERT, K., 1982, *Rising Ground Water-Problem or Resource?* (Louisville, KY), Ground Water, v. 20, no. 2, pp. 217-223.

HAWKINS, G. W., 1979, *Processing Groundwater-Level Data by Digital Computer,* Hydrological Sciences Bull., v. 24, no. 4, p. 529-538.

HEALY, J. H., and others, 1968, *The Denver Earthquakes,* Science, v. 161, pp. 1301-1310.

HEATH, R. C., 1976, *Design of Ground-Water Level Observations-Well Programs,* Ground Water, v. 14, no. 2, pp. 71-77.

INESON, J., 1963, *Form of Ground-water Fluctuations Due to Nuclear Explosion,* Nature, v. 198, pp. 22-23.

JACOB, C. E., 1939, *Fluctuations in Artesian Pressure Produced by Passing Railroad Trains as shown in a Well on Long Island, New York,* Trans. Am. Geophys. Union, v. 20, pp. 666-674.

JACOB, C. E., 1943, *Correlation of Ground-Water Levels and Precipitation on Long Island, New York,* Trans. Am. Geophys. Union, v. 24, pp. 564-573.

JACOB, C. E., 1944, *Correlation of Ground-Water Levels and Precipitation on Long Island, New York,* Trans. Am. Geophys. Union, v. 25, pp. 928-939.

JACOB, C. E., 1945, *Correlation of Groundwater Levels and Precipitation on Long Island, N. Y.,* New York Dept. Conserv. Water Power and Control Comm. Bull. GW-14.

KENNEDY, K.G., HOLLINGSHEAD, S.C., and LEECH, R.E.J., 1986, *Determining Head and Pressure Distribution in Low Transmissivity Formations and Soils,* Ground Water Monitoring Rev., v. 6, no. 2, pp. 92-98.

KERNODLE, J. M., and WHITESIDES, D. V., 1977, *Rising Ground-Water Level in Downtown Louisville, Kentucky, 1922-1977,* U. S. Geol. Survey Water-Resources Invest. 77-92.

KLEIN, M., and KASER, P., 1963, *A Statistical Analysis of Ground-water Levels in Twenty Selected Observation Wells in Ohio,* Tech. Rept. 5, Ohio Dept. Natural Resources, Div. of Water, Columbus, 124 pp.

KOHOUT, F. A., 1961, *Fluctuations of Ground-Water Levels Caused by Dispersion of Salts,* J. Geophys. Res., v. 66, pp. 2429-2434.

LAMAR, D. L., and MERIFIELD, P. M., 1980, *Water Level Monitoring along San Andreas and San Jacinto Faults, Southern California, During Fiscal Year 1979,* 80-1141 NTIS Springfield, Virginia 22161, 76 p.

La ROCQUE, G. A., Jr., 1941, *Fluctuations of Water Level in Wells in the Los Angeles Basin, California, during Five Strong Earthquakes, 1933-1940,* Trans. Am. Geophys. Union, v. 22, pp. 374-386.

LeGRAND, H. E., and STRINGFIELD, V. T., 1971, *Water Levels in Carbonate Rocks,* Ground Water, v. 9, no. 3, pp. 4-10.

LEWIS, M. R., 1932, *Flow of Ground Water as Applied to Drainage Wells,* Trans. Am. Soc. Civil Engrs., v. 96, pp. 1194-1211.

LUSCZYNSKI, N. J., 1952, *The Recovery of Ground-Water Levels in Brooklyn, N.Y. from 1947 to 1950,* U. S. Geol. Survey Circ. 167, 29 pp.

MAASLAND, M., 1959, *Water-table Fluctuations Induced by Intermittent Recharge,* J. Geophys. Res., v. 64, pp. 549-559.

MARINE, I. W., 1975, *Water Level Fluctuations Due to Earth Tides in a Well Pumping from Slightly Fractured Crystalline Rocks,* Water Resources Res., v. 11, no. 1, pp. 165-173.

MAUCHA, L., and SARVARY, I., 1970, *Tidal Phenomena in the Karstic Water Level,* Bull. Internat. Assoc. Sci. Hydrology, v. 15, no. 2, June.

McGINNIS, L. D., 1963, *Earthquakes and Crustal Movement as Related to Water Load in the Mississippi Valley Region,* Illinois State Geol. Survey Publ. C344.

MEYER, A. F., 1960, *Effect of Temperature on Ground-Water Levels,* J. Geophys. Res., v. 65, pp. 1747-1752.

MOYLE, W. R., Jr., 1980, *Ground-Water-Level Monitoring for Earthquake Prediction — A Progress Report Based on Data Collected in Southern California, 1976-79 U.S. Geol. Survey,* Sacramento, California 95825, PB80-413 NTIS Springfield, Virginia 22161, 64 pp.

NETHERLANDS STATE INSTITUTE FOR WATER SUPPLY, 1948, *The Effect of the Yearly Fluctuations in Rainfall on the Flow of Ground Water from an Extended Area of Recharge,* General Assembly Oslo, Internat. Assoc. Sci. Hydrology, v. 3, pp. 47-56.

PARKER, G. G., and STRINGFIELD, V. T., 1950, *Effects of Earthquakes, Trains, Tides, Winds, and Atmospheric Pressure Changes on Water in the Geologic Formations of Southern Florida,* Econ. Geol., v. 45.

PECK, A. J., 1960, *The Water Table as Affected by Atmospheric Pressure,* J. of Geophys. Res., v. 65, no. 8, pp. 2383-2388.

PIPER, A. M., 1933, *Fluctuations of Water-Surface in Observation-Wells and at Stream Gaging-Stations in the Mokelumne Area, California, during the Earthquake of December 20, 1932,* Trans. Am. Geophys. Union, v. 14, pp. 471-475.

REMSON, IRWIN, and RANDOLPH, J. R., 1958, *Application of Statistical Methods to the Analysis of Ground-Water Levels,* Trans. Am. Geophys. Union, v. 39, no. 1, pp. 75-83.

RICHARDSON, R. M., 1956, *Tidal Fluctuations of Water Level Observed in Wells in East Tennessee,* Trans. Am. Geophys. Union, v. 37, pp. 461-462.

RIGGS, H. C., 1953, *A Method of Forecasting Low Flow of Streams,* Trans. Am. Geophys. Union, v. 34, pp. 427-434.

ROBERTS, W. J., and ROMINE, H. E., 1947, *Effect of Train Loading on the Water Level in a Deep Glacial-Drift Well in Central Illinois,* Trans. Am. Geophys. Union, v. 28, pp. 912-917.

ROBINSON, E. S., and BELL, R. T., 1971, *Tides in Confined Well-Aquifer Systems,* J. Geophys. Res., v. 76, no. 8, March 10, pp. 1857-1869.

ROBINSON, T. W., 1939, *Earth-Tides Shown by Fluctuations of Water-Levels in Wells in New Mexico and Iowa,* Trans. Am. Geophys. Union, v. 20, pp. 656-666.

RORABAUGH, M. I., 1956, *Prediction of Ground-Water Levels on Basis of Rainfall and Temperature Correlations,* Trans. Am. Geophys. Union, v. 37, pp. 436-441.

ROSE, N. A., and ALEXANDER, W. H., Jr., 1945, *Relation of Phenomenal Rise of Water Levels to a Defective Gas Well, Harris County, Texas,* Bull. Am. Assoc. Petroleum Geologists, v. 29, pp. 253-279.

RUSSELL, R. R., 1963, *Ground-Water Levels in Illinois through 1961,* Rept. Inv. 45, Illinois State Water Survey, Urbana, 51 pp.

SAINES, MARVIN, 1981, *Errors in Interpretation of Ground-water Level Data,* Ground Water Monitoring Review, v. 1, no. 1, pp. 56-61.

SANTOS, C., 1973, *Prediction of the 1972 Managua, Nicaragua, Earthquake from Ground Water Changes, Inferred Probability of Earthquakes in the City of Managua, Nicaragua, During the Summer of 1973,* NTIS AD 762 134, Springfield, Virginia 22161.

SASMAN, R. T., and others, 1977, *Water-Level Decline and Pumpage in Deep Wells in Chicago Region, 1971-1975,* Illinois State Water Survey, Urbana, Illinois 61801, 35 pp.

SCHNEIDER, R., 1961, *Correlation of Ground-Water Levels and Air Temperatures in the Winter and Spring in Minnesota,* U. S. Geol. Survey Water-Supply Paper 1539-D, 14 pp.

SMEDEMA, L. B., and ZWERMAN, P. J., 1967, *Fluctuations of the Phreatic Surface: 1. Role of Entrapped Air Under a Temperature Gradient,* Soil Sci., v. 103, pp. 354-359.

SOKOL, DANIEL, 1963, *Position and Fluctuations of Water Level in Wells Perforated in more than one Aquifer,* J. Geophys. Res., v. 68, no. 4, pp. 1079-1080.

SOREN, JULIAN, 1976, *Basement Flooding and Foundation Damage from Watertable Rise in the East New York Section of Brooklyn, Long Island, NY,* NTIS Accession No. PB-261 190/AS.

STALLMAN, R. W., 1955, *Numerical Analysis of Regional Water Levels to Define Aquifer Hydrology* (abstract), Trans. Am. Geophys. Union, v. 36, no. 3.

STALLMAN, R. W., 1956, *Numerical Analysis of Regional Water Levels to Define Aquifer Hydrology,* Trans. Am. Geophys. Union, v. 37, no. 4.

STALLMAN, R. W., 1961, *Relation Between Storage Changes at the Water Table and Observed Water-Level Changes,* U. S. Geol. Survey Prof. Paper 424-B, pp. 39-40.

STALLMAN, R. W., 1965, *Effects of Water-Table Conditions on Water-Level Changes near Pumping Wells,* Water Resources Res., v. 1, no. 2.

STEGGEWENTZ, J. H., 1933, *De Invloed van de Getijbeweging van Zeeen en Getijrivieren op de Stijghoogte van Grondwater,* (tidal fluctuation) Thesis, Technische Hogeschool, Delft, The Netherlands.

STEWART, J. W., 1961, *Tidal Fluctuations of Water Levels in Wells in Crystalline Rocks in North Georgia,* U. S. Geol. Survey Prof. Paper 424-B, pp. 107-109.

SULAM, D. J., 1979, *Analysis of Changes in Ground-Water Levels in a Sewered and an Unsewered Area of Nassau County, Long Island, New York,* Ground Water, v. 17, no. 5, Sept.-Oct., pp. 446-455.

TASKER, G. D., and GUSWA, J. H., 1978, *Application of a Mathematical Model to Estimate Water Levels,* Ground Water, v. 16, no. 1, pp. 18-21.

THOMAS, H. E., 1940, *Fluctuations of Ground-Water Levels During the Earthquakes of November 10, 1938 and January 24, 1939,* Bull. Seismological Soc. Am. v. 30, pp. 93-97.

TROUSDELL, K. B., and HOOVER, M. D., 1955, *A Change in Ground-Water Level after Clearcutting of Loblolly Pine in the Coastal Plain,* J. Forestry, v. 53, pp. 493-498.

TROXELL, H. C., 1936, *The Diurnal Fluctuation in the Ground-Water and Flow of the Santa Ana River and its Meaning,* Trans. Am. Geophys. Union, v. 17, pp. 496-504.

TUINZAAD, H., 1954, *Influence of the Atmospheric Pressure on the Head of Artesian Water and Phreatic Water,* General Assembly Rome, Internat. Assoc. Sci. Hydrology, v. 2, pp. 32-37.

TURN, L. J., 1975, *Diurnal Fluctuations of Water Tables Induced by Atmospheric Pressure Changes,* J. Hydrology, v. 26, pp. 1-16.

U. S. GEOLOGICAL SURVEY, *Ground-Water Levels in the United States,* Water-Supply Papers, published intermittently.

VACHER, H. L., 1978, *Hydrology of Small Oceanic Islands — Influence of Atmospheric Pressure on the Water Table,* Ground Water, v. 16, no. 6, Nov.-Dec., pp. 417-424.

VANDENBERG, A., 1980, *Regional Ground-Water Motion in Response to an Oscillating Water Table,* J. Hydrology, v. 47, no. 3/4, July.

VAN DER KAMP, G. S. J. P., 1972, *Tidal Fluctuations in a Confined Aquifer Extending Under the Sea,* Internat. Geol. Congress, 24th Session, Section II — Hydrogeology, Montreal, pp. 101-106.

VEATCH, A. C., 1906, *Fluctuations of the Water Level in Wells with Special Reference to Long Island, New York,* U. S. Geol. Survey Water Supply Paper 155, 83 pp.

VENETIS, C., 1971, *Estimating Infiltration and/or the Parameters of Unconfined Aquifers from Ground Water-Level Observations,* J. Hydrology, v. 12, pp. 161-169. pp. 47-52.

VISWANATHAN, M. N., 1983, *The Rainfall/Water-Table Level Relationship of an Unconfined Aquifer,* Ground Water, v. 21, no. 1, pp. 49-56.

VORHIS, R. C., 1955, *Interpretation of Hydrologic Data Resulting from Earthquakes,* Geol. Rundschau, v. 43, pp. 47-52.

VORHIS, R. C., 1964, *Earthquake-Induced Water-Level Fluctuations from a Well in Dawson County, Georgia,* Seismological Soc. Am. Bull., v. 54, pp. 1023-1133.

VORHIS, R. C., 1967, *Hydrologic Effects of the Earthquake of March 27, 1964, Outside Alaska,* U. S. Geol. Survey Prof. Paper 544-C, 54 pp.

WEEKS, E. P., 1979, *Barometric Fluctuations in Wells Tapping Deep Unconfined Aquifers,* Water Resources Res., v. 15, no. 5, Oct. p. 1167.

WENZEL, L. K., 1936, *Several Methods of Studying Fluctuations of Ground-Water Levels,* Trans. Am. Geophys. Union, v. 17, pp. 400-405.

WERNER, P. W., 1946, *Notes on Flow-Time Effects in the Great Artesian Aquifers of the Earth,* Trans. Am. Geophys. Union, v. 27, pp. 687-708.

WERNER, P. W., and NOREN, D., 1951, *Progressive Waves in Non-Artesian Aquifers,* Trans. Am. Geophys. Union, v. 32, pp. 238-244.

WHITE, W. N., 1932, *A Method of Estimating Ground-Water Supplies Based on Discharge by Plants and Evaporation from Soil,* U. S. Geol. Survey Water-Supply Paper 659, pp. 1-105.

WILLIAMS, J., and LIU, T., 1971, *The Response to Tidal Fluctuations of Two Nonhomogeneous Coastal Aquifer Models,* Hawaii Water Resources Res. Center Tech. Rep. No. 51.

WILLIAMS, J., and LIU, T., 1973, *The Response to Tidal Fluctuations of a Leaky Aquifer System,* Hawaii Water Resources Res. Center Tech. Rep. No. 66.

WILLIAMSON, R. E., and CARREKER, J. R., 1970, *Effect of Water-Table Levels on Evapotranspiration and Crop Yield,* Trans. Am. Soc. Agr. Engrs., v. 13, no. 2, Mar.-Apr.

WINOGRAD, I. J., 1970, *Noninstrumental Factors Affecting Measurement of Static Water Levels in Deeply Buried Aquifers and Aquitards, Nevada Test Site,* Ground Water, v. 8, no. 2, pp. 19-29.

ZALTSBERG, E., 1984, *Application of Statistical Methods to Forecasting of Natural Groundwater Tables,* Canadian J. Earth Sci., v. 19, p. 1486.

ZAPOROZEC, ALEXANDER, 1980, *Drought and Ground-Water Levels in Northern Wisconsin,* Geoscience Wisconsin, v. 5, June. Univ. Wisconsin-Extension, Geological and Natural History Survey, Madison Wisconsin, 53706, 92 pp.

ZONES, C. P., 1957, *Changes in Hydraulic Conditions in the Dixie Valley Areas, Nevada, After the Earthquake of December 16, 1954,* Seismological Soc. Am. Bull, v. 47, pp. 387-396.

SPRINGS/SEEPS/GEOTHERMAL RESOURCES/ENERGY STORAGE/HEAT PUMP

BIBLIOGRAPHY

HEIKEN, G., and SAYER, S., 1980, *Bibliography of the Geological and Geophysical Aspects of Hot Dry Rock Geothermal Resources — 1979,* P. O. Box 1663, MS575, Univ. of Calif., Los Alamos Scientific Lab., Los Alamos, New Mexico 87545.

NATIONAL WATER WELL ASSOCIATION, 1980, *A Ground-Water Heat Pump Anthology,* 500 West Wilson Bridge Rd., Worthington, Ohio 43085.

SUMMERS, W. K., 1972, *Annotated and Indexed Bibliography of Geothermal Phenomena,* New Mexico Bur. of Mines and Mineral Resources, Socorro, New Mexico, 665 pp.

GENERAL WORKS

AIRCONDITIONING AND REFRIGERATION INSTITUTE, 1982, *Standard for Ground Water Source Heat Pumps,* ARI, 1815 Fort Myer Dr., Arlington, Virginia 22209. (definitions, testing requirements, specifications).

ANDERSON, D. N., and AXTELL, L. H., 1972, *Geothermal Overview of the Western United States,* Geothermal Resources Council, P. O. Box 1033, Davis, California 95616, Spec. Rept. No. 1, 200 pp.

ANDREWS, C. B., 1978, *The Impact of the Use of Heat Pumps on Ground-Water Temperatures,* Ground Water, v. 16, no. 6, Nov.-Dec., pp. 437-444.

ANONYMOUS, 1981, *Atlas of Subsurface Temperatures in the European Community,* (45-map atlas comprising terrestrial heat flow and groundwater circulation) Th. Schafer, Gmbh, Tivolistrasse 4, D-3000, Hanover 1, West Germany.

BACON, DOUG, 1980, *Environmental Implications of Widespread Use of the Ground-Water Geothermal Heat Pump,* Ground Water Heat Pump Journal, v. 1, no. 2, Summer.

BAHLS, L. L., and MILLER, M. R., 1973, *Saline Seeps in Montana,* in Second Annual Report, Montana Environmental Quality Council, pp. 35-44.

BAKER, C. H., Jr., 1968, *Thermal Springs Near Midway, Utah,* U. S. Geol. Survey Prof. Paper 600-D, pp. 63-70.

BEDINGER, M. S., and others, 1979, *The Waters of Hot Springs National Park, Arkansas — Their Nature and Origin,* U. S. Geol. Survey Prof. Paper 1044-C.

BENSEMAN, R. F., 1959, *Subsurface Discharge from Thermal Springs,* J. Geophys. Res. v. 64, no. 8.

BOUWER, HERMAN, 1979, *Geothermal Power Production with Irrigation Waste Water,* Ground Water, v. 17, no. 4, pp. 375-384.

BOWEN, ROBERT, 1980, *Geothermal Resources,* Halsted Press, John Wiley & Sons, Inc., One Wiley Dr., Somerset, New Jersey 08873.

BROWN, P.L., and others, 1983, *Saline-Seep Diagnosis, Control, and Reclamation,* Agric. Research Service, Beltsvile, MD., USDA/ARS-CRR-30, NTIS PB83-214213, 27 pp.

BROWNELL, D. H., Jr., GARG, S. K., and PRITCHETT, J. W., 1977, *Governing Equations for Geothermal Reservoirs,* Water Resour. Res., v. 13, no. 6, pp. 929-934.

BRUNE, GUNNAR, 1975, *Major and Historical Springs of Texas,* Texas Water Development Board Rep. 189., P. O. Box 13087, Austin, Texas 78711.

BRYAN, K., 1919, *Classification of Springs,* J. Geol., v. 27, pp. 522-561.

BURDON, D. J., and SAFADI, C., 1963, *Ras-el-Ain: The Great Karst Spring of Mesopotamia, an Hydrogeologic Study,* J. Hydrology, v. 1, pp. 58-95.

CALIFORNIA DEPT. OF CONSERVATION, DIV. OF MINES AND GEOLOGY, 1965, *California Laws for Conservation of Geothermal Energy.*

CALIFORNIA DEPT. OF CONSERVATION, DIV. OF MINES AND GEOLOGY, 1966, *Geothermal Resources in California,* Mineral Inf. Service, v. 20, no. 7, June.

CALIFORNIA DEPT. OF CONSERVATION, DIV. OF MINES AND GEOLOGY, 1968, *Geothermal Resources*, Mineral Inf. Service, v. 21, no. 2, Feb.

CALIFORNIA DEPT. OF WATER RESOURCES, 1967, *Investigation of Geothermal Waters in the Long Valley Area, Mono County*, Office Rept. July.

CALIFORNIA DEPT. OF WATER RESOURCES, 1970, *Geothermal Wastes and the Water Resources of the Salton Sea Area*, Bull. 143-7, 126 pp.

CALIFORNIA LEGISLATURE, 1967, *Geothermal Resources-Foundation for a Potentially Significant New Industry in California*, Senate Permanent Fact Finding Comm., 4th Progress Rept, Regular Session.

CALIFORNIA LEGISLATURE, 1967, *Senate Bill No. 169* — Geothermal Resources Legislation.

CLYDE, C. G., and others, 1980, *Ground-Water Heat Pump Equipment Selection Procedures for Architects, Designers, and Contractors*, Utah State University, Mechanical Engineering Dept., Water Research Lab., Logan 84322.

CRAIG, H., 1966, *Isotopic Composition and Origin of the Red Sea and Salton Sea Geothermal Brines*, Sci., v. 154, no. 3756.

CRAIG, H., 1969, *Discussion — Source Fluids for the Salton Sea Geothermal System*, J. Am. Sci., v. 267, Feb.

DEFFERDING, L. J., 1979, *State-of-the-Art of Liquid Waste Disposal for Geothermal Energy Systems*, Battelle Pacific Northwest Labs, Richland, Washington, NTIS Springfield, Virginia 22161, DOE/EV-0083, 252 pp.

DEXHEIMER, R. D., 1984, *Water Source Heat Pump Handbook*, National Water Well Assoc. (NWWA), Dublin, Ohio 43017, 241 pp.

DOERING, E. J., and SANDOVAL, F. M., 1976, *Hydrology of Saline Seeps in the Northern Great Plains*, Trans. Am. Soc. Agric. Engrs., v. 19, no. 5, pp. 856-861.

DUTCHER, L. C., 1972, *Preliminary Appraisal of Ground Water in Storage With Reference to Geothermal Resources in the Imperial Valley Area, California*, U. S. Geol. Survey Circ. 649, 57 pp.

FISHER, W. A., and others, 1966, *Freshwater Springs of Hawaii from Infrared Images*, U. S. Geol. Survey, Hydrologic Atlas 218.

FOURNIER, R. O., and MORGENSTERN, J. C., 1971, *A Device for Measuring Down-Hole Pressures and for Sampling Fluids in Geothermal Wells*, U. S. Geol. Survey Prof. Paper 750-C, pp. 146-150; *A Device for Collecting Down-Hole Water and Gas Samples in Geothermal Wells*, pp. 151-155.

GAITHER, B. E., 1977, *The Relation of Spring Discharge Behavior to the Hydrologic Properties of Carbonate Aquifers*, M. S. Thesis, Geology, Penn. State U., 210 pp.

HALVORSEN, A. D., and BLACK, A. L., 1974, *Saline-Seep Development in Dryland Soils of Northeastern Montana*, J. Soil and Water Conservation, v. 29, no. 2, Mar.-Apr.

HARBRIDGE HOUSE, INC., 1980, *Ground Water and Energy*, Proc. National Workshop May 1980, NTIS, Springfield, Virginia 22161, CONF-800137 (Sum.), 26 pp.

HELGESON, H. C., 1968, *Geologic and Thermodynamic Characteristics of the Salton Sea Geothermal System*, J. Am. Sci., v. 266.

HEMLEY, J. J., 1967, *Aqueous Solutions and Hydrothermal Activity*, Trans. Am. Geophys. Union, v. 48, no. 2, June.

HOBBA, W. A., JR., and others, 1979, *Hydrology and Geochemistry of Thermal Springs of the Appalachians*, U. S. Geol. Survey Prof. Paper 1044-E.

INGEBRITSEN, S. E., and SOREY, M. L., 1985, *A Quantitative Analysis of the Lassen Hydrothermal System, North Central California*, Water Resources Res., v. 21, no. 6, June.

JONES, J. W., Sr., 1980, *Engineering Report: A High Efficiency Ground-Water Heat Pump System*, Ground Water Heat Pump Journal, v. 1, no. 2, Summer.

KARANJAC, J. and GUNAY, G., 1980, *Dumauli Spring, Turkey — The Largest Karstic Spring in the World ?*, J. Hydrology, v. 45, no. 3/4, Feb.

KAZMANN, RAPHAEL, and WHITEHEAD, WALTER, 1980, *The Spacing of Heat Pump Supply and Discharge Wells*, Ground Water Heat Pump Journal, v. 1, no. 2, Summer.

KEEFER, W. R., 1971, *The Geologic Story of Yellowstone National Park*, U. S. Geol. Survey Bull. 1347, 92 pp.

KLEY, WOLFRAM, and HEEKMANN, WINFRIED, 1981, *Waste Heat Balance in Aquifers Calculated by a Computer Programme*, Ground Water, v. 19, no. 2, March-April, pp. 144-148.

KOENIG, J. B., 1969, *Geothermal Exploration in the Western United States*, United Nations Symp. on Development and Utilization of Geothermal Resources, New York.

KRIZ, H., 1973, *Processing of Results of Observations of Spring Discharge*, Ground Water, v. 11, no. 5, pp. 3-14.

KRON, A., and STIX, J., 1983, *The 1982 Geothermal Gradient Map of the United States,* Nat. Geophys. Data Center, NOAA, Code E/GC1, 325 Broadway, Boulder, Colorado 80303.

KRUGER, P., and OTTE, C., 1973, *Geothermal Energy*, Stanford Univ. Press, Stanford, California, 372 pp.

LYFORD, F. P., 1972, *The Nature and Extent of Peat Deposits and Possible Effects of Peat Mining on Manmade Features and Springs near Mescalero New Mexico*, U. S. Geol. Survey Open-file Rept., 24 pp.

McKAY, HAROLD, 1978, *McKay's Guide to Hot Mineral Springs and Spas in the United States and Canada*, McKay & Associates, Inc., 3719 Colonial Drive, Las Vegas, Nevada 89121.

McNITT, J. R., 1960, *Geothermal Power*, Calif. Div. Mines Info. Serv., v. 13, no. 3.

McNITT, J. R., 1968, *Worldwide Development of Geothermal Industry*, Symp. Geothermal Resources, Am. Assoc. Petroleum Geologists, Bakersfield, Calif., March.

MAJKRZAK, D. W., 1983, *Investigation of Thermal Impacts on a Shallow Aquifer from Recharge of a Ground Water Heat Pump System*, PB83-182063, NTIS, 5285 Port Royal Rd., Springfield, Virginia 22161, 60 pp.

MANN, J. A., and CHERRY, R. N., 1970, *Large Springs of Florida's "Sun Coast" — Citrus and Hernando Countries*, Bureau of Geology, Florida Dept. of Nat. Resources, Tallahassee, Leaflet No. 9.

MEINZER, O. E., 1927, *Large Springs in the United States*, U. S. Geol. Survey Water-Supply Paper 557, 94 pp.

MERCADO, G. S., 1960, *Aspectos Quimicos del Aprovechamiento de la Energia Geotermico, Campo Cerro Prieto B.C.*, Com. Federal Electricidad y Com. Energia Geotermica de Mexico, Mexicali, B.C., Oct.

MERCADO, G. S., 1969, *Cerro Prieto Geothermal Field, Baja California, Mexico*, Com. Energia Geotermico, Trans. Am. Geophys. Union, EOS v. 50, no. 2, Feb.

MEYER, C. F., and TODD, D. K., 1973, *Heat-Storage Wells*, Water Well Journal, v. 27, no. 10, pp. 35-41, Oct.

MEYER, C. F., and TODD, D. K., 1973, *Conserving Energy with Heat Storage Wells*, Environmental Sci. and Technology, v. 7, pp. 512-516.

MILLER, R. T., *Anisotropy in the Ironton and Galesville Sandstones Near a Thermal-Energy-Storage Well, St. Paul, Minnesota*, Ground Water, v. 22, no. 5, pp. 532-538.

MOLZ, F. J., and others, 1979, *Thermal Energy Storage in a Confined Aquifer: Experimental Results*, Water Resources Res., v. 15, p. 1509.

MUNDORF, J. C., 1971, *Nonthermal Springs of Utah*, Utah Geol. and Mineral Survey Bull. 16, Salt Lake City, Utah, 84108, 70 pp.

ORIO, CARL, 1980, *Normal and Extraordinary Maintenance of the Ground-Water Geothermal Heat Pump*, Ground Water Heat Pump Journal, v. 1, no. 2, Summer.

PARKER, J. D., and others, *The ASHRAE Design/Data Manual for Ground-Coupled Heat Pumps*, Am. Soc. Heating, Refrigerating and Air-Conditioning Engineers, 1791 Tullie Circle NE, Atlanta, Georgia 30329, 312 pp.

PARTIN, J. R., 1980, *Residential Heat Pumps: Some Alternatives*, Ground Water Heat Pump Journal, v. 1, no. 3, Fall.

PEARL, R. H., 1976, *Hydrologic Problems Associated with Developing Geothermal Energy Systems*, Ground Water, v. 14, no. 3, pp. 128-137.

PERRAULT, P., 1957, *On the Origin of Springs*, Transl. by A. LaRocque, Hafner, New York, 209 pp.

PRINZ, E., and KAMPE, R., 1934, *Handbuch der Hydrologie, Band II: Quellen*, J. Springer, Berlin, 290 pp.

RANDALL, W., and others, 1968, *Electrical Resistivity and Geochemistry of Aquifers in the Durmid Dome, Imperial Valley*, Trans. Am. Geophys. Union, v. 49, no. 4, Abstracts, Dec.

RAWLINGS, PHIL, 1985, *The State of the Art: Closed-Loop Earth-Coupled Heat Pumps*, Water Well J., v. 39, no. 7, Jul.

REX, R. W., 1966, *Heat Flow in the Imperial Valley of California*, Trans. Am. Geophys. Union, v. 47, no. 1, Abstracts.

REX, R. W., 1968, *Geochemical Water Facies in the Imperial Valley of California*, Trans. Am. Geophys. Union, v. 49, no. 4, Abstracts, Dec.

REX, R. W., 1968, *Investigation of the Geothermal Potential of the Lower Colorado River Basin, Phase 1 — The Imperial Valley Project*, Inst. of Geophysics and Planetary Physics, Univ. of California, Riverside, California, Oct. 30.

RINEHART, J. S., 1974, *Geysers*, EOS, Trans. Am. Geophys. Union, v. 56, no. 12, pp. 1052-1062.

RINEHART, J. S., 1980, *Geysers & Geothermal Energy*, Springer-Verlag New York, Inc., P. O. Box 2485, Secaucus, New Jersey 07094, 256 pp.

SABINS, F. F., 1967, *Infrared Imagery and Geologic Aspects, Indio Hills*, Photogrammetric Eng., July.

SAMMEL, E. A., 1984, *Analysis and Interpretation of Data Obtained in Tests of the Geothermal Aquifer at Klamath Falls, Oregon*, U. S. Geol. Survey, WRI 84-4216, 158 pp.

SAUTY, J. P., and others, 1982, *Sensible Energy Storage in Aquifers. 1. Theoretical Study, 2. Field Experiments and Comparison with Theoretical Results*, Water Resources Res., v. 18, no. 2, Apr.

SCHNEIDER, ROBERT, 1972, *Distortion of the Geothermal Field in Aquifers by Pumping*, U. S. Geol. Survey Prof. Paper 800-C, Geol. Survey Research 1972, pp. C267-C270.

SMITH, E., 1979, *Spring Discharge in Relation to Rapid Fissure Flow*, Ground Water, v. 17, no. 4, Jul.-Aug., pp. 346-350.

SOUTHER, J. G., and HALSTEAD, E. C., 1973, *Mineral and Thermal Waters of Canada*, Canada Geol. Survey Paper 73-18, Ottawa, Ontario.

SPICER, H. C., 1936, *Rock Temperatures and Depths to Normal Boiling Point of Water in the United States*, Bull. Am. Assoc. Petroleum Geologists, pp. 989-1021.

STEARNS, N. D., and others, 1937, *Thermal Springs in the United States,* U. S. Geol. Survey Water-Supply Paper 679-B, pp. 59-191.

SUN, P-C. P., and others, 1963, *Large Springs of East Tennessee,* U. S. Geol. Survey Water-Supply Paper 1755, 52 pp.

THOMPSON, J. M., 1985, *Chemistry of Thermal and Nonthermal Springs in the Vicinity of Lassen Volcanic National Park,* J. Volcanology and Geothermal Res., v. 25, no. 1-2, pp. 81-104.

TYBACH, L. and MUFFLER, L. J. P., 1981, *Geothermal Systems: Principles and Case Histories,* Wiley-Interscience, New York, New York 10016, 360 pp.

UNITED NATIONS, 1964, *Geothermal Energy,* Proc. Conf. New Sources of Energy, Rome, v. 2 and 3 Aug.

U. S. CONFERENCE OF MAYORS, 1984, *Ground Water and Geothermal: Urban District Heating Applications,* USCM, Office of Development Programs, 1620 Eye St., N. W., Washington, DC 20006, 91 pp.

U. S. DEPARTMENT OF ENERGY, 1980, *Ground Water and Energy,* Rept. National Workshop, Albuquerque, New Mexico, Jan 29-31, 1980, NTIS CONF-800137-Summary, Springfield, Virginia 22161.

U. S. DEPARTMENT OF ENERGY, DIV. OF GEOTHERMAL ENERGY, 1981, *Geothermal Resource Maps of Idaho, Colorado, Utah, New Mexico and California, as well as Thermal Springs List for the United States,* NOAA/NGSDC, Data Mapping Group, Code D64, Boulder, Colorado 80303.

VAN DER KAMP, G., 1982, *Interactions Between Heat Flow and Groundwater Flow - A Review,* Earth Physics Branch Open File No. 82-19, K. G. Campbell Corp., 880 Wellington St., Ottawa, Ontario, Canada K1R 6K7, 82 pp.

VAUX, H. J. Jr., 1977, *The Production of Water from Geothermal Reservoirs: Some Economic Considerations,* Water Resour. Bull., v. 13, no. 3, pp. 469-478.

WARING, G. A., 1951, *Summary of Literature on Thermal Springs,* General Assembly Brussels, Internat. Assoc. Sci. Hydrology, v. 2, pp. 289-293.

WARING, G. A., revised by BLANKENSHIP, R. R. and BENTALL, RAY, 1965, *Thermal Springs of the United States and other Countries of the World — A Summary,* U. S. Geol. Survey Prof. Paper 492, 383 pp.

WEISS, R. B., COFFEY, T. O., and WILLIAMS, T. L., 1980, *Geothermal Environmental Impact Assessment: Ground-Water Monitoring Guidelines for Geothermal Development*, PB80-144801, NTIS, 5285 Port Royal Rd., Springfield, Virginia 22161.

WHITE, D. E., 1957, *Thermal Waters of Volcanic Origin*, Bull. Geol. Soc. Am., v. 68, pp. 1637-1658.

WHITE, D. E., 1957, *Magmatic, Connate, and Metamorphic Waters*, Bull. Geol. Soc. Am., v. 68, pp. 1659-1682.

WHITE, D. E., 1965, *Geothermal Energy*, U. S. Geol. Survey Circ. 519, 17 pp.

WHITE, D. E., and BRANNOCK, W. W., 1950, *The Sources of Heat and Water Supply of Thermal Springs, with Particular Reference to Steamboat Springs, Nevada*, Trans. Am. Geophys. Union, v. 31, pp. 566-574.

WHITE, D. E., and others, 1963, *Geothermal Brine Well: Mile-Deep Drill Hole May Tap Ore-Bearing Magmatic Water and Rocks Undergoing Metamorphism*, Sci. v. 139, pp. 919-922.

WIBERG, N. E., 1983, *Heat Storage in Aquifers Analyzed by Finite Element Method,* Ground Water, v. 21, no. 2, pp. 178-187.

YOUNG, H. W., and MITCHELL, J. C., 1973, *Geothermal Investigations in Idaho, Part I, Geochemistry and Geologic Setting of Selected Thermal Waters*, Idaho Dept. of Water Resources Water Info. Bull, No. 30.

THEORY OF GROUNDWATER FLOW

AGUADO, E., SITAR, NICHOLAS, and REMSON, IRWIN, 1977, *Sensitivity Analysis in Aquifer Studies*, Water Resources Res., v. 13, no. 4, pp. 733-737.

AHMED, NAZEER, and SUNADA, D. K., 1969, *Nonlinear Flow in Porous Media*, J. Hydraulics Div. Am. Soc. Civil Engrs., v. 95, no. HY 6, Nov., pp. 1847-1857.

AMERICAN PETROLEUM INSTITUTE, 1957, *Microscopic Behavior of Fluids in Porous Systems*, API Research Proj. 47b, Dallas, Texas.

ARAVIN, V. L., and NUMEROV, S. N., 1953, *Theory of Fluid Flow in Undeformable Porous Media*, Jerusalem, Israel Program for Sci. Transl. 1965. (Russian original published in Moscow).

ARON, G., and others, *Cyclic Pumping for Drainage Purposes*, Ground Water, v. 5, no. 1, pp. 35-38, Jan.-Feb.

AVERY, S. B., 1953, *Analysis of Ground-Water Lowering Adjacent to Open Water*, Trans. Am. Soc. Civil Engrs., v. 118, pp. 178-208.

BABBITT, H. E., and CALDWELL, D. H., 1948, *The Free Surface Around, and the Interference Between, Gravity Wells*, Univ. Illinois Eng. Exp. Sta. Bull. 374, v. 45, no. 30, 60 pp.

BACHMAT, Y., and BEAR, J., 1964, *The General Equations of Hydrodynamic Dispersion in Homogeneous, Isotropic Porous Mediums*, J. Geophysical Research, v. 69, pp. 2561-2567.

BACHMAT, Y. and others, 1984, *Development of Single-Well Techniques for Quantitative Ground Water Studies,* Gesellschaft fuer Strahlen- und Umweltforschung m.b.H., Neuherberg bei Munich (West Germany), in German, DE85750072/WEP, NTIS, Springfield, Virginia 22161, 85 pp. (single-well pulse technique to measure dispersivity)

BACK, WILLIAM, and FREEZE, R. A., 1983, *Physical Hydrogeology,* Scientific and Academic Editions, 35 W. 50th St., New York, New York 10020, 448 pp. (papers emphasizing groundwater research in North America).

BAKHMETEFF, B. A., and FEODOROFF, N. V., 1937, *Flow Through Granular Media*, J. Appl. Mech., v. 4A, p. 97; Discussion, v. 5A, also 1937, pp. 86-90.

BAKR, A. A., and others, 1978, *Stochastic Analysis of Spatial Variability in Subsurface Flows, 1, Comparison of One- and Three-dimensional Flows*, Water Resour. Res., v. 14, no. 2, pp. 263-271.

BALDWIN, G. V., and McGUINNESS, C. L., 1963, *A Primer on Ground Water*, U. S. Geol. Survey, 26 pp.

BASAK, P., and MURTY, V. V. N., 1978, *Pollution of Ground Water Through Nonlinear Diffusion*, J. Hydrology, v. 38, no. 3/4, Aug.

BEAR, JACOB, and DAGAN, G., 1965, *The Relationship Between Solutions of Flow Problems in Isotropic and Anisotropic Soils*, J. Hydrology, v. 3, pp. 88-96.

BEAR, JACOB, and others, 1968, *Physical Principles of Water Percolation and Seepage,* Arid Zone Research, v. 29, UNIPUB, Inc., New York.

BEAR, JACOB, and PINDER, G. F., 1978, *Porous Medium Deformation in Multiphase Flow*, J. Engineering Mechanics Div. Am. Soc. Civil Engrs., v. 104, no. EM4, Aug.

BENNETT, G. D., 1976, *Introduction to Ground-Water Hydraulics — A Programmed Text for Self-Instruction*, Techniques of Water-Resources Investigations of the United States Geological Survey, Book 3, Chaper B2, 172 pp.

BIANCHI, W. C., and HASKELL, Jr., E. E., 1966, *Air in the Vadose Zone as it Affects Water Movements Beneath a Recharge Basin*, Water Resources Res., v. 2, no. 2, pp. 315-322.

BOAST, C. W., and KIRKHAM, D., 1971, *Auger Hole Seepage Theory*, Soil Sci. Soc. Amer. Proc., v. 35, pp. 365-373.

BOGOMOLOV, G. V., and others, 1972, *The Principles of Paleohydrogeological Reconstruction of Ground-Water Formation*, Internat. Geol. Congress, 24th Session, Montreal, Section II — Hydrogeology, pp. 205-207.

BOKHARI, S. M. H., and others, 1968, *Drawdowns due to Pumping from Strip Aquifers*, J. Irrig. and Drainage Div., Am. Soc. Civil Engrs., v. 94, no. IR2, June.

BORELI, M., 1955, *Free-Surface Flow Toward Partially Penetrating Wells*, Trans. Am. Geophys. Union, v. 36, pp. 664-672.

BOSTOCK, C. A., 1971, *Estimating Truncation Error in Image-Well Theory*, Water Resources Res., v. 7, no. 6. pp. 1658-1660.

BOULTON, N. S., 1951, *The Flow Pattern near a Gravity Well in a Uniform Water-Bearing Medium*, J. Inst. Civil Engrs., v. 10.

BOULTON, N. S., 1954, *The Drawdown of the Water-table Under Non-steady Conditions Near a Pumped Well in an Unconfined Formation*, Proc. Inst. Civil Engrs., v. 3, pt. III, pp. 564-579.

BOULTON, N. S., 1954, *Unsteady Radial Flow to a Pumped Well Allowing for Delayed Yield from Storage*, Internat. Assoc. Sci. Hydrology, Gen. Assembly Rome, v. 2, Pub. 37.

BOULTON, N. S., 1973, *The Influence of Delayed Drainage on Data from Pumping Tests in Unconfined Aquifers*, J. Hydrology, v. 19, no. 2, pp. 157-169, June.

BOULTON, N. S., and PONTIN, J. M. A., 1971, *An Extended Theory of Delayed Yield from Storage Applied to Pumping Tests in Uncofined Anisotropic Aquifers*, J. Hydrology, v. 14, no. 1, Oct.

BOUWER, HERMAN, 1964, *Unsaturated Flow in Ground-Water Hydraulics*, J. Hydraulics Div., Amer. Soc. Civil Engrs., v. 90, no. HY5, pp. 121-144.

BOUWER, HERMAN, and RICE, R. C., 1978, *Delayed Aquifer Yield as a Phenomenon of Delayed Air Entry*, Water Resources Res., v. 14, no. 6, Dec.

BREDEHOEFT, J. D., and PAPADOPULOS, I. S., 1965, *Rates of Vertical Groundwater Movement Estimated from the Earth's Thermal Profile*, Water Resources Res., v. 1, no. 2, pp. 325-328.

BRISCOE, JOHN, 1984, *The Optimal Spacing of Interfering Wells: An Analytical Solution*, Ground Water, v. 22, no. 5, pp. 573-578.

BROWNELL, L. E., and KATZ, D. L., 1947, *Flow of Fluids through Porous Media, I — Single Homogeneous Fluids*, Chem. Eng. Progress, v. 43, pp. 537-548.

BRUSTKERN, R. L., and MOREL-SEYTOUX, H. J., 1975, *Description of Water and Air Movement During Infiltration*, J. of Hydrology, v. 24, no. 1/2, Jan.

BRUTSAERT, W., and CORAPCIOGLU, M. Y., 1976, *Pumping of Aquifer with Viscoelastic Properties*, J. Hydraul. Div. Am. Soc. Civ. Eng., v. 102 (HY11), pp. 1663-1675.

BRUTSAERT, W. and others, 1961, *Predicted and Experimental Water Table Drawdown during Tile Drainage*, Hilgardia, v. 31, pp. 389-418, Nov.

CARNAHAN, C. L., *Non-Equilibrium Thermodynamics of Groundwater Flow Systems: Symmetry Properties of Phenomenological Coefficients and Consideration of Hydrodynamic Dispersion*, J. Hydrology, v. 31 (1/2), pp. 125-150.

CASAGRANDE, A., 1937 and 1940, *Seepage Through Dams*, J. New England Water Works Assoc., June 1937, also, J. of Boston Soc. Civ. Engrs. 1940, pp. 295-337.

CASE, C. M., 1971, *Projections of Darcy's Law as Related to the Ellipse of Direction*, Water Resources Res., v. 7, no. 5, pp. 1354-1356.

CASE, C. M., and COCHRAN, G. F., 1972, *Transformation of the Tensor Form of Darcy's Law in Inhomogenous and Anistropic Soils*, Water Resources Res., v. 8, no. 3, pp. 728-733.

CASE, C. M., and PECK, M. K., 1977, *Transform Approach to Solution of Groundwater Flow Equations*, J. Hydrology v. 32, (3/4), pp. 305-320.

CASTILLO, E., KARADI, G. M., and KRIZCK, R. J., 1972, *Unconfined Flow Through Jointed Rock*, Water Resources Bull., v. 8, no. 2, pp. 266-281.

CHANDLER, R. L., and McWHORTER, D. B., 1975, *Upconing of the Salt-Water Fresh-Water Interface Beneath a Pumping Well*, Ground Water, v. 13, no. 4, pp. 354-359.

CHILDS, E. C., 1945, *The Water Table, Equipotentials, and Streamlines in Drained Land*, Soil Sci., v. 59.

CHILDS, E. C., 1971, *Drainage of Groundwater Resting on a Sloping Bed*, Water Resources, Res., v. 7, no. 5, pp. 1256-1263.

CICCIOLI, PAOLO, and others, 1980, *Organic Solute-Mineral Surface Interactions: A New Method for the Determination of Ground-Water Velocities*, Water Resources Res., v. 16, no. 1.

COOLEY, R. L., and CASE, C. M., 1973, *Effect of a Water Table Aquitard on Drawdown in an Underlying Pumped Aquifer*, Water Resources Res., v. 9, no. 2, pp. 434-447.

COOLEY, R. L., FORDHAM, J. W., and WESTPHAL, J. A., 1973, *Some Applications of Statistical Methods to Ground-Water Flow Analysis*, Tech. Rep. Serv. H-W, Hydrology and Water Res. Pub. 14, Desert Research Institute, Univ. of Nevada.

COOLEY, R. L., and CUNNINGHAM, A. B., 1979, *Consideration of Total Energy Loss in Theory of Flow to Wells*, J. Hydrology, v. 43, no. 1/4, Oct.

COOPER, H. H., Jr., 1966, *The Equation of Groundwater Flow in Fixed and Deforming Coordinates*, J. Geohys. Res., v. 71, no. 20, pp. 4785-4790.

COOPER, H. H., Jr., and others, 1967, *Response of a Finite-Diameter Well to an Instantaneous Charge of Water*, Water Resources Res., v. 3, pp. 263-269.

CUSHMAN, J., and KIRKHAM, D., 1978, *Solute Travel Times to Wells in Single or Multiple Layered Aquifers*, J. Hydrol., v. 37, (1/2), pp. 169-184.

DAGAN, G., 1967, *A Method of Determining the Permeability and Effective Porosity of Unconfined Anisotropic Aquifers*, Water Resources Res., v. 3, no. 4, 1059-1071.

DAGAN, G., 1979, *The Generalization of Darcy's Law for Nonuniform Flows*, Water Resources Res., v. 15, no. 1, Feb.

DARCY, HENRI, 1856, *Les Fontaines Publiques de la Ville de Dijon*, Victor Dalmont, Paris, 647 pp.

DE GLEE, G. J., 1930, *Over Grondwaterstroomingen bij Wateronttrekking door Middel van Putten*, J. Waltman Jr., Delft, 175 pp.

DE JOSSELIN, DE JONG, G., 1958, *Longitudinal and Transverse Diffusion in Granular Deposits*, Trans. Am. Geophys. Union, v. 39, pp. 67-74.

DELORME, L. D., 1972, *Groundwater Flow Systems, Past and Present*, Internat. Geol. Congress, 24th Session, Montreal, Section II — Hydrogeology, pp. 222-226.

DEVER, R. J., Jr., and CLEARY, R. W., 1979, *Unsteady-State, Two-Dimensional Response of Leaky Aquifers to Stream Stage Fluctuations*, Advances in Water Resources, v. 2, no. 1, Mar.

DeVRIES, JAN, 1979, *Predictions of Non-Darcy Flow in Porous Media*, J. Irrigation and Drainage Div. Am. Soc. Civil Engrs., v. 105, no. IR2, June.

DeWIEST, R. J. M., 1961, *On the Theory of Leaky Aquifers*, J. Geophys. Res. v. 66, pp. 4257-4262.

DeWIEST, R. J. M., 1962, *Free Surface Flow in Homogeneous Porous Medium*, Trans. Am. Soc. Civil Engrs., v. 127, pt. 1, pp. 1045-1089.

DeWIEST, R. J. M., 1963, *Flow to an Eccentric Well in a Leaky Circular Aquifer with Varied Lateral Replenishment*, Geofisica Pura e Aplicata, v. 54, pp. 87-102.

DeWIEST, R. J. M., 1963, *Russian Contributions to the Theory of Ground-Water Flow*, Ground Water, v. 1, no. 1, pp. 44-48.

DeWIEST, R. J. M., 1964, *History of the Dupuit-Forchheimer Assumptions in Ground-Water Flow*, Paper presented at the Annual Winter Meeting of the Am. Soc. of Agric. Engrs., Dec., New Orleans.

DeWIEST, R. J. M., 1966, *On the Storage Coefficient and the Equations of Groundwater Flow*, Geophys. Res. v. 71, no. 4.

DOGRU, A. H., ALEXANDER, W., and PANTON, R. L., 1978, *Numerical Solution of Unsteady Flow Problems in Porous Media by Spline Functions*, J. Hydrol., v. 38 (1/2), pp. 179-195.

DUGUID, J. O., and LEE, P. C. Y., 1977, *Flow in Fractured Porous Media*, Water Resources Res., v. 13, no. 3, pp. 558-566.

DUPUIT, JULES, 1863, *Etudes Théoriques et Pratiques sur le Mouvement des Eaux dans les Canaux Découverts et à Travers les Terrains Perméables*, 2nd ed., Dunot, Paris, 304 pp.

EDELMAN, J. H., 1972, *Ground Water Hydraulics of Extensive Aquifers*, Bull. 13, Internat. Inst. for Land Reclamation and Improvement, Wageningen, The Netherlands, 216 pp.

EHLIG, C., and HALEPASKA, J. C., 1976, *A Numerical Study of Confined-Unconfined Aquifers Including Effects of Delayed Yield and Leakage*, Water Resources Res., v. 12, pp. 1175-1183.

ELDOR, M., and DAGAN, G., 1972, *Solutions of Hydrodynamic Dispersion in Porous Media*, Water Resources Res., v. 8, pp. 1316-1331.

ELIASON, O. L., and GARDNER, W., 1933, *Computing the Effective Diameter of a Well Battery by Means of Darcy's Law*, Agr. Eng., v. 14, pp. 53-54.

ENGELUND, F., 1951, *Mathematical Discussion of Drainage Problems*, Danish Acad. Tech. Sci. Trans. Bull. 3.

ENGELUND, F., 1953, *On the Laminar and Turbulent Flows of Groundwater through Homogeneous Sand*, Danish Tech. Sci. Trans. Bull. 4.

ERNST, L. F., 1962, *Ground-Water Flow in the Saturated Zone and its Calculation when Horizontal, Parallel Open Conduits are Present* (in Dutch), Centrum voor Landbouwpublikatie en Landbouwdocumentatie, Wageningen.

ERNST, L. F., 1978, *Drainage of Undulating Sandy Soils with High Ground-Water Tables. 1. A Drainage Formula Based on a Constant Hydraulic Head Ratio; 2. The Variable Hydraulic Head Ratio*, J. of Hydrology, v. 39, no. 1/2, Oct.

FAIR, G. M., and HATCH, L. P., 1933, *Fundamental Factors Governing the Streamline Flow of Water through Sand*, J. Am. Water Works Assoc., v. 25, pp. 1551-1565.

FERRIS, J. G., 1950, *A Quantitative Method for Determining Ground-Water Characteristics for Drainage Design*, Argic. Eng., v. 31, pp. 285-289.

FERRIS, J. G., 1959, *Ground Water*, in C. O. Wisler and E. F. Brater, *Hydrology*, Chap. 7, John Wiley & Sons, Inc, New York.

FERRIS, J. G., and others, 1962, *Theory of Aquifer Tests*, U. S. Geol. Survey Water-Supply Paper 1536-E, pp. 69-174.

FERRIS, J. G., and SAYRE, A. N., 1955, *The Quantitative Approach to Ground-Water Investigations*, Econ. Geology, 50th Anniv., pp. 714-747.

FOLEY, F. C., and others, 1953, *Ground-Water Conditions in the Milwaukee-Waukesha Area, Wisconsin*, U. S. Geol. Survey Water-Supply Paper 1229.

FORCHHEIMER, P., 1886, *Über die Ergiebigkeit von Brunnenanlagen und Sickerschlitzen*, Z. Architekt Ing-Verein, Hannover, v. 32, no. 7.

FRANKE, O. L., and COHEN, PHILIP, 1972, *Regional Rates of Ground-Water Movement on Long Island, New York*, U. S. Geol. Survey Prof. Paper 800-C, Geol. Survey Research 1972, pp. C271-C277.

FREEZE, R. A., 1971, *Influence of the Unsaturated Flow Domain on Seepage Through Earth Dams*, Water Resources Res., v. 7, no. 4, pp. 929-941.

FREEZE, R. A., 1972, *Role of Subsurface Flow in Generating Surface Runoff, 1. Base Flow Contributions to Channel Flow*, Water Resources Res., v. 8, no. 3, pp. 609-623; *2. Upstream Source Areas*, Water Resources Res., v. 8, no. 5, pp. 1272-1283.

FREEZE, R. A., 1975, *A Stochastic-Conceptual Analysis of One-Dimensional Groundwater Flow in Nonuniform Homogeneous Media*, Water Resources, Res., v. 11, no. 5, pp. 725-741.

FREEZE, R. A., and BANNER, J., 1969/1970, *The Mechanism of Natural Groundwater Recharge and Discharge*, Water Resources Res., v. 5, pp. 153-171, v. 6, pp. 138-155.

FREEZE, R. A., and WITHERSPOON, P. A., 1967, *Theoretical Analysis of Regional Groundwater Flow: 1. Analytical and Numerical Solutions to the Mathematical Model*, Water Resources Res., v. 2, no. 4, pp. 641-656.

FREEZE, R. A., and WITHERSPOON, P. A., 1967, *Theoretical Analysis of Regional Groundwater Flow, 2. Effect of Water-Table Configuration and Subsurface Permeability Variation*, Water Resources Res., v. 3, no. 2, 623-634.

FREEZE, R. A., and WITHERSPOON, P. A., 1968, *Theoretical Analysis of Regional Groundwater Flow, 3. Quantitative Interpretations*, Water Resources Res., v. 4, no. 3, pp. 581-590.

FREEZE, R. A., 1971, *Three-Dimensional, Transient, Saturated-Unsaturated Flow in a Groundwater Basin*, Water Resources Research, v. 7, pp. 347-366.

FRIND, E. O., 1979, *Exact Aquitard Response Functions for Multiple Aquifer Mechanics*, Advances in Water Resources, v. 2, no. 2, June.

FULLER, M. R., 1908, *Summary of the Controlling Factors of Artesian Flow*, U. S. Geol. Survey Bull. 319, 44 p.

GABRYSCH, R. K., 1968, *The Relation between Specific Capacity and Aquifer Transmissibility in the Houston Area, Texas*, Ground Water, v. 6, no. 4, pp. 9-14.

GAMBOLATI, GIUSEPPE, 1973, *Equation for One-Dimensional Vertical Flow of Groundwater, 1. The Rigorous Theory*, Water Resources Res., v. 9, no. 4, pp. 1022-1028.

GAMBOLATI, GIUSEPPE, 1973, *Equation for One-Dimensional Vertical Flow of Ground Water: 2, Validity Range of the Diffusion Equation*, Water Resources Res., v. 9, no. 5, pp. 1385-1395.

GAMBOLATI, GIUSEPPE, 1974, *Second-Order Theory of The Flow in Three-Dimensional Deforming Media*, Water Resources Res., v. 10, no. 6, p. 1217-1228.

GAMBOLATI, GIUSEPPE, 1976, *Transient Free Surface Flow to a Well: An Analysis of Theoretical Solutions*, Water Resources Res., v. 12, no. 1, pp. 27-39. (Formulas)

GAMBOLATI, GIUSEPPE, 1977, *Deviations from the Theis Solution in Aquifers Undergoing Three-Dimensional Consolidation*, Water Resources Res., v. 13, no. 1, pp. 62-68.

GARDNER, W., and others, 1928, *The Drainage of Land Overlying Artesian Basins*, Soil Sci., v. 26.

GELHAR, L. W., and COLLINS, M. A., 1971, *General Analysis of Longitudinal Dispersion in Nonuniform Flow*, Water Resources Research, v. 7, pp. 1511-1521.

GLOVER, R. E., 1966, *Ground-Water Movement*, U. S. Bureau of Reclamation Engrng. Monograph no. 31, Denver, 76 pp.

GLOVER, R. E., and BITTINGER, M. W., 1959, *Transient Ground Water Hydraulics*, Colorado State Univ. Pub. CER 59 REG 16, 57 pp.

GREEN, R. E., and COREY, J. C., 1971, *Calculation of Hydraulic Conductivity, a Further Evaluation of Some Predictive Methods*, Soil Sci. Soc. Am. Proc., 35, pp. 3-8.

GREEN, W. H., and AMPT, C. A., 1911, *Studies on Soil Physics, 1, Flow of Air and Water Through Soils*, J. Agric. Sci. v. 4, pp. 1-24.

GRAY, W. G., and PINDER, G. F., 1974, *Galerkin Approximation of the Time Derivative in the Finite Element Analysis of Groundwater Flow*, Water Resources Res., v. 10, no. 4, pp. 821-828.

GRIFFITHS, W. T., and IFE, D., 1984, *A Simplified Graphical Solution of the Theis Equation*, J. Hydrology, v. 73, no. 1/2.

GROVE, D. B., 1978, *The Use of Galerkin Finite-Element Methods to Solve Mass Transport Equations*, U. S. Geol. Survey Water-Resources Inv. 77-49, 61 pp.

GUITJENS, J. C., and LUTHIN, J. N., 1971, *Effect of Soil Moisture Hysteresis of the Water-table Profile Around a Gravity Well*, Water Resources Res., v 7, no. 2, pp. 334-346.

GUPTA, S. P., Sr., and GREENKORN, R. A., 1973, *Dispersion During Flow in Porous Media with Bilinear Absorption*, Water Resources Res., v. 9, no. 5, pp. 1357-1368.

GUPTA, S. P., Sr., and GREENKORN, R. A., 1976, *Solution for Radial Flow with Nonlinear Adsorption*, J. Environ. Eng. Div. Am. Soc. Civ. Eng., v. 102 (EE1), pp. 87-94.

GUYTON, W. F., 1941, *Applications of Coefficients of Transmissibility and Storage to Regional Problems in the Houston District*, Texas, Trans. Am. Geophys. Union, 21st Ann. Meeting, pt. 3.

HAGEN, G., 1939, *Bewegung des Wassers in Engen Cylindrischen Rohren*, Pogg. Ann. v. 47.

HALL, F. R., and MOENCH, A. F., 1972, *Application of the Convolution Equation to Stream Aquifer Relationships*, Water Resources Res., v. 8, no. 2, pp. 487-493.

HALL, H. P., 1955, *An Investigation of Steady Flow Toward a Gravity Well*, La Houille Blanche, v. 10, pp. 8-35.

HAMEL, G., 1934, *Über Grundwasserstromung*, Zeitschr. Angew, Math. Mech., v. 14.

HANSEN, V. E., 1953, *Unconfined Ground-Water Flow to Multiple Wells*, Trans. Am. Soc. Civil Engrs., v. 118, pp. 1098-1130.

HANSEN, V. E., 1955, *Infiltration and Water Movement During Irrigation*, Soil Sci., v. 79, pp. 93-105.

HANSHAW, B. B., and others, 1965, *Carbonate Equilibria and Radiocarbon Distribution Related to Groundwater Flow in the Floridan Limestone Aquifer, U. S. A.*, Intl. Assoc. Sci. Hydrology Publ. 74, pp. 601-614.

HANTUSH, M. S., 1957, *Nonsteady Flow to a Well Partially Penetrating an Infinite Leaky Aquifer*, Proc. Iraqi Sci., v. 1, pp. 10-19.

HANTUSH, M. S., 1959, *Non-Steady Flow to Flowing Wells in Leaky Aquifers*, J. Geophys. Res., v. 64, no. 8, p. 1043.

HANTUSH, M. S., 1960, *Modification of the Theory of Leaky Aquifers*, J. Geophys. Res., v. 65, no. 11, pp. 3713-3725.

HANTUSH, M. S., 1961, *Economical Spacing of Intefering Wells*, Internat. Assoc. Sci. Hydrology, Symposium of Athens, Publ. 56, Gentbrugge, Belgium, pp. 350-364.

HANTUSH, M. S., 1961, *Drawdown Around a Partially Penetrating Well*, Am. Soc. Civil Engrs. Hydraulics Div. pp. 83-98, July.

HANTUSH, M. S., 1962, *Drainage Wells in Leaky Water-Table Aquifers*, Am. Soc. Civil Engrs. J. Hydraulics Div. pp. 123-137, March.

HANTUSH, M. S., 1962, *Flow of Groundwater in Sands of Nonuniform Thickness*, J. Geophysical Res., v. 67, pp. 703-720 and 1527-1535.

HANTUSH, M. S., 1964, *Drawdown Around Wells of Variable Discharge*, J. Geophysical Res., v. 69, pp. 4221-4235.

HANTUSH, M. S., 1964, *Hydraulics of Wells*, in: V.T. Chow (ed.), *Advances in Hydroscience*, v. 1, pp. 281-432, Academic Press, New York.

HANTUSH, M. S., 1966, *A Method for Analyzing a Drawdown Test in Anisotropic Aquifers*, Water Resources Res., v. 2, no. 2, pp. 281-285.

HANTUSH, M. S., 1966, *Wells in Homogeneous Anisotropic Aquifers*, Water Resources Res., v. 2, no. 2, pp. 273-279.

HANTUSH, M. S., 1967, *Flow of Groundwater in Relatively Thick Leaky Aquifers*, Water Resources Res., v. 3, no. 2, p. 583.

HANTUSH, M. S., 1967, *Flow to Wells in Aquifers Separated by a Semi-Pervious Layer*, J. Geophysical Res., v. 72, no. 6, pp. 1709-20.

HANTUSH, M. S., and JACOB, C. E., 1954, *Plane Potential Flow of Groundwaters with Linear Leakage*, Trans. Am. Geophys. Union, v. 35, pp. 917-936.

HANTUSH, M. S., and JACOB, C. E., 1955, *Nonsteady Radial Flow in an Infinite Leaky Aquifer and Nonsteady Green's Functions for an Infinite Strip of Leaky Aquifer*, Trans. Am. Geophys. Union, v. 36, no. 1., pp. 95-112.

HANTUSH, M. S., and JACOB, C. E., 1955, *Steady Three-Dimesional Flow to a Well in a Two-Layered Aquifer*, Trans. Am. Geophys. Union, v. 36, pp. 286-292.

HANTUSH, M. S., and JACOB, C. E., 1960, *Flow to an Eccentric Well in a Leaky Circular Aquifer*, J. Geophys. Res., v. 65, pp. 3425-3431.

HANTUSH, M. S., and PAPADOPULOS, I. S., 1962, *Flow of Ground Water to Collector Wells*, J. Hydraulics Div., Amer. Soc. Civil Engrs., v. 88, no. HY5, pp. 221-244.

HATCH, L. P., 1940, *Flow Through Granular Media*, J. Appl. Mech., v. 7A, pp. 109-112.

HAUSHILD, W., and KRUSE, G., 1960, *Unsteady Flow of Ground Water into Surface Reservoir*, Proc. Am. Soc. Civil Engrs. no. HY7, pp. 13-20.

HERBERT, ROBIN, 1969, *A Design Method for Deep Well Dewatering Installations*, Ground Water, v. 7, no. 2, pp. 24-34.

HERRERA, ISMAEL, 1970, *Theory of Multiple Leaky Aquifers*, Water Resources Res., v. 6, no. 1, pp. 185-193.

HERRERA, ISMAEL, and RODARTE, LEOPOLDO, 1973, *Integrodifferential Equations for Systems of Leaky Aquifers and Applications*: 1, *The Nature of Approximate Theories*, Water Resources Res., v. 9, no. 4, pp. 995-1005.

HERRERA, ISMAEL, and YATES, ROBERT, 1977, *Integrodifferential Equations for Systems of Leaky Aquifers and Applications, 3, A Numerical Method of Unlimited Applicability*, Water Resources Res., v. 13, no. 4, pp. 725-732.

HIBSCH, GUNTER, and KREFT, ANDRZEJ, 1979, *Determination of Aquifer Transport Parameters*, J. Hydraulics Div. Am. Soc. Civil Engrs., v. 105, no. HY9, Sept.

HOOPES, J. A., and HARLEMAN, D. R. F., 1967, *Dispersion in Radial Flow from a Recharge Well*, J. Geophysical Res., v. 72, pp. 3595-3607.

HOPF, L., and TREFFTZ, E., 1921, *Grundwasserströmung in einen Abfallenden Gelande mit Abfanggraben*, Zeitschr. Angew. Math. Mech., v. 1.

HORTON, J. H., and HAWKINS, R. H., 1965, *Flow Path of Rain from the Soil Surface to the Water Table*, Soil Sci., 100, pp. 377-383.

HORTON, R. E., 1906, *Surface Drainage of Land by Tile*, Mich. Engr.

HUANG, Y. H., 1971, *Unsteady Flow Toward Partially Penetrating Artesian Wells*, Kentucky Water Resources Res. Inst. PB 202-206, 1971.

HUANG, Y. H., 1973, *Unsteady Flow Toward an Artesian Well*, Water Resources Res., v. 9, no. 2, pp. 426-433.

HUANG, Y. H., and WU, S-J, 1974, *Analysis of Unsteady Flow Toward Artesian Wells by Three-Dimensional Finite Elements*, Univ. Kentucky Water Resources Research Institute Rep. No. 75, 164 pp.

HUBBERT, M. K., 1940, *The Theory of Groundwater Motion*, J. Geol., v. 48, no. 8, pt. 1, pp. 785-944.

HUBBERT, M. K., 1956, *Darcy's Law and the Field Equations of the Flow of Underground Fluids*, Trans., Am. Inst. Mining and Metall. Engrs., v. 207, pp. 222-239, also in Bull. Internat. Assoc. Sci. Hydrology, no. 5, 1957 and Oct. 1956 issue of J. Petroleum Tech.

HUBBERT, M. K., 1969, *The Theory of Ground-Water Motion and Related Papers* (reprints of 3 papers with corrections and 1856 paper by Henry Darcy), Hafner Publishing Co., New York and London, 310 pp.

HUNT, B. W., 1970, *Unsteady Seepage Toward Narrow Ditch*, J. Hydraulics Div., Am. Soc. Civil Engrs., v. 96, no. HY10, Oct.

HUNT, B. W., 1971, *Vertical Recharge of Unconfined Aquifers*, J. Hydraulics Div. Am. Soc. Civil Engrs., v. 97, pp. 1017-1030.

HURR, T. R., 1966, *A New Approach for Estimating Transmissibility from Specific Capacity*, Water Resources Res., v. 2, no. 4, pp. 657-663.

INTERNAT. ASSOC. HYDRAULIC RESEARCH, 1972, *Fundamentals of Transport Phenomena in Porous Media*, Elsevier, Amsterdam, 392 pp.

IRMAY, S., 1956, *Extension of Darcy Law to Unsteady Unsaturated Flow Through Porous Media*, Internat., Assoc., Sci. Hydrology, Symposia Darcy, Pub. 41, pp. 57-66.

ISRAELSEN, O. W., and McLAUGHLIN, W. W., 1935, *Drainage of Land Overlying an Artesian Ground-Water Reservoir*, Utah Agr. Exp. Sta. Bull. 259.

JACOB, C. E., 1940, *On the Flow of Water in an Elastic Artesian Aquifer*, Trans. Am. Geophys. Union, v. 21, pp. 574-586.

JACOB, C. E., 1941, *Notes on the Elasticity of the Lloyd Sand on Long Island, New York*, Trans. Am. Geophys. Union, v. 22, pp. 783-787.

JACOB, C. E., 1946, *Radial Flow in a Leaky Artesian Aquifer*, Trans. Am. Geophys. Union, v. 27, no. 2, p. 198-208.

JACOB, C. E., 1947, *Drawdown Test to Determine Effective Radius of Artesian Well*, Trans. Am. Soc. Civil Engrs., v. 112.

JACOB, C. E., 1950, *Flow of Ground Water*, in Rouse, Hunter, *Engineering Hydraulics*, Chap. 5, John Wiley & Sons, Inc., New York, pp. 321-386.

JACOB, C. E., 1963, *Determining the Permeability of Water-Table Aquifers*, U. S. Geol. Survey Water Supply Paper 1536-I, pp. 245-271.

JACOB, C. E., 1963, *Recovery Method for Determining the Coefficient of Transmissibility*, U. S. Geol. Survey Water Supply Paper 1536-I.

JACOB, C. E., and LOHMAN, S. W., 1952, *Nonsteady Flow to a Well of a Constant Drawdown in an Extensive Aquifer*, Trans. Am. Geophys. Union, v. 33, no. 4, pp. 559-569.

JAVANDEL, I., and WITHERSPOON, P. A., 1969, *A Method of Analyzing Transient Fluid Flow in Multilayered Aquifers,* Water Resources Res., v. 5, no. 4, pp. 856-869.

JAVANDEL, I., and ZAGHI, N., 1975, *Analysis of Flow to an Extended Fully Penetrating Well,* Water Resour. Res., v. 11., no. 1, pp. 159-164.

JENKINS, D. N., and PRENTICE, J. K., 1982, *Theory of Aquifer Test Analysis in Fractured Rocks under Linear (Nonradial) Flow Conditions,* Ground Water, v. 20, no. 1, pp. 12-21.

JOHNSON, M. L. and W. F. BRUTSAERT, 1979, *The Transition Problem in Pumped Aquifers,* Water Resources Res., v. 15, no. 5, Oct.

JORGENSEN, D. G., 1980, *Relationships Between Basic Soils-Engineering Equations and Basic Ground-Water Flow Equations,* U. S. Geol. Survey Water-Supply Paper, W 2064, 40 p.

KAHN, M. Y., KIRKHAM, DON, and HANDY, R. L., 1976, *Shape of Steady State Perched Groundwater Mounds,* Water Resources Res., v. 12, no. 3, pp. 429-436.

KAHN, M. Y., KIRKHAM, DON, and TOKSOZ, S., 1971, *Steady State Flow Around a Well in a Two-Layer Aquifer,* Water Resources Res., v. 7, no. 1, pp. 155-165.

KARANJAC, JASMINKO, 1972, *Well Losses Due to Reduced Formation Permeability,* Ground Water, v. 10, no. 4, pp. 42-45, Jul.-Aug.

KASHEF, A. I., 1965, *Exact Free Surface of Gravity Wells,* J. Hydraul. Div. Am. Soc. Civil Engrs, v. 91, no. HY4, pp. 167-184.

KASHEF, A. I., 1969, *Ground-Water Movement Toward Artificial Cuts,* Water Resources Res., v. 5, no. 5, P. 1032.

KASHEF, A. I., 1970, *Interference Between Gravity Wells-Steady State Flow,* Water Resources Bull., v. 6, no. 4, pp. 617-630; also in Ground Water, v. 8, no. 6, pp. 25-33(1970).

KASHEF, A. I., and others, 1952, *Numerical Solutions of Steady-state and Transient Flow Problems,* Purdue Univ. Exp. Sta. Bull., v. 36, res. ser. 117.

KAZMANN, R. G., 1946, Notes on Determining the Effective Distance to a Line of Recharge, *Trans. Am. Geophys. Union, v. 27, pp. 854-859.*

KEELY, J. F., and TSANG, C. F., 1983, *Velocity Plots and Capture Zones of Pumping Centers for Ground-Water Investigations,* Ground Water, v. 21, no. 6, pp. 701-714.

KHADER, M. H. A., and RAMADURGAIAH, D., 1976, *Unsteady Flow to a Nonpenetrating Artesian Well,* Ground Water, v. 14, no. 4, pp. 200-204.

KHAN, M. Y., and KIRKHAM, DON, 1971, *Spacing of Drainage Wells in a Layered Aquifer,* Water Resources, Res., v. 7, no. 1, pp. 166-183.

KING, F. H., 1899, *Principles and Conditions of the Movements of Ground Water,* U. S. Geol. Survey 19th Ann. Rep., pt. 2, pp. 59-294.

KIPP, K. L., Jr., 1973, *Unsteady Flow to a Partially Penetrating, Finite Radius Well in an Unconfined Aquifer,* Water Resources Res., v. 9, pp. 448-462.

KIRKHAM, DON, 1945, *Artificial Drainage of Land, Streamline Experiments, the Artesian Basin-III,* Trans. Am. Geophys. Union, v. 26, no. 3.

KIRKHAM, DON, 1959, *Exact Theory of Flow into a Partially Penetrating Well,* J. Geophys. Res. v. 64, pp. 1317-1327.

KIRKHAM, DON, 1964, *Exact Theory of Shape of the Free Surface About a Well in a Semi-Confined Aquifer,* J. Geophys. Res., v. 69, pp. 2537-2549.

KIRKHAM, DON, 1969, *Ground-Water Seepage Patterns to Wells for Unconfined Flow,* Iowa State Water Resources Res. Inst. Rept., W70-02759.

KIRKHAM, DON, and AFFLECK, S. B., 1977, *Solute Travel Times to Wells,* Ground Water, v. 15, no. 3, pp. 231-242.

KIRKHAM, DON, and SOTRES, M. O., 1978, *Casing Depths and Solute Travel Times to Wells,* Water Resources Res., v. 14, no. 2, Apr.

KOZENY, J., 1933, *Theorie und Berechnung der Brunnen, Wasserkraft and Wasserwirtschaft,* v. 28, pp. 88-92, 101-105, and 113-116.

KRAIJENHOFF VAN DE LEUR, D. A., 1958, *A Study of Nonsteady Groundwater Flow With Special Reference to a Reservoir Coefficient,* Ingenieur, v. 70, pp 87-94.

KRAIJENHOFF VAN DE LEUR, D. A., 1962, *A Study of Nonsteady Ground-Water Flow, II, Computation Methods for Flow to Drains,* Ingenieur, v. 74, pp. 285-292, Utrecht.

KROSZYNSKI, U. I., and DAGAN, G., 1975, *Well Pumping in Unconfined Aquifers: the Influence of the Unsaturated Zone,* Water Resources Res., v. 11, no. 3, pp. 479-490.

KUIPER, L. K., 1972, *Drawdown in a Finite Circular Aquifer with Constant Well Discharge,* Water Resources Res., v. 8, no. 3, pp. 734-736.

KUIPER, L. K., 1972, *Groundwater Flow in an Inhomogeneous Aquifer,* Water Resources Res., v. 8, no. 3, pp. 722-724.

KUMAR, A., and KIMBLER, O. K., 1970, *Effect of Dispersion, Gravitational Segregation, and Formation Stratification on the Recovery of Freshwater Stored in Saline Aquifers,* Water Resources Res., v. 6, pp. 1689-1700.

LAWRENCE, F. E., and BRAUNWORTH, P. L., 1906, *Fountain Flow of Water in Vertical Pipe,* Trans. Am. Soc. Civil Engrs., v. 57, p. 264.

LEWIS, M. R., 1932, *Flow of Groundwater as Applied to Drainage Wells,* Trans. Am. Soc. Civil Engrs., v. 96.

LI, W. H., and YEH, G. T., 1968, *Dispersion at the Interface of Miscible Liquids in a Soil,* Water Resources Res., v. 4, pp. 369-378.

LINDQUIST, E., 1933, *On the Flow of Water through Porous Soil,* Repts. to the First Congress of Large Dams, Stockholm.

LIU, P. L. F., and LIGGETT, J. A., 1978, *An Efficient Numerical Method of Two-Dimensional Steady Groundwater Problems,* Water Resources Res., v. 14, no. 3, pp. 385-390.

LOHMAN, S. W., 1972, *Ground-water Hydraulics,* U. S. Geol. Survey Prof. Paper No. 708, 70 pp.

LUTHIN, J. N. (ed.), 1957, *Drainage of Agricultural Lands,* Am. Soc. Agronomy, Madison, Wis., 620 pp.; also Water Information Center, Plainview, New York 11803.

LUTHIN, J. N., and DAY, P. R., 1955, *Lateral Flow Above a Sloping Water Table,* Proc. Soil Sci. Soc. Am., v. 18, pp. 406-410.

LUTHIN, J. N., and SCOTT, V. H., 1952, *Numerical Analysis of Flow Through Aquifers Towards Wells,* Agr. Eng. v. 33, no. 5.

MAASLAND, D. E. L., and BITTINGER, M. W., (eds.), 1963, *Flow in Porous Media,* Symp. on Transient Water Hydraulics, Fort Collins, Colorado State Univ., 223 pp.

MARINO, M. A., 1974, *Growth and Decay of Ground-Water Mounds Induced by Percolation,* J. Hydrology, v. 22, no. 3-4.

MARINO, M. A., 1978, *Flow Against Dispersion in Adsorbing Porous Media,* J. Hydrology, v. 38 (1/2), pp. 197-205.

MARINO, M. A., 1978, *Flow Against Dispersion in Nonadsorbing Porous Media,* J. Hydrology, v. 37 (1/2), pp. 149-158.

McBRIDE, M. S., and PFANNKUCH, H. O., 1975, *The Distribution of Seepage Within Lakebeds,* U. S. Geol. Survey J. Res., v. 3, no. 5, pp. 505-512.

MEINZER, O. E., 1923 (reprinted 1969), *Outline of Ground-Water Hydrology, with Definitions,* U. S. Geol. Survey Water Supply Paper 494, 71 pp.

MEINZER, O. E., 1928, *Compressibility and Elasticity of Artesian Aquifers,* Economic Geology, v. 23, pp. 263-291.

MEINZER, O. E., and FISHEL, V. C., 1934, *Tests of Permeability with Low Hydraulic Gradients,* Trans. Am. Geophys. Union, v. 15, pp. 405-409.

MEINZER, O. E., and HARD, H. A., 1925, *The Artesian Water Supply of the Dakota Sandstone in North Dakota, with Special Reference to the Edgeley Quadrangle,* U. S. Geol. Survey Water Supply Paper 520, pp. 73-95.

MELESCHENKO, N. T., 1936, *Analysis of Groundwater Movement under Structures Equipped with Drainage Openings,* Izv. Nauchn. Issled, Inst. Gidrothn, v. 19.

MEYER, R., 1955, *A Few Recent Theoretical Results Concerning Ground-Water Flow,* La Houille Blanche, v. 10, pp. 86-108.

MJATIEV, A. N., 1947, *Pressure Complex of Underground Water and Wells,* Isvestiya Akademiya Nauk, U.S.S.R. Div. Tech. Sci., No. 9.

MOBASHERI, F., and SHAHBAZI, M., 1969, *Steady-State Lateral Movement of Water Through the Unsaturated Zone of an Unconfined Aquifer,* Ground Water, v. 7, no. 6, pp. 28-35.

MOENCH, A. F., 1971, *Ground-Water Fluctuations in Response to Arbitrary Pumpage,* Ground Water v. 9, no. 2, pp. 4-8.

MOENCH, A. F., 1973, *Analytical Solutions to the One-Dimensional Nonlinear Diffusion Equation for Flow Through Porous Media,* Water Resources Res., v. 9, no. 5, pp. 1378-1384.

MOENCH, A. F., and PRICKETT, T. A., 1972, *Radial Flow in an Infinite Auifer Undergoing Conversion for Artesian to Water-Table Conditions,* Water Resources Res., v. 8, no. 2, pp. 494-499.

MOGG, J. L., 1959, *The Effect of Aquifer Turbulence on Well Drawdown,* Proc. Am. Soc. Civil Engnrs. Hydraulics Div., pp. 99-122, Nov.

MONGAN, C. E., 1985, *Validity of Darcy's Law Under Transient Conditions,* U.S. Geol. Survey Prof. Paper 1331, 16 pp.

MOREL-SEYTOUX, H. J., and KHANJI, J., 1974, *Derivation of an Equation of Infiltration,* Water Resources Res., v. 10, no. 4, pp. 795-800.

MOTZ, L. H., 1978, *Steady-State Drawdowns in Coupled Aquifers,* J. Hydraulics Div. Am. Soc. Civil Engrs., v. 104, no. HY7, July, pp. 1061-1074.

MUALEM, Y., and BEAR, JACOB, 1974, *The Shape of the Interface in Steady Flow in a Stratified Aquifer,* Water Resources Res., v. 10, no. 6, pp. 1207-1216.

MUSKAT, M., 1932, *Potential Distributions in Large Cylindrical Disks with Partially Penetrating Electrodes,* Physics, v. 2.

NAHRGANG, G., 1954, *Zur Theorie des Volkommen and Unvolkommen Brunnens,* J. Springer, Berlin, 43 pp.

NEUMAN, S. P., 1972, *Theory of Flow in Unconfined Aquifers Considering Delayed Response of the Water Table,* Water Resources Res., v. 8, no. 4, pp. 1031-1045.

NEUMAN, S. P., 1974, *Effect of Partial Penetration on Flow in Unconfined Aquifers Considering Delayed Gravity Response,* Water Resources Res., v. 10, no. 2, pp. 303-312.

NEUMAN, S.P., and others, 1984, *Determination of Horizontal Aquifer Anisotropy with Three Wells,* Ground Water, v. 22, no. 1, pp. 66-72.

NEUMAN, S. P., and WITHERSPOON, P. A., 1968, *Theory of Flow in Aquicludes Adjacent to Slightly Leaky Aquifers,* Water Resources Res., v. 4, no. 1, pp. 103-112.

NEUMAN, S. P., and WITHERSPOON, P. A., 1969, *Applicability of Current Theories of Flow in Leaky Aquifers,* Water Resources Res., v. 5, no. 4, pp. 817-829.

NEUMAN, S. P., and WITHERSPOON, P. A., 1969, *Theory of Flow in a Confined Two Aquifer System,* Water Resources Res., v. 5, no. 4, pp. 803-816.

NEUMAN, S. P., and WITHERSPOON, P. A., 1969, *Transient Flow of Ground Water to Wells in Multiple Aquifer Systems,* Geotech. Eng. Rep. 69-1, Univ. of Calif., Berkeley, California.

NEUMAN, S. P., and WITHERSPOON, P. A., 1970, *Variational Principles for Confined and Unconfined Flow of Ground Water,* Water Resources Res, v. 6, no. 5, Oct.

NEUZIL, C. E., and TRACY, J. V., 1981, *Flow Through Fractures,* Water Resources Res., v. 17, no. 1, Feb.

NIWAS, S., AND SINGHAL, D.C., 1981, *Estimation of Aquifer Transmisssivity from Dar-Zarrouk Parameters in Porous Media,* Hydrology, v. 50, pp. 393-399.

OGATA, A., 1976, *Two-Dimensional Steady-State Dispersion in a Saturated Porous Medium,* U. S. Geol. Surv. J. Res., v. 4, no. 3, pp. 277-284.

ORTIZ, N. V., and others, 1978, *Growth of Ground-Water Mounds Affected by In-Transit Water,* Water Resources Res. v. 14, no. 6, Dec.

ORTIZ, N. V., and others, 1979, *Effects of In-Transit Water on Ground-Water Mounds Beneath Circular and Rectangular Recharge Areas,* Water Resources Res. v. 15, no. 3, Jun.

OVERPECK, A.C., and HOLDEN, W.R., 1970, *Wells Imaging and Fault Detection in Anisotropic Reservoirs,* J. of Petroleum Tech., pp. 1317-1325.

PAPADOPULOS, I. S., 1965, *Nonsteady Flow to a Well in an Infinite Anisotropic Aquifer,* Symp. Internat. Assoc. Sci. Hydrology Dubrovnik.

PAPADOPULOS, I.S., and COOPER, H. H., Jr., 1967, *Drawdown in a Well of Large Diameter,* Water Resources Res., v. 3, no. 1, pp. 241-244.

PETERSON, D. F., Jr., 1955, *Hydraulics of Wells,* Proc. Am. Soc. Civil Engrs., v. 81, sep. 708, 23 pp.

PHILIP, J.R., 1968, *Steady Infiltration from Buried Point Sources and Spherical Cavities,* Water Resources Res., v. 4, no. 5, pp. 1039-1047.

PHILIP, J.R., 1969, *Theory of Infiltration,* Advances in Hydroscience, v. 5, pp. 215-296.

PHILIP, J.R., 1973, *On Solving the Unsaturated Flow Equation: 1. The Flux-Concentration Relation,* Soil Sci., v. 116, no. 5, pp. 328-335.

PHILIP, J.R., 1984, *Travel Times from Buried and Surface Infiltration Point Sources,* Water Resources Res., v. 20, no. 7, Jul.

POISEVILLE, J. M., 1846, *Experimental Investigations on the Flow of Liquids in Tubes of Very Small Diameters,* Proc. Roy. Acad. Sci. Inst., France, Math. Phys. Sci. Mem.

RAO, D. B., and others, 1971, *Drawdown in a Well Group Along a Straight Line,* Ground Water, v. 9, no. 4, pp. 12-18, Jul.-Aug.

RAO, D. B., KARADI, G. M. and KRIZEK, R. J., 1973, *Unsteady Drawdown at a Partially Penetrating Well in a Transversely Isotropic Artesian Aquifer,* Ground Water, v. 11,

REILLY, T.E., FRANKE, O.L., and BENNETT, G.D., *The Principle of Superposition and its Application in Ground-Water Hydraulics,* U.S. Geol. Survey Open-File Rept. 84-459, 36 pp.

REMSON, IRWIN, and others, 1959, *Zone of Aeration and its Relationship to Ground Water Recharge,* J. Am. Water Works Assoc., v. 51, p. 371, Mar.

RIJTEMA, P. E., and WASSINK, H. (eds.), 1968, *Water in the Unsaturated Zone,* Proc. Symp. Wageningen, June 19-25, 1966, Internat. Assoc. Sci. Hydrology, Pub. nos. 82 and 83.

RODARTE, LEOPOLDO, 1976, *Theory of Multiple Leaky Aquifers, 1, The Integrodifferential and Differential Equations for Small and Large Values of Time,* Water Resources Res., v. 12, no. 2, pp. 163-170.

ROGERS, W. E., 1954, *Introduction to Electric Fields,* McGraw-Hill, 333 pp. (Viscous fluid)

ROSE, H. E., 1945, *An Investigation into the Laws of Flow of Fluids Through Beds of Granular Materials,* Proc. Inst. Mech. Engrs., v. 153, pp. 141-148.

ROSE, H. E., 1945, *On the Resistance Coefficient — Reynolds Number Relationship for Fluid Flow Through a Bed of Granular Material,* Proc. Inst. Mech. Engrs., v. 153, pp. 154-168.

RUBIN, HILLEL, 1971, *Effect on Nonlinear Stabilizing Salinity Profiles on Thermal Convection in a Porous Medium Layer,* Water Resources Res., v. 9, no. 1, pp. 211-221.

RUBIN, HILLEL, 1973, *Effect of Solute Dispersion on Thermal Convection in a Porous Medium Layer,* Water Resources Res., v. 9, no. 4, pp. 968-974.

RUBIN, HILLEL, 1975, *Effect of Solute Dispersion on Thermal Convection in a Porous Medium Layer, 2,* Water Resources Res., v. 11, no. 1, pp. 154-158.

RUBIN, HILLEL, 1976, *Onset of Thermohaline Convection in a Cavernous Aquifer,* Water Resources Res., v. 12, no. 2, pp. 141-147.

RUBIN, HILLEL, 1977, *Thermal Convection in a Cavernous Aquifer,* Water Resources Res., v. 13, no. 1, pp. 34-40.

RUBIN, JACOB and JAMES, R. V., 1973, *Dispersion Affected Transport of Reacting Solutes in Saturated Porous Media,* Water Resources Res. v. 9, no. 5, pp. 1332-1356.

RUMER, R. R., 1962, *Longitudinal Dispersion in Steady and Unsteady Flow,* J. Hydraulics Div. Amer. Soc. Civil Engrs., v. 88, no. HY4, pp. 147-172.

RUMER, R. R. and DRINKER, P. A., 1966, *Resistance to Laminer Flow Through Porous Media,* J. Hydraulics Div., Amer. Soc. Civil Engrs., v. 92, no. 11Y5, pp. 155-163.

RUSHTON, K. R., and RATHOD, K. S., 1980, *Flow in Aquifers When the Permeability Varies with Depth,* Hydrological Sciences Bull., v. 25, no. 4., Dec.

RUSSELL, W. L., 1928, *The Origin of Artesian Pressure,* Econ. Geol., v. 23, pp. 132-157.

SAFFMAN, P. G., 1959, *A Theory of Dispersion in a Porous Medium,* J. Fluid Mech., v. 6, pp. 321-349.

SAHNI, B. M., 1973, *Physics of Brine Coming Beneath Skimming Wells,* Ground Water, v. 11, no. 1, pp. 19-24.

SALEEM, M., 1969, *An Inexpensive Method of Determining the Direction of Natural Flow of Ground-Water,* J. Hydrology, v. 9, pp. 73-89.

SALEEM, Z. A., and JACOB, C. E., 1971, *Optimal Use of Coupled Leaky Aquifers,* Water Resources Res., v. 7, no. 2, pp. 382-393.

SALEEM, Z. A., and JACOB, C. E., 1973, *Drawdown Distribution due to Well Fields in Coupled Leaky Aquifers,* 1, *Infinite Aquifer System,* Water Resources Res., v. 9, no. 6, pp. 1671-1678, 2. *Finite Aquifer Systems,* Water Resources Res., v. 10, no. 2, pp. 336-342.

SAYRE, A. N., 1950, *Ground Water,* Sci. Am., v. 183, no. 5, pp. 14-19.

SCHEIDEGGER, A. E., 1957, *On the Theory of Flow of Miscible Phases in Porous Media,* Internat. Assoc., Sci. Hydrology, General Assembly Toronto, Pub. 44, v. 2, pp. 236-242.

SCHEIDEGGER, A. E., 1961, *General Theory of Dispersion in Porous Media,* J. Geophys. Res., v. 66, no. 10, pp. 3273-3278.

SCHNEEBELI, G., 1955, *Experiences sur la Limite de Validité de la Loi de Darcy et l'Apparition de la Turbulence dans un Ecoulement de Filtration,* La Houille Blanche, v. 10, no. 2 pp. 141-149.

SCHWARTZ, F. W., 1977, *Macroscopic Dispersion in Porous Media: The Controlling Factors,* Water Resources Res., v. 13, no. 4, pp. 743-752.

SELIM, S. M., and KIRKHAM, DON, 1974, *Screen Theory for Wells and Soil Drainpipes,* Water Resources Res., v. 10, no. 5, pp. 1019-1030.

SHAHBAZI, M., and others, 1968, *Effect of Topography on Ground Water Flow,* Intl. Assoc. Sci. Hydrology Publ. 77, pp. 314-319.

SHAMIR, U. Y., and HARLEMAN, D. R. F., 1967, *Numerical Solutions for Dispersion in Porous Mediums,* Water Resources Res., v. 3, pp. 557-581.

SHARP, J. M., Jr., and DOMENICO, P. A., 1976, *Energy Transport in Thick Sequences of Compacting Sediment,* Geol. Soc. Am. Bull., v. 87, no. 3, pp. 390-400.

SHAW, F. S., and SOUTHWELL, R. V., 1941, *Relaxation Methods Applied to Engineering Problems, Pt. VII, Problems Relating to the Percolation of Fluids Through Porous Materials,* Proc. Roy. Soc. London, ser. A, v. 178.

SHEN, H. T., 1976 *Transient Dispersion in Uniform Media Flow,* J. Hydraul. Div. Am. Soc. Civ. Eng., v. 102 (HY6), pp. 707-716.

SIKKEMA, P. C., and VAN DAM, J. C., 1982, *Analytical Formulae for the Shape of the Interface in a Semiconfined Aquifer,* J. Hydrology, v. 56, no. 3/4, Apr.

SINGH, RAMESHWAR, 1976, *Prediction of Mound Geometry under Recharge Basins,* Water Resources Res., v. 12, no. 4, pp. 775-780.

SKIBITZKE, H. E., and ROBINSON, G. M., 1963, *Dispersion in Ground Water Flowing Through Heterogeneous Materials*, U. S. Geol. Survey Prof. Paper 386-B, 3 pp.

SLICHTER, C. S., 1898, *Theoretical Investigations of the Motion of Groundwaters*, U. S. Geol. Surv. 19th Ann. Rept. pt. 2, pp. 301-384.

SLICHTER, C. S., 1902, *The Motion of Underground Waters*, U. S. Geol. Survey Water Supply Paper 67.

SLICHTER, C. S., 1905, *Field Measurements of the Rate of Movement of Underground Water*, U. S. Geol. Survey Water Supply Paper 140.

SMITH, I. W., and others, 1973, *Rayleigh-Ritz and Garlerkin Finite Elements for Diffusion-Convection Problems*, Water Resources Res., v. 9, no. 3, pp. 593-606.

SMITH, W. O., 1961, *Mechanism of Gravity Drainage and its Relation to Sepcific Yield of Uniform Sands*, U. S. Geol. Survey Prof. Paper 402-A 1961.

SMITH, W. O., and SAYRE, A. N., 1964, *Turbulence in Ground-Water Flow*, U. S. Geol. Survey Prof. Paper 402-E, 9 pp.

SMITH, W. O., 1967, *Infiltration in Sands and its Relation to Groundwater Recharge*, Water Resources Res., v. 3, no. 2, pp. 539-555.

STALLMAN, R. W., 1963, *Computation of Ground-Water Velocity from Temperature Data*, U. S. Geol. Survey Water-Supply Paper 1544-H, pp. 36-46.

STALLMAN, R. W., 1964, *Multiphase Fluids in Porous Media*, U. S. Geol. Survey Prof. Paper 411-E, pp. 1-54.

STALLMAN, R. W., 1965, *Effects of Water Table Conditions on Water Level Changes near Pumping Wells*, Water Resources Res., v. 1, pp. 295-312.

STALLMAN, R. W., 1967, *Flow in the Zone of Aeration*, in Chow, V. T. (ed.), *Advances in Hydroscience*, v. 4, pp. 151-195. Academic Press, New York.

STALLMAN, R. W., and PAPADOPULOS, I. S., 1966, *Measurement of Hydraulic Diffusivity of Wedge-Shaped Aquifers Drained by Streams*, U. S. Geol. Survey Prof. Paper 514, 50 pp.

STEGGEWENTZ, J. H., and VAN NES, B. A., 1939, *Calculating the Yield of a Well Taking Account of Replenishment of the Groundwater from Above*, Water and Water Eng., v. 41, pp. 561-563.

STERNBERG, Y. M., 1969, *Flow to Wells in the Presence of Radial Discontinuities,* Ground Water, v. 7, no. 6, pp. 17-21.

STERNBERG, Y. M., 1973, *Efficiency of Partially Penetrating Wells,* Ground Water, v. 11, no. 3, pp. 5-8.

STRACK, O. D. l., 1976, *A Single-Potential Solution for Regional Interface Problems in Coastal Aquifers,* Water Resources Res., v. 12, no. 6, pp. 1165-1174.

STRELTSOVA, T. D., 1972, *Unsteady Radial Flow in an Unconfined Aquifer,* Water Resources Res., v. 8, no. 4, pp. 1059-1066.

STRELTSOVA, T. D., 1973, *Flow near a Pumped Well in an Unconfined Aquifer under Nonsteady Conditions,* Water Resource Res., v. 9, no. 1, pp. 227-235.

STRELTSOVA, T. D., 1973, *On the Leakage Assumption Applied to Equations of Groundwater Flow,* J. Hydrology, v. 20, pp. 237-253.

STRELTSOVA, T. D., 1974, *Drawdown in Compressible Unconfined Aquifer,* J. Hydraulics Div. Am. Soc. Civil Engrs., v. 100, no. HY11, Nov.

STRELTSOVA, T. D., 1976, *Analysis of Aquifer-Aquitard Flow,* Water Resources Res., v. 12, no. 3, pp. 415-422.

STRELTSOVA, T. D., 1976, *Hydronamics of Groundwater Flow in a Fractured Formation,* Water Resources Res., v. 12, no. 3, pp. 405-414.

STRELTSOVA, T. D., and KASHEF, A. A. I., 1974, *Critical State of Salt-Water Upconing Beneath Artesian Discharge Wells,* Water Resources Bull., v. 10, no. 5, Oct.

STRELTSOVA, T. D., and RUSHTON, K. R., 1973, *Water Table Drawdown Due to a Pumped Well in an Unconfined Aquifer,* Water Resources Res., v. 9, no. 1, pp. 236-242.

SWARTZENDRUBER, D., 1962, *Non-Darcy Flow Behavior in Liquid Saturated Porous Media,* J. Geophys. Res., v. 67, pp. 5205-5213, Dec.

TANG, D. H., and PINDER, G. F., 1979, *A Direct Solution to the Inverse Problem in Ground-Water Flow,* Advances in Water Resources, v. 2, no. 2, Jun.

THEIS, C. V., 1932, *Equations for Lines of Flow in Vicinity of Discharging Artesian Well,* Trans. Am. Geophys. Union, v. 13, pp. 317-320.

THEIS, C. V., 1935, *The Relation Between the Lowering of the Piezometric Surface and the Rate and Duration of Discharge of a Well Using Groundwater Storage,* Trans. Am. Geophys. Union, v. 16, pp. 519-524.

THEIS, C. V., 1937, *Amount of Ground-Water Recharge in the Southern High Plains,* Trans. Am. Geophys. Union, 18th Ann. Meeting, pt. 2.

THEIS, C. V., 1938, *The Significance and Nature of the Cone of Depression in Ground-Water Bodies,* Econ. Geol., v. 33, pp. 889-902.

THEIS, C. V., and others, 1954, *Estimating Transmissibilities from Specific Capacity,* U. S. Geol. Survey Groundwater Note no. 24, 11 pp.

THEIS, C.V., 1963, *Spacing of Wells,* U.S. Geol. Survey Water-Supply Paper 1545-C, pp. C113-C117.

THIEM, A., 1870, *Die Ergiebigkeit Artesischer Bohrlocher, Schacht-brunnen und Filtergallerien,* J. Gasbeleuchtung Wasserversorgung, v. 14, Munich.

THIEM, GUNTER, 1906, *Hydrologische Methoden,* J. M. Gebhardt, Leipzig, 56 p.

TODD, D. K., and BEAR, J., 1961, *Seepage Through Layered Anisotropic Porous Media,* J. Hydraulics Div., Amer. Soc. Civil Engrs., v. 87, no. HY3, pp. 31-57.

TODOROVIC, P. A., 1970, *A Stochastic Model of Longitudinal Diffusion in Porous Media,* Water Resources Res., v. 6, pp. 211-222.

TÖTH, J. A., 1963, *A Theoretical Analysis of Groundwater Flow in Small Drainage Basins,* J. Geophysical Res., v. 68, pp. 4795-4812.

TOWNER, G. D., 1975, *Drainage of Groundwater Resting on a Sloping Bed with Uniform Rainfall,* Water Resources Res., v. 11, no. 1, pp. 144-147.

TSENG, M. T., and RAGAN, R. M., 1973, *Behavior of Groundwater Flow Subject to Time-Varying Recharge,* Water Resources Res., v. 9, no. 3, pp. 734-742.

U. S. GEOLOGICAL SURVEY, 1961-1970, *Fluid Movement in Earth Materials,* U. S., Prof. Paper 411, Chaps. A to I.

VACHAUD, G., and THONY, J. L., 1971, *Hysteresis During Infiltration and Redistribution in a Soil Column at Different Water Contents,* Water Resources Res., v. 7, pp. 111-127.

VANDEN BERG, A., and LENNOZ, D. H., 1973, *Determining Wedge Angle for a Wedge Aquifer,* Ground Water, v. 11, no. 1, pp. 25-31, Jan.-Feb.

VAN DER KAMP, GARTH, 1976, *Determining Aquifer Transmissivity by Means of Well Response Tests: the Underdamped Case,* Water Resources Res., v. 12, no. 1, pp. 71-77.

VAN DER PLOEG, R. R., and others, 1971, *Steady State Well Flow Theory for a Confined Elliptical Aquifer,* Water Resources Res., v. 7, no. 4, pp. 942-954.

VAN VOAST, W. A., and NOVITZKI, R. P., 1966, *Ground Water Flow Related to Streamflow and Water Quality,* Water Resources Res., v. 4, no. 4, Aug.

VERDERNIKOV, V. V., 1934, *Versickerung aus Kanalen,* Wasserkraft Wasserwirtsch., vs. 11-13.

VERMA, R. D., and BRUTSAERT, W., 1971, *Unsteady Free-Surface Ground-Water Seepage,* Am. Soc. Civil Eng. Hydraul. Div. Jour., 97 (HY8), pp. 1213-1299.

VREEDENBURGH, C. G. J., and STEVENS, O. M., 1936, *Electric Investigation of Underground Water Flow Nets,* Proc. Internat. Conf. Soil Mech. Found. Eng., v. 1.

WALTON, W. C., 1955, *Ground-water Hydraulics as an Aid to Geologic Interpretation,* Ohio J. Sci., v. 55, no. 1.

WALTON, W. C., 1979, *Review of Leaky Artesian Aquifer Test Evaluation Methods,* Ground Water, v. 17, no. 3, pp. 270-283, May-June.

WARD, J. C., 1964, *Turbulent Flow in Porous Media,* J. Hydraulics Div. Am. Soc. Civil Engrs., v. 90, no. HY5, pp. 1-12, Sept.

WATSON, K. K., 1974, *Some Applications of Unsaturated Flow Theory in Drainage for Agriculture,* J. van Schilfgaarde (ed.), Agronomy, No. 17, American Society of Agronomy, Madison, Wisconsin.

WAY, SHAO-CHIN, and McKEE, C.R., 1984, *Detection of an Impermeable Boundary in an Anisotropic Formation: A Case Study,* Ground Water, v. 22, no. 5, pp. 579-583.

WEINBERGER, Z., and MANDEL, S., 1973, *The Role of Molecular Diffusion in Dispersion Theory,* J. Hydrology, v. 19, no. 4, Aug.

WENZEL, L. K., 1933, *Specific Yield Determined from a Thiem's Pumping Test,* Trans. Am. Geophys. Union. v. 14, pp. 475-477.

WENZEL, L. K., 1936, *The Thiem Method for Determining Permeability of Water-Bearing Materials and its Application to the Determination of Specified Yield, Results of Investigations in the Platte River Valley, Nebr.,* U. S., Geol. Survey Water-Supply Paper 679-A.

WENZEL, L. K., 1942, *Methods for Determining Permeability of Water-Bearing Materials with Special Reference to Discharging-Well Methods,* U. S. Geol. Survey Water-Supply Paper 887, 192 pp.

WENZEL, L. K., and GREENLEE, A. L., 1943, *A Method for Determining Transmissibility and Storage Coefficients by Tests of Multiple Well Systems,* Trans. Am. Geophys. Union, v. 24, pp. 547-564.

WENZEL, L. K., and SAND, H. H., 1942 *Water Supply of the Dakota Sandstone in the Ellentown-Jamestown Area, North Dakota,* U. S. Geol. Survey Water-Supply Paper 889-A.

WERNER, P. W., 1957, *Some Problems in Non-Artesian Groundwater Flow,* Trans. Am. Geophys. Union, v. 38, pp. 511-518.

WHITEHEAD, W. R., and LANGHETEE, E. J., 1978, *Use of Bounding Wells to Counteract the Effects of Preexisting Groundwater Movement,* Water Resources Res., v. 14, no. 2, pp. 273-280.

WIGLEY, T. M. L., 1968, *Flow into a Finite Well with Arbitrary Discharge,* J. Hydrology, v. 6, no. 2, pp. 209-213.

WILLIAMS, R. A., LAI, R. Y., and KARADI, G. M., 1972, *Nonsteady Flow to a Well with Time-Dependent Drawdown,* Water Resources Bull., v. 9, no. 2, pp. 294-303.

WILSON, C. R., and WITHERSPOON, P. A., 1976, *Flow Interference Effects at Fracture Intersections,* Water Resources Res., v. 12, no. 1, pp. 102-104.

WITHERSPOON, P. A., and NEUMAN, S. P., 1967, *Evaluating a Slightly Permeable Caprock in Aquifer Gas Storage, 1. Caprock of Infinite Thickness,* Trans. AIME, 240, p. 949.

WOLANSKI, E. J., and WOODING, R. A., 1973, *Steady Seepage Flow to Sink Pairs Symetrically Situated Above and Below a Horizontal Diffusing Interface, 1, Parallel Line Sinks,* Water Resources Res., v. 9, no. 2, pp. 415-425.

WOLFF, R. G., 1970, *Field and Laboratory Determination of the Hydraulic Diffusivity of a Confining Bed,* Water Resources Res., v. 6, no. 1, pp. 194-203, Feb.

WOLFF, R. G., and PAPADOUPULOS, S. S., 1972, *Determination of the Hydraulic Diffusivity of Heterogeneous Confining Bed,* Water Resources Res., v. 8, no. 4, pp. 1051-1058.

WOODING, R. A., and CHAPMAN, T. G., 1966, *Groundwater Flow over a Sloping Impermeable Layer,* J. Geophys. Res., v. 71, no. 12.

WYLLIE, M. R. J., and SPANGLER, M. B., 1952, *Application of Electrical Resistivity Measurements to Problem of Fluid Flow in Porous Media,* Bull. Am. Assoc. Petroleum Geologists, v. 36, pp. 359-403.

YANG, S. T., 1949, *Seepage Toward a Well Analyzed by the Relaxation Method,* Doctoral Thesis, Harvard Univ.

YEH, W. W-G., and TAUXE, G. W., 1971, *Optimal Identification of Aquifer Diffusivity using Quasilinearization,* Water Resour. Res., v. 7, no. 4, pp. 955-962.

YOUNGS, E. G., 1969, *Unconfined Aquifers and the Concept of the Specific Yield,* Bull. Internat. Assoc. Sci. Hydrology, v. 14, no. 2, June.

YOUNGS, E. G., 1971, *Seepage Through Unconfined Aquifers with Lower Boundaries of Any Shape,* Water Resources Res., v. 7, no. 3, pp. 624-631.

YOUNGS, E. G, 1980, *The Analysis of Ground-Water Seepage in Heterogeneous Aquifers,* Hydrological Sciences Bull., v. 25, no. 2, June.

ZANGER, C. N., 1953, *Flow in Porous Media — Theory and Problems of Water Percolation,* U. S. Bur. Reclam., Eng. Monograph 8, Denver, 76 p.

ZUCKER, W. B., and others, *Hydrologic Studies Using the Boussinesq Equation with a Recharge Term,* Water Resources Res., v. 9, no. 3, pp. 586-592.

PUMPING TESTS/FIELD OBSERVATIONS/ INSTRUMENTATION

BIBLIOGRAPHY

CALIFORNIA DEPT. OF WATER RESOURCES, 1963, *Permeability, Coefficients of Transmissibility, Coefficients of Storage, Methods of Determination and Application to Ground-Water Problems-Annotated Bibliography through 1961,* 75 pp.

GENERAL WORKS

BARKER, J.A. and BLACK, J.H., 1983, *Slug Tests in Fissured Aquifers,* Water Resources Res., v. 19, no. 6.

BARKER, J.A. and HERBERT, R., 1982, *Pumping Tests in Patchy Aquifers,* Ground Water, v. 20, no. 2, pp. 150-155.

BENNETT, G. D., and PATTEN, E. P., Jr., 1962, *Constant-Head Pumping Test of a Multiaquifer Well to Determine Characteristics of Individual Aquifers,* U. S. Geol. Survey Water-Supply Paper 1536-G, pp. 181-203.

BENTALL, RAY, 1963, *Methods of Determining Permeability, Transmissibility and Drawdown,* U. S. Geol. Survey Water-Supply Paper 1536-I, pp. 243-341.

BENTALL, RAY, (compiler), 1963, *Shortcuts and Special Problems in Aquifer Tests,* U. S. Geol. Survey Water-Supply Paper 1545-C, 117 pp.

BIANCHI, W. C., and HASKELL, E. E., Jr., 1968, *Field Observations Compared with Dupuit-Forchheimer Theory for Mound Heights under a Recharge Basin,* Water Resources Res, v. 4, no. 5, pp. 1049-1059.

BIERSCHENK, W. H., 1964, *Determining Well Efficiency by Multiple Step-Drawdown Tests,* Internat. Assoc. Sci. Hydrology, Pub. 64, pp. 493-507.

✓BIRSOY, Y. K., and SUMMERS, W. K., 1980, *Determination of Aquifer Parameters from Step Tests and Intermittent Pumping Data,* Ground Water, v. 18, no. 2, pp. 137-146, Mar.-Apr.

BIRTLES, A. B., and MOREL, E. H., 1979, *Calculation of Aquifer Parameters from Sparse Data,* Water Resources Res., v. 15, no. 4, Aug. P90

BLACK, J. H., and KIPP, K. L., Jr., 1977, *The Significance and Prediction of Observation Well Response Delay in Semiconfined Aquifer-Test Analysis,* Ground Water, v. 15, no. 6, pp. 446-451.

BOULTON, N. S., 1963, *Analysis of Data from Nonequilibrium Pumping Tests Allowing for Delayed Yield from Storage,* Proc. Inst. Civil Engrs., v. 26, pp. 469-482, London.

BOULTON, N. S., 1964, *Discussion on the Analysis of Data from Nonequilibrium Pumping Tests Allowing for Delayed Yield from Storage,* Proc. Inst. Civil Engrs., v. 28, London.

BOULTON, N. S., 1970, *Analysis of Data from Pumping Tests in Unconfined Anisotropic Aquifers,* J. Hydrology, v. 10, no. 4, June.

BOULTON, N. S., and STRELTSOVA, T. D., 1975, *New Equations for Determining the Formation Constant of an Aquifer from Pumping Test Data,* Water Resources Res., v. 11, no. 1, pp. 148-153.

✓BOUWER, HERMAN, and RICE, R. C., 1976, *A Slug Test for Determining Hydraulic Conductivity of Unconfined Aquifers with Completely or Partially Penetrating Wells,* Water Resources Res., v. 12, no. 3, pp. 423-428.

BRERETON, N. R., 1979, *Step-Drawdown Pumping Tests for the Determination of Aquifer and Borehole Characteristics,* Tech. Rep. 103, Water Resources Center, Marlow, Great Britain.

BROWN, R. H., 1953, *Selected Procedures for Analyzing Aquifer Test Data,* J. Am. Water Works Assoc., v. 45, no. 8, pp. 844-866.

BRUIN, J., and HUDSON, H. E., Jr., 1955, *Selected Methods for Pumping Test Analyses,* Illinois State Water Surv. Rept. of Invest. 25.

CARR, P. A., and VAN DER KAMP, G., 1969, *Determining Aquifer Characteristics by the Tidal Method,* Water Resources Res., v. 5, no. 5, pp. 1023-1031.

CHOW, V. T., 1951, *Drawdown in Artesian Wells Computed by Nomograph,* Civil Eng., v. 21, no. 10, pp. 48-49.

Pumping Tests

CHOW, V. T., 1952, *On the Determination of Transmissibility and Storage Coefficients from Pumping Test Data,* Trans. Am. Geophys. Union, v. 33, pp. 397-404.

CLARK, J.E., GERMOND, B.J., and BENNETT, K.C., 1983, *Aquifer Test Monitoring by Electrical Pressure Transducer in Comparison with the Hand-Held Tape Method,* DE8300649, NTIS, 5285 Port Royal Rd., Springfield, Virginia 22161, 28 pp.

COBB, P. M., McELWEE, C. E., and BUTT, M. A., 1982, *Analysis of Leaky Aquifer Pumping Test Data: An Automated Numerical Solution Using Sensitivity Analysis,* Ground Water v. 20, no. 3, pp. 325-333.

COOPER, H. H., Jr., and JACOB, C. E., 1946, *A Generalized Graphical Method for Evaluating Formation Constants and Summarizing Well-Field History,* Trans. Am. Geophys. Union, v. 27, pp. 526-534.

CROLEY, T. E., II, 1977, *Hydrologic and Hydraulic Computations on Small Programmable Calculators,* Iowa Institute of Hydraulic Research, 837 pp.

CSALLANY, S., 1966, *Graphical Method for Determining Coefficient of Transmissibility,* J. Am. Water Works Assoc., v. 58, pp. 628-634.

DEAL, C. D., 1979, *Water Well Drawdown Monitoring System,* Off. Gaz. U. S. Patent Office, v. 980, p. 58. (pressure sensors)

DUMBLE, J.P., and CULLEN, K.T., 1983, *The Application of a Microcomputer in the Analysis of Pumping Test Data in Confined Aquifers,* Ground Water, v. 21, no. 1, pp. 79-83.

EAGON, H. B., Jr., and JOHE, D. E., 1972, *Practical Solutions for Pumping Tests in Carbonate-Rock Aquifers,* Ground Water, v. 10, no. 4, pp. 6-13, Jul.-Aug.

EARLOUGHER, R. C., Jr., 1977, *Advances in Well Test Analysis,* Society of Petroleum Engineers AIME, 264 pp. (finite difference)

FERRIS, J. G., 1948, *Groundwater Hydraulics as a Geophysical Aid,* Mich. Dept. Conserv., Geol. Surv. Div. Tech. Rept. 1.

FERRIS, J. G., and KNOWLES, D. B., 1954, *Slug Test for Estimating Transmissibility,* U. S. Geol. Survey Groundwater Note 26.

FIGUEROA, G. D., 1971, *Influence Chart for Regional Pumping Effects,* Water Resources Res., v. 7, no. 1, pp. 209-210.

FOOSE, R. M., 1969, *Mine Dewatering and Recharge in Carbonate Rocks near Hershey, Pennsylvania,* in *Legal Aspects of Geology in Engineering Practice,* Geol. Soc. Am. Eng. Case Histories, no. 7, pp. 45-60.

GARBER, M. S., and KOOPMAN, F. C., 1968, *Methods of Measuring Water Levels in Deep Wells,* U. S. Geol. Survey Techniques of Water Resources-Inv., Bk. 8, Chap. A1, 23 pp.

GERAGHTY, J. J., and others, 1967, *The Status of Ground-Water Resources of Nansemond County and Isle of Wight County, Virginia,* Geraghty & Miller, Inc., Plainview, New York, 61 pp.

GILLILAND, J. A., 1968, *Digitizing, Storing and Recovering Observation Well Hydrographs,* J. Hydrology, v. 6, no. 2, Apr.

GILLILAND, J. A., 1968, *Groundwater Instrumentation and Observation Techniques,* Proc. 7th Canadian Hydrology Symp., pp. 37-57.

GRIMESTEAD, GARRY, 1981, *Inverse Solutions of the Theis Equation Determined with Programmable Calculators: I — Single Parameter Calculations,* pp. 382-386; *II — Determination of Aquifer Hydraulic Constants from Numerical Test Data,* pp. 387-391, Ground Water v. 19, no. 4.

GRINGARTEN, A.C., and WITHERSPOON, P.A., 1972, *A Method of Analyzing Pump Test Data from Fractured Aquifers,* Proc. Symp. Percolation through Fissured Rock, Sept. 18-19, 1971, Stuttgart, pp. T3-B-1 to T3-B-8, Internat. Society for Rock Mechanics and Internat. Assoc. of Engineering Geology.

GUPTA, C.P., and SINGH, V.S., 1985, *Effect of Incomplete Recovery Prior to a Pump Test in a Large-Diameter Well,* J. Hydrology, v. 79, no. 1/2, July 10.

GUTHRIE, MARILYN, 1986, *Use of a Geoflowmeter for the Determination of Ground Water Flow Direction,* Ground Water Monitoring Rev., v. 6, no. 2, pp. 81-86.

GUYTON, W. F., 1942, *Results of Pumping Tests of the Carrizo Sand in the Lufkin Area, Texas,* Trans. Am. Geophys. Union, v. 22, pt. 1, pp. 40-48.

HACKBARTH, D. A., 1978, *Application of the Drill-Stem Test to Hydrogeology,* Ground Water, v. 16, no. 1, pp. 5-11.

HANTUSH, M. S., 1956, *Analysis of Data from Pumping Tests in Leaky Aquifers,* Trans. Am. Geophys. Union, v. 37, pp. 702-714.

HANTUSH, M. S., 1961, *Aquifer Tests on Partially Penetrating Wells,* J. Hydraulics Div. Proc. Am. Soc. Civil Engrs., v. 87, no. HY5, Paper 2943, pp. 171-195.

HANTUSH, M. .S, 1961, *Economical Spacing of Interfering Wells,* Internat. Assoc. Sci. Hydrology, Publ. no. 57, pp. 350-364.

HANTUSH, M. S., 1964, *Depletion of Storage, Leakage, and River Flow by Gravity Wells in Sloping Sands,* J. of Geophys. Res., v. 69, pp. 2551-60.

HARRILL, J. R., 1970, *Determining Transmissivity from Water-Level Recovery of a Step-Drawdown Test,* U. S. Geol. Survey Prof. Paper 700-C, pp. 212-213.

HERBERT, ROBIN, and KITCHING, RON, 1981, *Determination of Aquifer Parameters from Large-Diameter Dug Well Pumping Tests,* Ground Water, v. 19, no. 6, pp. 593-599.

HICKEY, J.J., 1984, *Field Testing the Hypothesis of Darcian Flow Through a Carbonate Aquifer,* Ground Water, v. 22, no. 5, pp. 544-548.

HIRD, J. M., 1969, *Control of Artesian Ground Water in Strip Mining Phosphate Ores in Eastern North Carolina* (abs), Mining Eng., v. 21, no. 12, p. 55.

HOLZCHUH III, J. C., 1976, *A Simple Computer Program for the Determination of Aquifer Characteristics from Pump Test Data,* Ground Water, v. 14, no. 5, pp. 283-285.

HURR, R. T., 1966, *A New Approach for Estimating Transmissibility from Specific Capacity,* Water Resources Res., v. 2, pp. 657-664.

INESON, J., 1963, *Application and Limitations of Pumping Tests, Hydrogeological Significance,* J. Inst. Water Engnrs., v. 17, pp. 200-215.

JACOB, C. E., 1944, *Notes on Determining Permeability by Pumping Tests under Water-Table Conditions,* U. S. Geol. Survey mimeographed report.

JACOB, C. E., 1946, *Drawdown Test to Determine Effective Radius of Artesian Well,* Proc. Am. Soc. Civil Engrs., v. 112, pp. 1047-1070.

JOHNSON, C. R., and GREENKORN, R. A., 1960, *Comparison of Core Analysis and Drawdown Test Results from a Water-Bearing Upper Pennsylvanian Sandstone of Central Oklahoma,* Bull. Geol. Soc. America, v. 71, pp. 1898.

KANO, T., 1939, *Frictional Loss of Head in the Wall of a Well*, Japan J. Astron. Geophys. v. 17, no. 1.

KARANJAC, J., 1972, *Well Losses Due to Reduced Formation Permeability*, Ground Water, v. 10, no. 4, pp. 42-49.

KELLY, J. E., ANDERSON, K. E., and BURNHAM, W. L., 1980, *The "Cheat Sheet": A New Tool for the Field Evaluation of Wells by Step-Testing*, Ground Water, v. 18, no. 3, pp. 294-298, May-Jun.

KOHUT, A.P.J., and others, 1983, *Pumping Effects of Wells in Fractured Granitic Terrain*, Ground Water, v. 21, pp. 564-572.

KRUSEMAN, G. P., and DE RIDDER, N. A., 1970, *Analysis and Evaluation of Pumping Test Data*, Internat. Inst. for Land Reclamation and Improvement, Bull. 11, Wageningen, The Netherlands, 200 pp.

LABADIE, J. W., and HELWEG, O. J., 1975, *Step-Drawdown Test Analysis by Computer*, Ground Water, v. 13, no. 5, pp. 438-444.

LAKSHMINARAYANA, V., and RAJAGOPALAN, S. P., 1978, *Type-Curve Analysis of Time-Drawdown Data for Partially Penetrating Wells in Unconfined Anisotropic Aquifers*, Ground Water, v. 16, pp. 328-333.

LANG, S. M., 1960, *Interpretation of Boundary Effects from Pumping Test Data*, J. Am. Water Works Assoc., v. 52, no. 3, pp. 356-364.

LANG, S. M., 1963, *Drawdown Patterns in Aquifers Having a Straight-Line Boundary*, U. S. Geol. Survey Water-Supply Paper 1545-C.

LEAP, DARRELL, 1985, *A Simple Pneumatic Device and Technique for Performing Rising Water Level Slug Tests*, Ground Water Monitoring Rev., v. 4, no. 4, pp. 141-146.

LEE, C. H., 1934, *The Interpretation of Water-Levels in Wells and Test-Holes*, Trans. Am. Geophys. Union, v. 15, pp. 540-554.

LEE, D. R., and CHERRY, J. A., 1980, *A Field Exercise on Ground-Water Flow Using Seepage Meters and Minipiezometers*, J. of Geol. Education, v. 27, no. 1, January.

LENNOX, D. H., 1966, *The Analysis and Application of the Step-Drawdown Test*, J. Hydraulics Div. Am. Soc. Civil Engrs., v. 92, no. HY6, pp. 25-48.

LENNOX, D. H., and VANDENBERG, ALBERT, 1967, *Drawdowns due to Cyclic Pumping*, J. Hydraulics Div. Proc. Am. Soc. Civil Engrs., v. 93, no. HY6, pp. 35-51.

LI, W. H., 1954, *Interaction Between Well and Aquifer,* Proc. Am. Soc. Civil Engrs., v. 80, sep. 578, 14 pp.

LOGAN, J., 1964, *Estimating Transmissibility from Routine Production Tests of Water Wells,* Ground Water, v. 2, no. 2, pp. 35-37.

MARINELLI, F., and ROWE, J.W., 1985, *Performance and Analysis of Drillstem Tests in Small-Diameter Boreholes,* Ground Water, v. 23, no. 3, pp. 367-376.

MATTHEWS, C. S. and RUSSELL, D. G., 1967, *Pressure Buildup and Flow Tests in Wells,* Am. Institute of Mining, Metal and Petroleum Engrs., Society of Petroleum Engrs., Monograph v. 1.

MOGG, J. L., 1969, *Step-Drawdown Test Needs Critical Review,* Ground Water, v. 7, no. 1.

NAWROCKI, M. A., 1971, *Comparison of Methods for Evaluating Aquifer Characteristics,* Ground Water, v. 9, no. 4, pp. 6-11.

NEUMAN, S. P., and WITHERSPOON, P. A., 1972, *Field Determination of the Hydraulic Properties of Leaky Multiple Aquifer Systems,* Water Resources Res., v. 8, no. 5, pp. 1284-1298.

NEUMAN, S. P., 1975, *Analysis of Pumping Test Data from Anisotropic Unconfined Aquifers Considering Delayed Gravity Response,* Water Resources Res., v. 11, no. 2, pp. 329-342.

NEUMAN, S. P., 1975, *Analysis of Pumping Test Data From Anisotropic Unconfined Aquifers Considering Delayed Gravity Response,* J. Water Resources Res., v. 11, pp. 329-342.

NEUZIL, C. E., 1982, *On Conducting the Modified "Slug" Test in Tight Formations,* Water Resources Res., v. 18, no. 3, Jun.

NORRIS, S. E., 1976, *Change in Drawdown Caused by Enlarging a Well in a Dolomite Aquifer,* Ground Water v. 14, no. 4, pp. 191-193.

PAPADOPULOS, S. S., BREDEHOEFT, J. D., and COOPER, H. H., 1973, *On the Analysis of "Slug Test" Data,* Water Resources Res., v. 9, no. 4, pp. 1087-1089.

PASCHETTO, J., and McELWEE, C. D., 1982, *Hand Calculator Program for Evaluating Theis Parameters from a Pumping Test,* Ground Water, v. 20, no. 5, pp. 551-555.

PARIZEK, R. R., and SIDDIQUI, S. H., 1970, *Determining the Sustained Yields of Wells in Carbonate and Fractured Aquifers,* Ground Water, v. 8, no. 5, pp. 12-21.

PATCHICK, P. F., 1967, *Estimated Water Well Specific Capacity Utilizing Permeability of Disturbed Samples,* J. Am. Water Works Assoc., v. 59, p. 1292, Oct.

PETERSEN, J. S., and others, 1955, *Effect of Well Screens on Flow into Wells,* Trans. Am. Soc. Civil Engrs., v. 120, pp. 563-607.

PICKENS, J. F., and others, 1978, *A Multilevel Device for Ground-Water Sampling and Piezometric Monitoring,* Ground Water, v. 16, no. 5, p. 322, Sept.-Oct.

POWERS, J.P. (ed.), 1984, *Dewatering: Avoiding the Unwanted Side Effects,* Am. Soc. Civil Engrs, 345 E. 47th St., New York, New York 10017.

PRICKETT, T. A., 1965, *Type-Curve Solution to Aquifer Tests under Water-Table Conditions,* Ground Water, v. 3, no. 3, pp. 5-14.

PROSSER, D. W., 1981, *A Method of Performing Response Tests on Highly Permeable Aquifers,* Ground Water, v. 19, no. 6, pp. 588-592.

RANDOLPH, R.B., KRAUS, R.E., and MASLIA, M.L., 1985, *Comparison of Aquifer Characteristics Derived from Local and Regional Aquifer Tests,* Ground Water, v. 23, no. 3, pp. 309-316.

RAYNER, F. A., 1980, *Pumping Test Analysis with a Handheld Calculator,* Ground Water, v. 18, no. 6, pp. 562-568, Nov.-Dec.

REED, J.E., 1980, *Type Curves for Selected Problems of Flow to Wells in Confined Aquifers,* Scientific Publications Co., P.O. Box 23041, Washington, DC., 106 pp.

REMSON, I., and LANG, S. M., 1955, *A Pumping-Test Method for the Determination of Specific Yield,* Trans. Am. Geophys. Union, v. 36, pp. 321-325.

REMSON, I., and VAN HYLCKEMA, T. E. A., 1956, *Nomographs for the Rapid Analysis of Aquifer Tests,* J. Am. Water Works Assoc., v. 48, pp. 511-516.

RIHA, MIRKO, 1979, *A Method of Aquifer Testing and its Application to an Anisotropic Inhomogeneous Multiple Aquifer System,* Ground Water, v. 17, no. 5, pp. 423-429.

RORABAUGH, M. I., 1953, *Graphical and Theoretical Analysis of Step Drawdown Test of Artesian Well,* Proc. Am. Soc. Civil Engrs., v. 79, sep. 362, 23 pp.

RUSHTON, K.R., 1985, *Interference Due to Neighboring Wells During Pumping Tests,* Ground Water, v. 23, no. 3, pp. 361-366.

RUSHTON, K. R., and CHAN, Y. K., 1976, -Pumping Test Analysis When Parameters Vary with Depth, Ground Water v. 14, no. 2, pp. 82-87.

RUSHTON, K. R., and CHAN, Y. K., 1977, *Numerical Pumping Test Analysis in Unconfined Aquifers,* J. Irrig. and Drainage Div. Am. Soc. Civ. Eng., vol. 103 (1R1), pp 1-12.

RUSHTON, K.R., and HOWARD, K. W. F., 1982, *The Unreliability of Open Observation Boreholes in Unconfined Aquifer Pumping Tests,* Ground Water, v. 20, no. 5, pp. 546-550.

RUSHTON, K. R., and RATHOD, K. S., 1980, *Overflow Tests Analysed by Theoretical and Numerical Methods,* Ground Water, v. 18, no. 1, pp. 61-69, Jan-Feb.

RUSHTON, K.R., and RATHOD, K.S., 1981, *Aquifer Response Due to Zones of Higher Permeability and Storage Coefficient,* J. Hydrology, v. 50, pp. 299-316.

RUSHTON, K.R., and WELLER, JANE, 1985, *Response to Pumping of a Weathered-Fractured Granite Aquifer,* J. Hydrology, v. 80, no. 3/4, Oct. 15.

SAGEEV, ABRAHAM, HORNE, R.N., and RAMEY, H.J., Jr., 1985, *Detection of Linear Boundaries by Drawdown Tests: A Semilog Type Curve Matching Approach,* Water Resources Res., v. 21, no. 3, pp. 305-310.

SAMMEL, E. A., 1974, *Aquifer Tests in Large-Diameter Wells in India,* Ground Water, v. 12, no. 5, pp. 265-272.

SANDERS, P.J., 1985, *New Tape For Ground Water Measurements,* Ground Water Monitoring Rev., v. 4, no. 1, pp. 39-42.

SCHAFER, D. C., 1978, *Casing Storage Can Affect Pumping Test Data,* The Johnson Drillers Journal, v. 50, no. 1, pp. 1-5.

SCHICHT, R. J., 1972, *Selected Methods of Aquifer Test Analysis,* Water Resources Bull., v. 8, no. 1, pp. 175-187.

SCHRALE, G. and BRANDWYK, J. F., 1979, *An Acoustic Probe for Precise Determination of Deep Water Levels in Boreholes,* Ground Water, v. 17, no. 1, p. 110, Jan.-Feb.

SEN, ZEKAI, 1982, *Type Curves for Large-Diameter Wells Near Barriers,* Ground Water, v. 20, no. 3, pp. 274-277.

SEN, ZEKAI, 1986, *Aquifer Test Analysis in Fractured Rocks with Linear Flow Pattern,* Ground Water, v. 24, no. 1, pp. 72-78.

SEN, ZEKAI., 1986, *Determination of Aquifer Parameters by the Slope-Matching Method,* Ground Water, v. 24, no. 2, pp. 217-223.

SHEAHAN, N. T., 1966, *Determining Transmissibility From Cyclic Discharge,* Ground Water, v. 4, no. 3, pp. 33-34.

SHEAHAN, THOMAS, 1971, *Type-Curve Solution of Step-Drawdown Test,* Ground Water, v. 9, no. 1, pp. 25-29.

SHUTER, E., and JOHNSON, A. I., 1961, *Evaluation of Equipment for Measurement of Water Level in Wells of Small Diameter,* U. S. Geol. Survey Cir. 453, 12 pp.

SLICHTER, C. S., 1904, *Approximate Methods of Measuring the Yield of Flowing Wells,* U. S. Geol. Survey Water Supply Paper 110, pp. 37-42.

SMITH, E.D., and VAUGHAN, N.D., 1985, *Aquifer Test Analysis in Nonradial Flow Regimes: A Case Study,* Ground Water, v. 23, no. 2, pp. 167-175.

STALLMAN, R. W., 1971, *Aquifer-Test Design, Observation and Data Analysis,* U. S. Geol. Survey Tech. of Water Resources Inv., 3 (BI), pp. 1-26.

STERNBERG, Y. M., 1967, *Transmissibility Determination From Variable Discharge Pumping Tests,* Ground Water, v. 5, no. 4, pp. 27-29.

STERNBERG, Y. M., 1968, *Simplified Solution for Variable Rate Pumping Test,* J. Hydraulics Div. Am. Soc. Civil Engrs., v. 94, no. HY1, pp. 177-180.

STEWART, D. M., 1970, *The Rock and Bong Techniques of Measuring Water Levels in Wells,* Ground Water, v. 8, no. 6, pp. 14-19.

SUMMERS, W. K., 1972, *Application of Harrill's Equation to a Limestone Aquifer,* Ground Water, v. 10, no. 4, pp. 21-23.

SUMMERS, W. K., 1972, *Specific Capacity of Wells in Crystalline Rocks,* Ground Water, v. 10, no. 6, pp. 37-47.

SUTCLIFFE, H., Jr., and JOYNER, B. F., 1966, *Packer Testing in Water Wells near Sarasota, Florida,* Ground Water, v. 4, no. 2, pp. 23-27, Jan.

THEIS, C. V., 1963, *Drawdowns Caused by a Well Discharging under Equilibrium Conditions from an Aquifer, Bounded on a Finite Straight-Line Source,* in *Shortcuts and Special Problems in Aquifer Tests,* U. S. Geol. Survey Water Supply Paper 1545-C.

TOMSON, M. B., and others, 1980, *A Nitrogen Powered Continuous Delivery, All-Glass-Teflon Pumping System for Ground-Water Sampling from Below 10 Meters,* Ground Water, v. 18, no. 5, pp. 444-446, Sept.-Oct.

TURCAN, A. N., Jr., 1963, *Estimating the Specific Capacity of a Well,* U. S. Geol. Survey Prof. Paper 450-E.

VANDENBERG, A., 1976, *Tables and Type Curves for Analysis of Pump Tests in Leaky Parallel-Channel Aquifers,* Inland Waters Directorate, Water Resources Branch, Ottawa, Canada, Tech. Bull. no. 96.

VANDENBERG, A., 1977, *Type Curves for Analysis of Pump Tests in Leaky Strip Aquifers,* J. Hydrology, v. 33, no. 1/2, pp. 15-26.

VAN POOLLEN, H. K., 1961, *Status of Drill-Stem Testing Techniques and Analysis,* J. Petroleum Tech., pp. 333-339, Apr.

WALTER, G. R., and THOMPSON, G. M., 1982, *A Repeated Pulse Technique for Determining the Hydraulic Properties of Tight Formations,* Ground Water, v. 20, no. 2, pp. 186-193.

WALTON, W. C., 1960, *Leaky Artesian Aquifer Conditions in Illinois,* Illinois State Water Survey Rept. of Invest. 39. Urbana, Illinois, 27 pp.

WALTON, W. C., 1962, *Selected Analytical Methods for Well and Aquifer Evaluation,* Illinois State Water Survey Bull. 49, Urbana Illinois, 81 pp.

WALTON, W. C., and NEILL, J. C., 1963, *Statistical Analysis of Specific Capacity Data for a Dolomite Aquifer,* J. Geophys. Res., v. 68, pp. 2251-2262.

WALTON, W. C., and STEWART, J. W., 1961, *Aquifer Tests in the Snake River Basalt,* Trans. Am. Soc. Civil Engnrs., v. 126, pp. 612-632.

WALTON, W.C., 1978, *Comprehensive Analysis of Water-table Aquifer Test Data,* Ground Water, v. 16, no. 5, pp. 311-317.

WARNER, D. L., and YOW, M. G., 1979, *Programmable Hand Calculator Programs for Pumping and Injection Wells: I — Constant or Variable Pumping (Injection) Rate, Single or Multiple Fully Penetrating Wells,* Ground Water, v. 17, no. 6, pp. 532-537, Nov-Dec.

WARNER, D. L., and YOW, M. G., 1980, *Programmable Hand Calculator Programs for Pumping and Injection Wells: II — Constant Pumping (Injection) Rate, Single Fully Penetrating Well, Semiconfined Aquifer,* Ground Water, v. 18, no. 2, pp. 126-133, Mar.-Apr.

WARNER, D. L, and YOW, M. G., 1980, *Programmable Hand Calculator Programs for Pumping and Injection Wells: III — Constant Pumping (Injection) Rate, Fully Confined Aquifer, Partially Penetrating Well,* Ground-Water, v. 18, no. 5, pp. 438-443, Sept.-Oct.

WENZEL, L. K., and GREENLEE, A. L., 1943, *A Method for Determining Transmissibility and Storage Coefficients by Tests of Multiple Well-Systems,* Trans. Am. Geophys. Union, 24th Ann. Meeting, pt. 2.

WIKRAMARATNA, R.S., 1985, *A New Method for the Analysis of Pumping Tests in Large- Diameter Wells,* Water Resources Res., v. 21, no. 2.

WILLIAMS, J. A., and SOROOS, R. L, 1973, *Evaluation of Methods of Pumping Test Analyses for Application to Hawaiian Aquifers,* Univ. of Hawaii Water Resources Res. Center Tech. Rept. 70, Honolulu, Hawaii 96822.

YEH, WILLIAM W-G., and SUN, NE-ZHENG, 1984, *An Extended Indentifiability in Aquifer Parameter Identification and Optimal Pumping Test Design,* Water Resources Res., v. 16, no. 12, pp. 1837-1847.

TRACERS AND GROUNDWATER DATING

BIBLIOGRAPHY

MATHER, J. D., 1968, *A Literature Survey of the Use of Radioisotopes in Ground-Water Studies,* Great Britain Inst. of Geol. Sciences, Tech. Communication no. 1, London.

GENERAL WORKS

ALEY, T.J., 1985, *Optical Brightener Sampling; A Reconnaissance Tool for Detecting Sewage in Karst Groundwater,* Hydrological Sci. and Techn.: Short Papers, v. 1, no. 1, 45-48 pp.

ALLISON, G.B. and HUGHES, M.W., 1983, *The Use of Natural Tracers as Indicators of Soil- Water Movement in a Temperate Semi-arid Region.* J. Hydrology, v. 60, no. 1/4.

AMBROSE, A. W., 1921, *Use of Detectors for Tracing Movement of Underground Water,* U. S. Bureau Mines Bull. 195, Washington, D. C., pp. 106-120.

BACK, W., and others, 1970, *Carbon-14 Ages Related to Occurrence of Salt Water,* J. Hydraulics Div., Amer. Soc. Civil Engrs., v. 96, no. HY11, pp. 2325-2336.

BAETSLE, L. H., and SOUFFRIA, J., 1966, *Fundamentals of the Dispersion of Radionuclides in Sandy Aquifers,* Isotopes in Hydrology Symp. Vienna, Internat. Atomic Energy Agency, pp. 617-628.

CARLSTON, C. W., and others, 1960, *Tritium as a Hydrologic Tool, the Wharton Tract Study,* Internat. Assoc. Sci. Hydrol. Pub. no. 52, pp. 503-512.

CARTER, R. C., and others, 1959, *Helium as a Ground-Water Tracer,* J. Geophys. Res., v. 64, pp. 2433-2439.

COLE, D. R., 1982, *Tracing Fluid Sources in the East Shore Area, Utah,* Ground Water, v. 20, no. 5, pp. 586-593. (stable isotope and fluid chemistry)

DANEL, P., 1953, *The Measurement of Ground-Water Flow,* Proc. Ankara Symp. on Arid Zone Hydrology, pp. 99-107, UNESCO, Paris.

DAVIS, N., and others, 1980, *Ground-Water Tracers — A Short Review,* Ground Water, v. 18, no. 1, pp. 14-23, Jan-Feb.

DAVIS, S.N., and others, 1985, *An Introduction to Ground-Water Tracers,* EPA/600/2-85/ 022, PB 86-100591, NTIS, Springfield, Virginia 22161, also NWWA, 6375 Riverside Dr., Dublin, Ohio 43017.

DE LAGUNA, WALLACE, 1970, *Tracer Aids Interpretation of Pumping Test,* Water Resources Res., v. 6, no. 1, p. 172, Feb.

DOLE, R. B., 1906, *Use of Fluorescein in the Study of Underground Water,* U. S. Geol. Survey Water Supply Paper 160, pp. 73-86.

DROST, W., and others, 1968, *Point Dilution Methods of Investigating Ground Water Flow by Means of Radioisotopes,* Water Resources Research, v. 4, pp. 125-146.

FOX, C. S., 1952, *Using Radioactive Isotopes to Trace Movement of Underground Waters,* Municipal Utilities, v. 90, no. 4, pp. 30-32.

GASPAR, E., and ONCESCU, M., 1972, *Radioactive Tracers in Hydrology,* Am. Elsevier Publ. Co., Inc., New York, 352 pp.

GAT, J. R., 1971, *Comments on the Stable Isotope Method in Regional Groundwater Investigations,* Water Resources Res., v. 7, no. 4, pp. 980-993.

GROVE, D. B., and BEETEM, W. A., 1971, *Porosity and Dispersion Constant Calculations for a Fractured Carbonate Aquifer Using the Two-Well Tracer Method,* Water Resources Res., v. 7, no. 1, pp. 128-134.

HALEVY, E., and NIR, A., 1962, *The Determination of Aquifer Parameters with the Aid of Radioactive Tracers,* J. Geophys. Res., v. 67, no. 6, pp. 2403-2409.

HALEVY, E., and others, 1967, *Borehole Dilution Techniques: A Critical Review, Isotopes in Hydrology,* Internat. Atomic Energy Agency, Vienna, pp. 531-564.

HANSHAW, B. B., BACK, W, and RUBIN, M., 1965, *Radiocarbon Determinations for Estimating Groundwater Flow Velocities in Central Florida,* Science, v. 148, pp. 494-495.

HARPAZ, Y., and others, 1963, *The Place of Isotope Methods in Ground Water Research, Radioisotopes in Hydrology,* Symp. Tokyo, Internat. Atomic Energy Agency, pp. 175-191.

HOURS, R., 1955, *Radioactive Tracers in Hydrology,* La Houille Blanche, v. 10, no. A, pp. 14-24.

INTERNATIONAL ATOMIC ENERGY AGENCY, 1963, *Radioisotopes in Hydrology,* Proc. Symp. on the Application of Radioisotopes in Hydrology, Tokyo, March 5-9, 1963, IAEA Vienna, 449 pp.

INTERNATIONAL ATOMIC ENERGY AGENCY, 1978, *Nuclear Techniques in Groundwater Pollution Research,* Symp. Cracow 1976, Doc. ISP518 UNIPUB, New York, New York 10010, 285 pp.

JESTER, W. A., and UHLER, K. A., 1974, *Identification and Evaluation of Water Tracers Amenable to Post-Sampling Neutron Activation Analysis,* Inst. for Research on Land and Water Resources, Research Publ. no. 85, The Pennsylvania State University, 92 pp.

KAUFMAN, W. J., 1960, *The Use of Radioactive Tracers in Hydrologic Studies,* Univ. of California Proc. Conf. on Water Research, Water Resources Center Rept. 2, pp. 6-14.

KAUFMAN, W. J., 1961, *Tritium as a Ground Water Tracer,* Trans. Am. Soc. Civil Engnrs. Paper 3203, pp. 436-446.

KAUFMAN, W. J., and ORLOB, G. T., 1956, *An Evaluation of Ground-Water Tracers,* Trans. Am. Geophys. Union, v. 37, pp. 297-306.

KAUFMAN, W. J., and ORLOB, G. T., 1956, *Measuring Ground Water Movement with Radioactive and Chemical Tracers,* J. Am. Water Works Assoc., v. 48, pp. 559-572.

KAUFMAN, W. J., and TODD, D. K., 1955, *Methods of Detecting and Tracing the Movement of Ground Water,* Inst. Eng. Research Rep. 93-1, Univ. of California, Berkeley, 130 pp.

KAUFMAN, W. J., and TODD, D. K., 1962, *Application of Tritium Tracer to Canal Seepage Measurements, Tritium in the Physical and Biological Sciences,* Internat. Atomic Energy Agency, Vienna, pp. 83-94.

KEELEY, J. W., and SCALF, M. R., 1969, *Aquifer Storage Determination by Radio-Tracer Techniques,* Ground Water, v. 7, pp. 17-22.

KESWICK, B. H., WANG, D., and GERBA, C. P., 1982, *The Use of Microorganisms as Ground-Water Tracers: A Review,* Ground Water, v. 20, no. 2, pp. 142-149.

KEYS, N. S., 1966, *The Application of Radiation Logs to Groundwater Hydrology,* Isotopes in Hydrology Symp. Vienna, Internat. Atomic Energy Agency, SM 83/33, pp. 477-486.

LEWIS, D. C., and others, 1966, *Tracer Dilution Sampling Technique to Determine Hydraulic Conductivity of Fractured Rock,* Water Resources Res., v. 2, pp. 533-542.

LLAMAS, M. R., SIMPSON, E. S., and MARTINEZ ALFARO, P. E., 1982, *Ground-Water Age Distribution in Madrid Basin, Spain.* Ground Water, v. 20, no. 6, pp. 688-695.

MALOSZEWSKI, P. and ZUBER, A., 1982, *Determining the Turnover Time of Ground-Water Systems with the Aid of Environmental Tracers,* 1. Models and Their Applicability. J. Hydrology, v. 57, no. 3/4, Jun.

MARINE, I. W., 1980, *Determination of the Location and Connectivity of Fractures in Metamorphic Rock with In-Hole Tracers,* Ground Water, v. 18, no. 3, pp. 252-261 May-Jun.

NAYMIK, T.G., and SIEVERS, M.E., 1985, *Characterization of Dye Tracer Plumes: In Situ Field Experiments,* Ground Water, v. 23, no. 6, pp. 746-752.

NELSON, R. W., and REISENAUER, A. E., 1963, *Application of Radioactive Tracers in Scientific Ground Water Hydrology,* Radioisotopes in Hydrology, Symp. Tokyo, Internat. Atomic Energy Agency, pp. 207-230.

PEARSON, F. J., Jr., and WHITE, D. E., 1967, *Carbon-14 Ages and Flow Rates of Water in Carrizo Sand, Atascosa County, Texas,* Water Resources Res., v. 3, pp. 251-261.

PEARSON, F. J., Jr., 1972, *The Evaluation and Application of C-14 Dating of Groundwater,* U. S. Geol. Survey, NTIS, Springfield, Virginia 22151, AD 748 877, 70 pp.

POLAND, J. F., and STEWART, G. L., 1975, *New Tritium Data on Movement of Groundwater in Western Fresno County, California,* Water Resources Res., v. 11, no. 5, pp. 716-724.

RABINOWITZ, D. D., GROSS, G. W., and HOLMES, C. R., 1977, *Environmental Tritium as a Hydrometeorological Tool in the Roswell Basin, New Mexico,* J. Hydrology, v. 32, pp. 3-46.

RIFAI, M. N. E., and others, 1956, *Dispersion Phenomena in Laminar Flow through Porous Media,* Inst. Eng. Res. Rep. 93-2, Univ. Calif. Berkeley, 157 pp.

SCHOFF, S. L, and MOORE, J. E., 1968, *Sodium as a Clue to Direction of Ground-Water Movement,* Nevada Test Site, U. S. Geol. Survey Prof. Paper 600-D, pp. 30-33.

SCHULTZ, T. R., and others, 1977, *Tracing Sewage Effluent Recharge — Tuscon, Arizona,* Proc. Third National Ground Water Quality Symposium, EPA-600/9-77-014, Robert S. Kerr Environmental Research Laboratory, Ada, Oklahoma, pp. 93-98.

SMITH, D. B., 1974, *Flow Tracing Using Isotopes,* in *Groundwater Pollution in Europe,* J. A. Cole (ed.), Water Information Center, Plainview, New York, pp. 377-387.

SMITH, D. B., and others, 1976, *The Age of Groundwater in the Chalk of London Basin,* Water Resources Res., v. 12, no. 3, pp. 392-404.

SPIRIDONOV, A. I., and others, 1973, *Some Problems in the Computation of the Age of Ground Waters,* Soviet Hydrology: Selected Papers no. 3.

STOUT, G. E. (ed.), 1967, *Isotope Techniques in the Hydrologic Cycle,* Geophysical Monograph Ser. no. 11, Amer. Geophysical Union, 199 pp.

SUDICKY, E. A., and FRIND, E. O., 1981, *Carbon 14 Dating of Ground Water in Confined Aquifers: Implications of Aquitard Diffusion,* Water Resources Res., v. 17, no. 4, Aug.

SUGISAKI, R., 1961, *Measurement of Effective Flow Velocity of Ground Water by Means of Dissolved Gases,* Am. Jour. Sci., v. 259, pp. 144-153.

THATCHER, L., and others, 1961, *Dating Desert Ground Water,* Science, v. 134, no. 3472, pp. 105-106.

THEIS, C. V., 1963, *Hydrologic Phenomena Affecting the Use of Tracers in Timing Ground Water Flow,* Radioisotopes in Hydrology, Symp. Tokyo, Internat. Atomic Energy Agency, pp. 193-206.

THRAILKILL, JOHN, and others, 1982, *Groundwater in the Inner Bluegrass Karst Region, Kentucky,* Kentucky Water Resources Research Inst., Lexington, PB83-108126, NTIS, Springfield, Virginia 22161, 147 pp. (dye tracing).

TORAN, LAURA, 1982, *Isotopes in Ground-Water Investigations,* Ground Water, v. 20, no. 6, p. 740-745.

VOGEL, J. C., and EHHALT, D., 1963, *The Use of Carbon Isotopes in Groundwater Studies,* Radioisotopes in Hydrology, Internat. Atomic Energy Agency, Vienna, pp. 383-395.

VOGEL, J. C., 1970, *Carbon-14 Dating of Groundwater,* Isotope Hydrology 1970, Internat. Atomic Energy Agency, Vienna, pp. 225-239.

VON BUTTLAR, H. 1959, *Ground-Water Studies in New Mexico Using Tritium as a Tracer, Part II,* J. Geophys. Res., v. 64, pp. 1031-1038.

VON BUTTLAR, H., and WENDT, I., 1958, *Ground-Water Studies in New Mexico Using Tritium as a Tracer,* Trans. Am. Geophys. Union, v. 39, pp. 660-668.

WEBSTER, D. S., and others, 1970, *Two-Well Tracer Test in Fractured Crystalline Rock,* U. S. Geol. Survey Water-Supply Paper 1544-I, 22 pp.

WIEBENGA, W. A., and others, 1967, *Radioisotopes as Groundwater Tracers,* J. Geophys. Res., v. 72, no. 15, pp. 4081-91.

WIGLEY, T. M. L., 1975, *Carbon 14 Dating of Groundwater from Closed and Open Systems,* Water Resources Res., v. 11, no. 2, pp. 324-328.

WOGMAN, N. A., 1976, *Nuclear Techniques Applicable to Studies of Pollutants in Ground Water,* Battelle Pacific Northwest Labs. Richland, Washington, 36 pp.

WOOD, W. W., and EHRLICH, G. G., 1978, *Use of Baker's Yeast to Trace Microbial Movement in Ground Water,* Ground Water, v. 16, no. 6, pp. 398-404, Nov.-Dec.

WORKING GROUP ON NUCLEAR TECHNIQUES IN HYDROLOGY, 1968, *Guidebook on Nuclear Techniques in Hydrology,* Tech. Rept. Ser. no. 91, Internat. Atomic Energy Agency, Vienna, 214 pp.

NATURAL WATER QUALITY

BIBLIOGRAPHY

CASE, L. C., and others, 1942, *Selected Annotated Bibliography on Oil Field Waters,* Bull. Am. Assoc. Petroleum Geologists, v. 26, no. 5, p. 865-881.

FETH, J. H., 1965, *Selected References on Saline Ground Water Resources of the United States,* U. S. Geol. Survey Circ. 499, 30 pp.

WATER STANDARDS AND CRITERIA

ADAMS, O. H., 1977, *The Safe Drinking Water Act and the Water Utility Owner,* J. Am. Water Works Assoc., pp. 229-233, May.

AMERICAN SOCIETY FOR TESTING MATERIALS, 1967, *Water Quality Criteria,* ASTM Spec. Tech. Publ. 416, Philadelphia, 120 pp.

AMERICAN SOCIETY FOR TESTING MATERIALS, 1969, *Manual on Water,* ASTM Spec. Tech. Publ. no. 442, Philadelphia, 360 pp.

AMERICAN WATER WORKS ASSOCIATION, 1967, *Sources of Nitrogen and Phosphorus in Water Supplies,* J. Am. Water Works Association, Task Group Report, pp. 344-366.

ANONYMOUS, 1940, *Progress Report of the Committee on Quality Tolerances of Water for Industrial Uses,* J. New Engl. Water Works Assoc., v. 54.

GORRELL, H. A., 1958, *Classification of Formation Waters based on Sodium Chloride Content,* Bull. Am. Assoc. Petroleum Geologists, v. 42, pp. 2513.

HAMMER, H. J., 1981, *An Assessment of Current Standards for Selenium in Drinking Water,* Ground Water, v. 19, no. 4, pp. 366-369.

KELLEY, W. P., 1941, *Permissible Composition and Concentration of Irrigation Water,* Trans. Am. Soc. Civil Engrs., v. 106, pp. 849-861.

McKEE, J. E., and WOLF, H. W. (eds.), 1963, *Water Quality Criteria,* Calif. State Water Quality Control Bd., Pub. no. 3-A.

NATIONAL ACADEMY OF SCIENCES, NATIONAL ACADEMY OF ENGINEERING, 1972, *Water Quality Criteria 1972,* Washington, D. C., 594 pp.

NATIONAL ACADEMY OF SCIENCES, 1977, 1980, 1983, *Drinking Water and Health,* U. S. Environmental Protection Agency, PB 269 519/5WP, NTIS Springfield, Virginia 22161. (scientific basis for revision of primary drinking water standards)

POWELL, S. T., 1948, *Some Aspects of the Requirements for the Quality of Water for Industrial Purposes,* J. Am. Water Works Assoc., v. 40, pp. 8-23.

U. S. DEPARTMENT OF HEALTH, EDUCATION AND WELFARE, 1962, *Public Health Service Drinking Water Standards,* U. S. Public Health Service Pub. 956 Washington, D. C., 61 pp.

U. S. ENVIRONMENTAL PROTECTION AGENCY, 1976, *Quality Criteria for Water,* Washington, D. C. 504 pp.

U. S. ENVIRONMENTAL PROTECTION AGENCY, 1977, *National Interim Primary Drinking Water Regulations,* PB-267 630/2WP, NTIS, Springfield, Virginia 22161.

U. S. FEDERAL WATER POLLUTION CONTROL ADMINISTRATION, 1968, *Water Quality Criteria,* Report of the Nat. Tech. Advisory Committee to Secretary of Interior, Washington, D. C., GPO, 234 pp.

WILCOX, L. V., 1948, *The Quality of Water for Irrigation Use,* U. S. Dept. Agric. Tech. Bull. 962, Washington, D. C., 40 pp.

WILCOX, L. V., 1955, *Classification and Use of Irrigation Waters,* U. S. Dept. Agric. Circ. 969, Washington, D. C., 19 pp.

WORLD HEALTH ORGANIZATION, 1961, *European Standards for Drinking Water,* Geneva, 52 pp.

WORLD HEALTH ORGANIZATION, 1963, *International Standards for Drinking Water,* Geneva, 206 pp.

LABORATORY AND FIELD METHODOLOGY/ WATER SAMPLING PROCEDURES/SAMPLING DEVICES

ALLISON, L. E., 1971, *A Simple Device for Sampling Groundwaters in*

Auger Holes, Soil Science Society of America Proceedings, v. 35, pp. 844-845.

AMERICAN PUBLIC HEALTH ASSOCIATION, AMERICAN WATER WORKS ASSOCIATION, AND WATER POLLUTION CONTROL FEDERATION, 1975, *Standard Methods for the Examination of Water and Wastewater, 14th ed.,* American Public Health Assoc., Washington, D. C., 1200 pp.

AMERICAN SOCIETY FOR TESTING MATERIALS, 1966, *Manual on Industrial Water and Industrial Waste Water,* 2nd ed., Philadelphia, 992 pp.

AMERICAN SOCIETY FOR TESTING MATERIALS, 1970, *Water and Atmospheric Analysis,* Philadelphia, Pennsylvania, 1,072 pp.

BANERJEE, S., YALKOWSKY, S.H., and VALVANI, S.C., 1980, *Water Solubility and Octanol/ Water Partition Coefficients of Organics Limitations of the Solubility-Partition Coefficient Correlation,* Env. Science and Technology, v. 14, no. 10, pp. 1227-1229.

BARCELONA, M.J., 1984, *TOC Determinations in Ground Water,* Ground Water, v. 22, no. 1, pp. 18-24.

BARCELONA, M.J., and others, 1985, *Practical Guide for Ground-Water Sampling,* Illinois State Water Survey, 2200 Griffith Dr., Champaign, Illinois 61820; PB 86-137304 NTIS Springfield, Virginia 22161, 184 pp.

BARCELONA, M.J., HELFRICH, J.A., GARSKE, E.E., 1984, *A Laboratory Evaluation of Ground Water Sampling Mechanisms,* Ground Water Monitoring Rev., v. 4, no. 2, pp. 32-41.

BARCELONA, M.J., HELFRICH, J.A., and GARSKE, E.E., 1985, *Sampling Tubing Effects on Groundwater Samples,* Illinois State Water Survey Reprint Ser. No. 656, 2204 Griffith Drive, Champaign, Illinois 61820.

BARNETT, P. R., and MALLORY, E. C., Jr., 1971, *Determination of Minor Elements in Water by Emission Spectroscopy,* Chap. A2, Techniques of Water-Resources Investigations, Book 5, Laboratory Analysis, U. S. Geol. Survey, 31 pp.

BARVENIK, M.J. and CADWGAN, R.M., 1983, *Multilevel Gas-Drive Sampling of Deep Fractured Rock Aquifers in Virginia,* Ground Water Monitoring Rev., v. 3, no. 4, pp. 26-33.

BROWN, E., SKOUGSTAD, M. W., and FISHMAN, M. J., 1970, *Methods for Collection and Analysis of Water Samples for Dissolved Minerals and Gases,* Chap. A1, Techniques of Water- Resources Investigations, Book 5, U. S. Geol. Survey, 160 pp.

CHERRY, R. N., 1965, *Portable Sampler for Collecting Water Samples from Specific Zones in Uncased or Screened Wells,* U. S. Geol. Survey Prof. Paper 525-C, pp. 214-216.

DUNLAP, W. J., and others, 1977, *Sampling for Organic Chemicals and Microorganisms in the Subsurface,* U. S. EPA, Robert S. Kerr Environmental Res. Lab. EPA-600/2-77-176, Ada, Oklahoma 74820, 27 pp.

FELDMANN, CHARLES, 1981, *Organic Analyses in Water Quality Control Programs: Training Manual,* PB81-124414, NTIS Springfield, Virginia 22161.

GIBB, J. P., SCHULLER, R. M., and GRIFFIN, R. A., 1981, *Procedures for the Collection of Representative Water Quality Data from Monitoring Wells,* Illinois State Water Survey and State Geol. Survey Cooperative Ground-Water Rep. No. 7, Champaign, Illinois 61820.

GILHAM, R. W., and JOHNSON, P. E., 1981, *A Positive Displacement Ground-Water Sampling Device,* Ground Water Monitoring Review, v. 1, no. 2, pp. 33-35.

GOERLITZ, D. F., and BROWN, E., 1972, *Methods for Analysis of Organic Substances in Water,* Chap. A3, Techniques of Water- Resources Investigations, Book 5, Laboratory Analysis, U. S. Geological Survey, 40 pp.

HARDER, A., H., and HOLDEN, W. R., 1965, *Measurement of Gas in Groundwater,* Water Resources Res., v. 1, no. 1, pp. 75-82.

HARRISON, S.S., 1986, *Low-Cost Apparatus for OnSite Monitoring of Methane in Ground Water,* Ground Water Monitoring Rev., v. 6, no. 2, pp. 73-76.

HEM, J. D., 1961, *Calculation and Use of Ion Activity,* U. S. Geol. Survey Water- Supply Paper 1535-C, 17 pp.

HORNBY, W.J., ZABCIK, J.D., and CRAWLEY, 1986 *Factors Which Affect Soil-Pore Liquid: A Comparison of Currently Available Samplers with Two New Designs,* Ground Water Monitoring Rev., v. 6, no. 2, p. 61-66.

HUTTON, L.G., 1984, *Field Testing of Water in Developing Countries*, Water Research Centre, Henley Road, Medmenham, P.O. Box 16, Marlow, Bucks, SL7 2HD, England, (water quality guidelines).

INTERNATIONAL HYDROLOGICAL DECADE/WORLD HEALTH ORGANIZATION, 1979, *Water Quality Surveys — A Guide for the Collection and Interpretation of Water Quality Data*, Reports in Hydrology, 23, UNIPUB, 345 Park Ave. South, New York, New York 10010.

KEELEY, J.F., 1982, *Chemical Time-Series Sampling*, Ground Water Monitoring Rev., v. 3, no. 4, pp. 29-38.

KEELY, J.F., and WOLF, FRED, 1983, *Field Applications of Chemical Time-Series Sampling*, Ground Water Monitoring Rev., v. 3, no. 4, pp. 17-25.

KERFOOT, W.B., 1985, *A Portable Well Point Sampler for Plume Tracking*, Ground Water Monitoring Rev., v. 4, no. 4, pp. 38-42.

KORTE, N., and EALEY, D., 1984, *Procedures for Field Chemical Analyses of Water Samples*, DE84004369, NTIS, 5285 Port Royal Rd., Springfield, Virginia 22161, 60 pp.

KORTE, N., and KEARL, P., 1984, *Procedures for the Collection and Preservation of Ground-water and Surface Water Samples and for the Installation of Monitoring Wells*, DE84-997264, NTIS, Springfield, Virginia 22161, 62 pp.

LEE, G.F., and JONES, R.A., 1983, *Guidelines for Sampling Ground Water*, Water Pollution Control Fed., v. 55, no. 1.

McMILLION, L. G., and KEELEY, J. W., 1968, *Sampling Equipment for Ground-Water Investigations*, Ground Water, v. 6, no. 2, pp. 9-11.

MORRISON, R. D., and BREWER, P. E., *Air-Lift Samplers for Zone of Saturation Monitoring*, Ground Water Monitoring Review, v. 1, no. 1, pp. 52-55.

NATIONAL TRAINING AND OPERATIONAL TECHNOLOGY CENTER, 1980, *Methods for the Determination of Chemical Contaminants in Drinking Water — Participants Handbook*, NTIS, 5285 Port Royal Rd., Springfield, Virginia 22161.

NIGHTINGALE, H. I., and BIANCHI, W. C., 1979, *Influence of Well Water Quality Variability on Sampling Decisions and Monitoring*, Water Resources Bull. 15, no. 5, Oct.

PETTYJOHN, W. A., and others, 1981, *Sampling Ground Water for Organic Contaminants,* Ground Water, v. 19, no. 2, pp. 180-189, Mar.-Apr.

PICKENS, J. F., and others, 1981, *A Multi-Level Device for Ground-Water Sampling,* Ground Water Monitoring Review, v. 1, no. 1, pp. 48-51.

RAINWATER, F. H., and THATCHER, L. L., 1960, *Methods for Collection and Analysis of Water Samples,* U. S. Geol. Survey Water-Supply Paper 1454, 301 pp.

SCALF, M. R., and others, 1981, *Manual of Ground-Water Quality Sampling Procedures,* PP82-103045, NTIS, Springfield, Virginia 22161, 105 pp.

SCALF, M.R., 1984, *Sampling Procedures for Groundwater Quality Investigations,* Robert S. Kerr Env. Research Lab., Ada, OK, EPA-600/D-84-137, NTIS, PB84- 194844. 37 pp.

SCHULLER, R. M., GIBB, J. P., and GRIFFIN, R. A., 1981, *Recommended Sampling Procedures for Monitoring Wells,* Ground Water Monitoring Review, v. 1, no. 1, pp. 42-47.

U. S. ENVIRONMENTAL PROTECTION AGENCY, 1972, *Handbook for Analytical Quality Control in Water and Wastewater Laboratories,* Analytical Quality Control Laboratory Nat. Environmental Res. Center, Cincinnati, Ohio, U. S. Government Printing Office, no. 479-971, 54 pp.

U. S. ENVIRONMENTAL PROTECTION AGENCY, 1974, *Manual of Methods for Chemical Analysis of Water and Wastes,* EPA-625-15-75-003, Methods Development and Quality Assurance Research Laboratory, National Environmental Research Center, Cincinnati, Ohio, 298 pp.

U. S. ENVIRONMENTAL PROTECTION AGENCY, 1982, *The Handbook for Sampling and Sample Preservation of Water and Wastewater,* EPA 600/4-82-029, Office of Research and Development, USEPA Cincinnati, Ohio 45268, 402 pp.

U. S. GEOLOGICAL SURVEY, 1973, *Laboratory Analysis,* Techniques of Water Resources Investigations, Book 5.

WARD, C.H., and others (eds.), 1985, *Ground Water Quality*; Comprehensive analysis based on papers Presented at 1st Internat. Conf. on Ground Water Quality Research, Oct. 1981, Rice Univ., Houston, Texas, John Wiley & Sons, Somerset, New Jersey 08873, 547 pp.

WILLARDSON, L. S., MEEK, B. D., and HUBER, M. J., 1973, *A Flow Path Ground Water Sampler,* Soil Science Society of America Proceedings, v. 36, pp. 965-966.

WILSON, L.G., DWORKIN, J.M., and HOLDEN, P.W., 1984, *Development of a Primer on Well Sampling for Volatile Organic Substances,* Arizona Water Resources Res. Center, Tucson, PB85-230043/WEP, NTIS, Springfield, Virginia 22161, 66 pp.

WOOD, W. W., 1973, *A Technique Using Porous Cups for Water Sampling at any Depth in the Unsaturated Zone,* Water Resources Res., v. 9, no. 2, pp. 486-488.

HEALTH ASPECTS

ANONYMOUS, 1971, *Virus and Water Quality: Occurrence and Control,* Proc. 13th Water Quality Conference, Univ. of Illinois, College of Engineering and Illinois Env. Protection Agency, 224 pp.

BITTON, G., and others, 1983, *Survival of Pathogenic and Indicator Organisms in Ground Water,* Ground Water v. 21, no. 4, pp. 405-410.

CRAUN, G. F., 1979, *Waterborne Disease Outbreaks in the United States,* J. Environ. Health, v. 4, p. 259.

CRAUN, G.F., 1985, *Summary of Waterborne Illness Transmitted Through Contaminated Groundwater,* Health Effects Research Lab., Cincinnati, OH, PB85-176857/WEP, NTIS, Springfield, Virginia 22161, 31 pp.

CRUMP, K. S., and GUESS, H. A., 1980, *Drinking Water and Cancer: Review of Recent Findings and Assessment of Risks,* Science Research Systems, Inc., Ruston, Louisiana, NTIS, Springfield, Virginia 22161, PB81-128167, 106 pp.

DREWRY, W.A., and ELIASSEN, R., 1968, *Virus Movement in Ground Water,* J. Water Poll. Control Fed., v. 40, pp. R257-R272, Aug.

DUFOUR, A.P., 1985, *Diseases Caused by Water Contact*, Health Effects Research Lab., Research Triangle Park, NC, PB85-164929/WEP, NTIS, Springfield, Virginia 22161, 52 pp.

FERLAND, R.K., 1985 *Using Cancer Risk Assessments to Determine "How Clean is Clean?"*, Proc. Symp. Groundwater Contamination and Reclamation, AWRA, Tucson, AZ, August 1985, pp. 73-79.

GERBA, C. P. and McNABB, J. F. 1982, *Microbial Aspects of Groundwater Pollution,* PB82-249343 NTIS, Springfield, Virginia 22161, 6 pp.

KASPER, D. R., and KNICKERBOCKER, K. S., 1980, *Organic Qualities of Ground Waters,* PB80-147 267, NTIS, Springfield, Virginia 22161.

KESWICK, B. H., and GERBA, C. P., 1980, *Viruses in Groundwater,* Environ. Sci. & Technology, v. 14, p. 1290; also PB83-188961, NTIS, Springfield, Virgina, 22161, 10 pp. (literature review).

LACEY, R.F., 1982, *Changes in Water Hardness and Cardiovascular Death Rates,* PB82-248923, NTIS, 5285 Port Royal Rd., Springfield, Virginia 22161, 36 pp.

LEWIS, R. J., Sr. and TATKEN, R. L., 1980, *Registry of Toxic Effects of Chemical Substances*: in 2 vols.; v. 1., 828 pp., v. 2, 770 pp. Prepared by Nat. Inst. for Occupational Safety in Health, Supt. of Doc. U. S. Govt. Printing Office No. 017-033-00366, Washington, D. C. 20402.

LYMAN, G.H. LYMAN, C.G., and JOHNSON. W., *Association of Leukemia With Radium Groundwater Contamination,* J. Am. Med. Assoc., v. 254, no. 5, pp.621-627.

MARZOUK, YOSEF, GOYAL, S. M. and GERBA, C. P., 1979, *Prevalence of Enteroviruses in Ground Water in Israel,* Ground Water, v. 17, no. 5, Sep.-Oct., pp. 487-491.

MERONEK, G. E., 1980, *Poliovirus Type I Removal by Montmorillonite Clay,* PB80-127210, NTIS, 5285 Port Royal Rd., Springfield, Virginia 22161.

NATIONAL ACADEMY OF SCIENCES, 1980, *Geochemistry of Water in Relation to Cardiovascular Disease,* National Academy of Sciences, Washington, D. C. 20418, 98 pp.

PIET, G. J., and ZOETEMAN, B. C. J., 1980, *Organic Water Quality Changes During Sand Bank and Dune Filtration of Surface Waters in The Netherlands,* J. Am. Water Works Assoc., Jul., pp. 400-404.

ROBERTS, P. V., and others, 1978, *Organic Contaminant Behavior During Groundwater Recharge,* 51st Annual Conf. Water Pollution Control Federation, Anaheim, California, Oct. 1978.

SAGIK, B. P., and others, 1980, *Assessment of the Potential Health Risks Associated with the Injection of Residual Domestic Wastewater Sludges into Soils,* Final Report 1, National Conference, PB80-169204, NTIS, 5285 Port Royal Rd., Springfield, Virginia 22161.

SCHEUERMAN, P. R., and others, 1979, *Transport of Viruses Through Organic Soils and Sediments,* J. Environmental Engineering Div., Am. Soc. Civil Engrs., v. 105, no. EE4, Aug.

SITTIG, M., 1983, *Priority Toxic Pollutants: Health Impacts and Allowable Limits,* Noyes Data Corp., Park Ridge, New Jersey 07656, 370 pp.

STATE OF CALIFORNIA, 1978, *Health Aspects of Wastewater Recharge: A State-of-the-Art Review,* Water Information Center, Plainview, New York 11803, 240 pp.

THOMAS, H. E., 1949, *Sanitary Quality of Ground-Water Supplies,* The Sanitarian, v. 11, pp. 147-151.

U. S. DEPARTMENT OF HEALTH, EDUCATION and WELFARE, 1973. *The Toxic Substances List,* National Institute for Occupational Safety and Health, Rockville, Maryland, 1001 pp.

VAUGHN, J. M. and LANDRY, E. F., 1980, *The Fate of Human Viruses in Ground-Water Recharge Systems,* Brookhaven National Lab. Rep. No. 51214, Upton, New York; also BNL-51214, NTIS, Springfield, Virginia 22161, 77 pp.

VAUGHN, J.M. and others, 1984, *Virus Entrainment in a Glacial Aquifer,* Brookhaven Natl. Lab., Upton, NY, DE85001344/WEP, NTIS, Springfield, Virginia 22161, 18 pp.

WATSON, A.P., and ZEIGHAMI, E.A., 1985, *Water Chemistry and Cardiovascular Disease Risk,* Oak Ridge Natl. Lab., TN, DE85006364/WEP, NTIS, Springfield, Virginia 22161, 108 pp.

WILLIS, C. J., and others, 1975, *Bacterial Flora of Saline Aquifers,* Ground Water, v. 13, no. 5, pp. 406-409.

MAPS AND GRAPHIC PROCEDURES

BACK, WILLIAM, 1961, *Techniques for Mapping Hydrochemical Facies,* U. S. Geol. Surv. Prof. Paper 424-D.

COLLINS, W. D., 1923, *Graphical Representation of Water Analyses,* Ind. Eng. Chem., v. 15.

DUROV, S. A., 1948, *Classification of Natural Waters and Graphic Representation of their Composition,* Akad. Nauk., USSR, v. 59, pp. 87-90.

FETH, J. H. and others, 1965, *Preliminary Map of the Conterminous United States Showing Depth to and Quality of Shallowest Ground Water Containing More Than 1000 Parts per Million Dissolved Solids,* U. S. Geol. Survey Hydrologic Inv. Atlas IIA-199, 31 pp.

HEM, J. D., 1960, *Chemical Equilibrium Diagrams for Ground-Water Systems,* Bull. Internat. Assoc. Sci. Hydrology, no. 19, Sept., pp. 45-53.

LANGELIER, W. F., and LUDWIG, H. F., 1942, *Graphical Methods for Indicating the Mineral Character of Natural Waters,* J. Am. Water Works Assoc., v. 34.

MORGAN, C. O., and McNELLIS, J. M., 1969, *Stiff Diagrams of Water-Quality Data Programmed for the Digital Computer,* Kansas State Geol. Survey, Spec. Distrib. Pub. 43.

NATIONAL WATER WELL ASSOCIATION, 1980, *Ground-Water Quality Atlas of the United States,* NWWA, Dublin, Ohio 43017, 272 pp.

PIPER, A. M., 1944, *A Graphic Procedure in the Geochemical Interpretation of Water Analyses,* Trans. Am. Geophys. Union, 25th Annual Meeting; pt. 6, Discussion by R. A. Hill, W. F. Langelier, and A. M. Piper.

PIPER, A. M., 1953, *A Graphic Procedure in the Geochemical Interpretation of Water Analyses,* U. S. Geol. Survey Ground Water Note 12.

SILIN-BEKCURIN, A. I., 1957, *Types of Hydrochemical Maps in Hydrology,* Internat. Assoc. Sci. Hydrology, General Assembly Toronto, Pub. 44, v. 2, p. 85.

STIFF, H. A., JR., 1951, *The Interpretation of Chemical Water Analysis by Means of Patterns,* J. Petroleum Tech., v. 3, no. 10, pp. 15-17.

UNESCO, 1975, *Legends for Geohydrochemical Maps,* UNIPUB, Box 433 New York 10016 New York, 62 pp.

ZAPOROZEC, ALEXANDER, 1972, *Graphical Interpretation of Water Quality Data,* Ground Water, v. 10, no. 2, pp. 32-43, Mar.-Apr.

SALINITY

ANONYMOUS, 1982, *Chemical Quality of Irrigation Waters in the Equus Beds Area* (salinity problem), Kansas Geol. Survey, Lawrence, Kansas, 66044.

BAKER, R. C. and others, 1964, *Natural Sources of Salinity in the Brazos River, Texas — with Particular Reference to the Croton and Salt Croton Creek Basins,* U. S. Geol. Survey Water-Supply Paper 1669-CC, 81 pp.

BREDEHOEFT, J. D., and others, 1963, *Possible Mechanism for Concentration of Brines in Subsurface Formations,* Bull. Am. Assoc. Petroleum Geologists v. 47, pp. 257-269.

BROWN, D. L., 1971, *Techniques for Quality-of-Water Interpretations from Calibrated Geophysical Logs, Atlantic Coastal Area,* Ground Water, v. 9, no. 4, pp. 25-38, Jul.-Aug.

FETH, J. H., and others 1965, *Preliminary Map of the Conterminous United States Showing Depth to and Quality of Shallowest Ground Water Containing More Than 1000 ppm Dissolved Solids,* U. S. Geol. Survey Hydrol. Inv. Atlas HA-199, 31 pp.

FOSTER, M. D., 1942, *Base Exchange and Sulfate Reduction in Salty Ground Waters Along Atlantic and Gulf Coasts,* Bull. Am. Assoc. Petroleum Geologists, v. 26, pp. 838-851.

HILL, R. A., 1942, *Salts in Irrigation Water,* Trans. Am. Soc. Civil Engrs. v. 107.

HISS, W. L., 1970, *Acquisition and Machine Processing of Saline Water Data from Southeastern New Mexico and Western Texas,* J. Water Resources Res., v. 6, no. 5, pp. 1471-1477, Oct.

JONES, B. F., 1966, *Geochemical Evolution of Closed Basin Water in the Western Great Basin,* Northern Ohio Geol. Soc., Symp. on Salt, 2nd Proc. v. 1, pp. 181-199.

JUNGE, C. E., and GUSTAFSON, P. E., 1957, *On the Distribution of Sea Salt over the United States and its Removal by Precipitation,* Tellas, v. 9, pp. 164-173.

KELLY, T. E., 1972, *Reconnaissance Investigation of Ground Water in the Rio Grande Drainage Basin — with Special Emphasis on Saline Ground-Water Resources,* U. S. Geol. Survey Hydr. Inv. Atlas, 46 pp.

KOHOUT, F. A. (ed.), 1972, *Saline Water — A Valuable Resource,* Papers of April 24, 1969 Symp. on Saline Groundwater, Am. Geophysical Union, Washington D. C., 90 pp.

KOHOUT, F. A. and HOY, N. D., 1963, *Some Aspects of Sampling Salty Ground Water in Coastal Aquifers,* Ground Water, v. 1, pp. 28-32, 43.

KRIEGER, R. A., and others, 1957, *Preliminary Survey of the Saline-Water Resources of the United States,* U. S. Geol. Survey Water-Supply Paper 1374, 172 pp.

KRIEGER, R. A., 1963, *The Chemistry of Saline Waters,* Ground Water, v. 1, no. 4, pp. 7-12.

LOEWENGART, S., 1961, *Airborne Salts — The Major Source of the Salinity of Waters in Israel,* Bull. Research Council Israel, v. 10G, pp. 183-206.

LOGAN, J., 1961, *Estimation of Electrical Conductivity from Chemical Analyses of Natural Waters,* J. Geophys. Res., v. 66, no. 8, pp. 2479-2483.

LOVE, S. K., 1944, *Cation Exchange in Groundwater Contaminated with Seawater near Miami, Florida,* Trans. Am. Geophys. Union, 25th Ann. Meeting, pt. 6.

MATTOX, R. B., 1970, *Groundwater Salinity,* Contrib. No. 13 of Committee on Desert and Arid Zones Research, Southwestern and Rocky Mountain Div., A.A.A.S., New Mexico Highlands Univ. Las Vegas, New Mexico, 150 pp.

PRYOR, W. A., 1956, *Quality of Water Estimated from Electric Resistivity Logs,* Illinois State Geol. Survey Circ. 215.

REVELLE, ROGER, 1941, *Criteria for Recognition of Sea Water in Ground Water,* Trans., Am. Geophys. Union v. 22, pp. 593-597.

RICHARDS, L. A., (ed.), 1954, *Diagnosis and Improvement of Saline and Alkali Soils,* U. S. Dept. of Agric. Handbook no. 60, 160 pp.

SCHOFIELD, J. C., 1955, *Methods of Distinguishing Sea-Ground-Water From Hydrothermal Water,* New Zealand J. Sci. and Technlogy, v. 37, pp. 597-602.

SCOFIELD, C. S., 1940, *Salt Balance in Irrigated Areas,* J. Agric. Res., v. 61, pp. 17-39.

SILIN-BEKCURIN, A. I., 1961, *Conditions of Formation of Saline Waters in Arid Zones;* in *Salinity Problems in the Arid Zones,* pp. 43-47, UNESCO, Paris.

U. S. SALINITY LABORATORY, 1954, *Diagnosis and Improvement of Saline and Alkali Soils,* U. S. Dept. Agric. Handbook 60.

WADLEIGH, C. H., and FIREMAN, M., 1948, *Salt Distribution Under Furrow and Basin Irrigated Cotton and its Effect on Water Removal,* Soil Sci. Soc. Amer. Proc., v. 13, pp. 527-530.

WHITE, D. E., 1965, *Saline Waters of Sedimentary Rocks,* in *Fluids in Subsurface Environments,* Am. Assoc. Petroleum Geol., Mem. 4, pp. 342-367.

WILCOX, L. V., 1962, *Salinity Caused by Irrigation,* J. Am. Water Works Assoc., v. 54, p. 217, Feb.

WULKOWICZ, G. M., and SALEEM, Z. A., 1974, *Chloride Balance of an Urban Basin in the Chicago Area,* Water Resources Res., v. 10, no. 5, pp. 974-982.

YAALON, D. H., 1963, *On the Origin and Accumulation of Salts in Groundwater and in Soils of Israel,* Res. Council of Israel Bull., v. 11G, no. 3, pp. 105-131.

TEMPERATURE

BALKE, K.D., and KLEY, W., 1981, *Ground Water Temperatures in Densely Populated Areas,* Bundesministerium für Forschung und Technologie, Bonn-Bad Godesberg (Germany, F.R.), BMFT-FB-T-81-028, (in German) DE82902321, NTIS, Springfield, Virginia 22161, 91 pp.

BODVARSSON, GUNNAR, 1969, *On the Temperature of Water Flowing Through Fractures,* J. Geophys. Res., v. 74, no. 8, Apr.

BRASHEARS, M. L., Jr., 1941, *Ground-Water Temperature on Long Island, New York, as Affected by Recharge of Warm Water,* Econ. Geol., v. 36, pp. 811-828.

COLLINS, W. D., 1925, *Temperature of Water Available for Industrial Use in the United States,* U. S. Geol. Survey Water-Supply Paper 520-F, pp. 97-104.

FINK, J. F., 1964, *Groundwater Temperatures in a Tropical Island Environment,* J. Geophys. Res., v. 69, no. 24.

HEATH, R. C., 1964, *Seasonal Temperature Fluctuations in Surficial Sand near Albany, N.Y.*, U. S. Geol. Survey Prof. Paper 475-D, Art. 168, pp. D204-D208.

LOVERING, T. S., and GOODE, H. D., 1963, *Measuring Geothermal Gradients in Drill Holes Less Than 60 Feet Deep, East Tintic District, Utah*, U. S. Geol. Survey Bull. 1172.

MINK, J. F., 1964, *Groundwater Temperatures in a Tropical Island Environment*, J. Geophys. Research, v. 69, pp. 5225-5230.

PLUHOWSKI, E. J., and KANTROWITZ, I. H., 1963, *Influence of Land-Surface Conditions on Ground-Water Temperatures in Southwestern Suffolk County, Long Island, N.Y.*, U. S. Geol. Survey Prof. Paper 475-B, Art. 51, pp. B186-B188.

SCHNEIDER, ROBERT, 1962, *An Application of Thermometry to the Study of Ground Water*, U. S. Geol. Survey Water-Supply Paper 1544-B, 16 pp.

TRAINER, F. W., 1968, *Temperature Profiles in Water Wells as Indicators of Bedrock Fractures*, U. S. Geol. Survey Research Prof. Paper 600-B, pp. B210-214.

WINSLOW, J. D., 1962, *Effect of Stream Infiltration on Ground-Water Temperatures near Schenectady, N.Y.*, U. S. Geol. Survey Prof. Paper 450-C, Art. 111, pp. C125-C128.

GENERAL WORKS/WATER TREATMENT/ RAINWATER

AMERICAN WATER WORKS ASSOC., 1971, *Water Quality and Treatment*, McGraw-Hill, New York, 654 pp.

ANONYMOUS, 1969, *Gas in Ground Water*, J. Am. Water Works Assoc., v. 61, no. 8, pp. 413-414, Aug.

ANONYMOUS, 1984, *Acid Precipitation and Drinking Water Quality in the Eastern United States*, New England Water Works Assoc., Dedham, Massachusetts, PB84-157932/CAG, NTIS, Springfield, Virginia 22161, 195 pp.

BACK, WILLIAM, 1960, *Origin of Hydrochemical Facies of Ground Water in the Atlantic Coastal Plain*, Internat. Geol. Cong., 21 Session, Part 1, Geochemical Cycles, pp. 87-95.

BACK, WILLIAM, 1963, *Preliminary Results of a Study of Calcium Carbonate Saturation of Ground Water in Central Florida*, Internat. Assoc. Sci. Hydrology, 8th year, no. 3, pp. 43-51.

BACK, WILLIAM, 1966, *Hydrochemical Facies and Ground-Water Flow Patterns in Northern Part of Atlantic Coastal Plain*, U. S. Geol. Survey Prof. Paper 498-A, 42 pp.

BACK, WILLIAM, and FREEZE, R. A., 1983, *Chemical Hydrogeology*, Scientific and Academic Editions, 35 West 50th St., New York, New York 10020, 432 pp.; also Hutchinson Ross Publ. Co., Stroudsburg, Pennsylvania.

BACK, WILLIAM, and HANSHAW, B. B., 1965, *Chemical Geohydrology*, in *Advances in Hydroscience* (V. T. Chow, ed.), v. 2, Academic Press, New York pp. 49-109.

BACK, WILLIAM, and HANSHAW, B. B., 1970, *Comparison of Chemical Hydrogeology of the Carbonate Peninsulas of Florida and Yucatan (Mexico)*, J. Hydrol., v. 10, no. 4, pp. 330-368.

BACK, WILLIAM, and HANSHAW, B. B., 1971, *Geochemical Interpretation of Groundwater Flow System*, Water Resources Bull., v. 7, no. 5, pp. 1008-1016.

BACK, WILLIAM, and LETOLLE, R. (eds.), 1980, *Geochemistry of Groundwater*, Papers presented at July 1980 Paris Symp., Elsevier North-Holland, New York, New York 10017, 370 pp.

BACK, WILLIAM, and others, 1979, *Geochemical Significance of Groundwater Discharge and Carbonate Solutions to the Formation of Caleta Xel Ha, Quintana Roo, Mexico*, Water Resources Res., v. 15, pp. 1521-1535.

BACK, WILLIAM, HANSHAW, B.B., and VAN DRIEL, J.N., 1984, *Role of Groundwater in Shaping the Eastern Coastline of the Yucatan Penninsula, Mexico*, in LaFleur, R.G., (ed.), *Groundwater as a Geomorphic Agent*: Boston, Allen and Unwin, pp. 281-293.

BAKER, M. N., 1948, *The Quest for Pure Water*, Am. Water Works Assoc., New York 527 pp.

BARNES, IVAN, and CLARKE, F. E., 1964, *Geochemistry of Ground Water in Mine Drainage Problems*, U. S. Geol. Survey Prof. Paper 473-A, 6 pp.

BARR, D. E., and NEWLAND, L. W., 1977, *Hydrogeochemical Relationships Using Partial Correlation Coefficients*, Water Resources Bull., v. 13, no. 4, pp. 843-846.

BLASZYK, TADEUSZ, and GORSKI, JOSEF, 1981, *Ground-Water Quality Changes During Exploitation*, Ground Water, v. 19, no. 1, Jan.-Feb., pp. 28-33.

BÖHLER, U. G., and others, 1968, *On the Treatment of a Ground Water Containing High Percentages of Iron and Manganese*, Wasserwirtschaft-Wassertechnik, v. 18, no. 4.

CALLAHAN, M., and others, 1979, *Water-Related Environmental Fate of 129 Priority Pollutants — Introduction and Technical Background, Metals and Inorganics, Pesticides and PCBs*, U. S. Environmental Protection Agency, 2 vols, v. 1-PB80-204373, v. 2-PB80-204381, NTIS, Springfield VA, 22161.

CARROLL, D., 1962, *Rainwater as a Chemical Agent of Geologic Processes — A Review*, U. S. Geol. Survey Water-Supply Paper 1535-G, 18 pp.

CLARKE, F. E., 1979, *The Corrosive Well Waters of Egypt's Western Desert*, U. S. Geol. Survey Water-Supply Paper 1757-0.

CLARKE, F. W., 1924, *The Data of Geochemistry*, U. S. Geol. Survey Bull. no. 770, 841 pp.

COGBILL, C. V., and LIKENS, G. E., 1974, *Acid Precipitation in the Northeastern United States*, Water Resources Res., v. 10, no. 6, pp. 1133-1137.

COWART, J. B., KAUFMAN, M. I., and OSMOND, J. K., 1978, *Uranium-Isotope Variations in Groundwater of the Floridan Aquifer and Boulder Zone of South Florida*, J. Hydrology, v. 36, (1/2), pp. 161-172.

DALTON, M. G., and UPCHURCH, S. B., 1978, *Interpretation of Hydrochemical Facies by Factor Analysis*, Ground Water, v. 16, no. 3, pp. 228-233.

DAY, B. A., and NIGHTINGALE, H. I., 1984, *Relationships Between Ground-Water, Silica, Total Dissolved Solids, and Specific Electrical Conductivity*, Ground Water, v. 22, no. 1, pp. 80-85.

DE GEOFFROY, J. G., and others, 1967, *Geochemical Coverage by Spring Sampling Method in Southwest Wisconsin*, Econ. Geol., v. 62, pp. 679-697.

DOCKINS, W. S., and others, 1980, *Sulfate Reduction in Ground Water of Southeastern Montana*, U. S. Geol. Survey, Helena, Montana; NTIS, Springfield, Virginia 22161, PB80-221971, 18 pp.

Water Quality

DRAKE, J. J., and WIGLEY, T. M. L., 1975, *The Effect of Climate on the Chemistry of Carbonate Groundwater*, Water Resources Res., v. 11, no. 6, pp. 958-962.

DRURY, J. S., and others, 1981, *Uranium in U. S. Surface, Ground, and Domestic Waters, Volume 1*, Oak Ridge National Lab, TN, ORNL/EIS 192/V1, EPA-570/9-81-001, NTIS, PB82-258740, 344 pp.

DUNIN, F. X., MATTHES, G., and GRAS, R. A., (eds.), 1984, *Relation of Groundwater Quantity and Quality*, Proc. Symp. XVIIIth General Assembly Internat. Union Geodesy and Geophysics, Hamburg, Germany - Aug. 1983, IAHS Pub. No. 146, IAHS, 2000 Florida Ave. NW, Washington, D.C. 20009, 316 pp.

DURUM, W. H., and HAFFTY, J., 1961, *Occurrence of Minor Elements in Water*, U. S. Geol. Survey Circ. 445, 11 pp.

EATON, F. M., 1935, *Boron in Soils and Irrigation Waters and its Effect on Plants, with Particular Reference to the San Joaquin Valley of California*, U. S. Dept. Agric. Tech. Bull. 448, 131 pp.

EISENBERG, D., and KAUZMANN, 1969, *The Structure and Properties of Water*, Oxford University Press, New York, 296 pp.

ENGBERG, R. A., 1973, *Selenium in Nebraska's Ground Water and Streams*, Nebraska Conserv. and Survey Paper 35.

ERICKSSON, E., 1959, *Atmospheric Transport of Oceanic Constituents in Their Circulation in Nature*, Tellus, v. 11, pp. 1-72.

FETH, J. H., and others, 1964, *Sources of Mineral Constituents in Water from Granite Rocks, Sierra Nevada, California and Nevada*, U. S. Geol. Survey Water-Supply Paper 1535-1, 70 pp.

FEULNER, A. J., and SCHUPP, R. G., 1963, *Seasonal Changes in the Chemical Quality of Shallow Ground Water in Northwestern Alaska*, U. S. Geol. Survey Prof. Paper 475 B, pp. 189-191.

FIREMAN, M., and MAGISTAD, O. C., 1945, *Permeability of Five Western Soils as Affected by the Percentage of Sodium of the Irrigation Water*, Trans. Am. Geophys. Union, v. 26, pp. 91-94.

FLEISCHER, MICHAEL, 1962, *Fluoride Content of Ground Water in the Conterminous United States*, U. S. Geol. Survey Misc. Geol. Investigations Map I-387.

FOSTER, M. D., 1950, *The Origin of High Sodium Bicarbonate Waters in the Atlantic and Gulf Coastal Plains*, Geochim. et Cosmochim. Acta, v. 1, pp. 33-48.

GINTER, R. L., 1934, *Sulphate Reduction in Deep Subsurface Waters*, in W. E. Wrather and F. H. Lahee (eds.), *Problems of Petroleum Geology*, Am. Assoc. Petroleum Geologists, Tulsa, Oklahoma, pp. 907-925.

GORHAM, E., 1955, *On the Acidity and Salinity of Rain*, Geochim. et Cosmochim. Acta, v. VII, pp. 231-239.

GROWITZ, D. J., and LLOYD, O. B., Jr., 1971, *Relationship Between Groundwater Levels and Quality in Shallow Observation Wells, Muddy Creek Basin, Southeastern York County, Pennsylvania*, U. S. Geol. Survey Prof. Paper 750-D, pp. 178-181.

HALLBERG, R. O., and MARTINELL, RUDOLF, *Vyredox — In Situ Purification of Ground Water*, Ground Water, v. 14, no. 2, pp. 88-93.

HANSHAW, B. B., and BACK, WILLIAM, 1979, *Major Geochemical Processes in the Evolution of Carbonate-Aquifer Systems*, J. Hydrology, v. 43, no. 1/4, Oct.

HARDCASTLE, J. H., and MITCHELL, J. K., 1976, *Water Quality and Aquitard Permeability*, J. Irrigation Drainage Div. Am. Soc. Civ. Eng., v. 102 (IR2), pp. 205-220.

HARMON, R. S., and others, 1975, *Regional Hydrochemistry of North American Carbonate Terrains*, Water Resources Res., v. 11, no. 6, pp. 963-967.

HASSAN, A. A., 1974, *Water Quality Cycle — Reflection of Activities of Nature and Man*, Ground Water, v. 12, no. 1, pp. 16-21.

HEM, J. D., 1950, *Geochemistry of Ground Water*, Econ. Geol., v. 45, pp. 72-81.

HEM, J. D., 1963, *Some Aspects of Chemical Equilibrium in Ground Water*, Ground Water, v. 1, pp. 30-34.

HEM, J. D., 1970, *Study and Interpretation of the Chemical Characteristics of Natural Water*, U. S. Geol. Survey Water-Supply Paper 1473, 363 pp.

HEM, J. D., 1972, *Chemistry and Occurrence of Cadmium and Zinc in Surface Water and Ground Water*, Water Resources Res., v. 8, no. 3, pp. 661-679.

HEM, J. D., and CROPPER, W. H, 1959, *Survey of Ferrous-Ferric Chemical Equilibria and Redox Potentials*, U. S. Geol. Survey Water-Supply Paper 1459-A, 31 pp.

HEM, J. D., and others, 1962, *Chemistry of Iron in Natural Water*, U. S. Geol. Survey Water-Supply Paper 1459, 269 pp.

HITCHON, BRIAN, and WALLACK, E. I., (eds.), 1985, *Practical Applications of Ground Water Geochemistry*, Proc. First Canadian/ American Conf. on Hydrogeology, June, 1984, Banff, Alberta, National Water Well Assoc., Dublin, Ohio 43017, 323 pp.

HOUNSLOW, A. W., 1980, *Ground-Water Geochemistry: Arsenic in Landfills*, Ground Water, v. 18, no. 4, Jul.-Aug., pp. 331-333. (fate of arsenic)

HUBERT, J. S., and CANTER, L. W., 1980, *Effects of Acid Rain on Ground-Water Quality*, National Center for Ground-Water Research Rep. 80-7, Norman, Oklahoma 73019.

HUBERTY, M. R., 1941, *Chemical Composition of Ground Waters*, Civil Eng., v. 11, pp. 494-495.

HUFEN, T. H., BUDDEMEIER, R. W., and LAU, L. S., 1972, *Tritium and Radiocarbon in Hawaiian Natural Waters*, Hawaii Water Resources Res. Center Tech. Report No. 53.

HUFEN, T. H., BUDDEMEIER, R. W., and LAU, L. S., 1974, *Isotopic and Chemical Characteristics of High-Level Groundwaters on Oahu, Hawaii,* Water Resources Res., v. 10, no. 2, pp. 366-370.

HUFEN, T. H., BUDDEMEIER, R. W., and LAU, L. S., 1974, *Radiocarbon, Carbon-13 and Tritium in Water Samples from Basaltic Aquifers and Carbonate Aquifers on the Island of Oahu, Hawaii,* Proc. Internat. Atomic Energy Agency, Vienna.

JONES, E. E., Jr., 1971, *Where Does Water Quality Improvement Begin?* Ground Water, v. 9, no. 3, pp. 24-28.

KHAN, R. A., FERRELL, R. E., and BILLINGS, G. K., 1972, *The Genesis of Selected Hydrochemical Facies in Baton Rouge, Louisiana, Ground Waters*, Ground Water, v. 10, no. 4, pp. 14-20.

KHAN, R. A., and others, 1972, *Geochemical Hydrology of the Baton Rouge Aquifers*, Louisiana State University, Bull. 8, March, 63 pp.

KHARAKA, Y. K., 1974, *Retention of Dissolved Constituents of Waste by Geologic Membranes*, in Jules Braunstein (ed.), *Underground Waste Management and Artificial Recharge,* preprint volume Am. Assoc. Petrol. Geol., pp. 420-435.

KHARAKA, Y. K., and BERRY, F. A. F., 1974, *The Influence of Geological Membranes on the Geochemistry of Subsurface Waters from Miocene Sediments at Kettleman North Dome in California*, Water Resources Res., v. 10, no. 2, pp. 313-327.

KLUSMAN, R. W., and EDWARDS, K. W., 1977, *Toxic Metals in Ground Water of the Front Range, Colorado*, Ground Water, v. 15, no. 2, pp. 160-169. (water quality in metal mining area)

KREITLER, C. W., 1975, *Determining the Source of Nitrate in Ground Water by Nitrogen Isotope Studies*, Rep. of Investigations 83, Bureau of Economic Geology, Univ. of Texas at Austin, Texas.

LARSON, T. E., 1949, *Geologic Correlation and Hydrologic Interpretation of Water Analyses*, Water and Sewage Works, v. 96, pp. 67-74.

LARSON, T. E., and HENLEY, L. M., 1966, *Occurrence of Nitrate in Well Waters*, Univ. of Illinois Water Resources Center, Res. Rept. No. 1.

LAU, L. S., and HUFEN, T., 1973, *Tritium Measurement of Natural Waters on Oahu, Hawaii, A Preliminary Interpretation*, Hawaii Water Resources Res. Center Tech. Rep. No. 65.

LAWRENCE, A. R., and others, 1976, *Hydrochemistry and Ground-Water Mixing in Part of the Lincolnshire Limestone Aquifer, England*, Ground Water, v. 14, no. 5, pp. 320-327.

LAWRENCE, F. W., and UPCHURCH, S. B., 1982, *Identification of Recharge Areas Using Geochemical Factor Analysis*, Ground Water, v. 20, no. 6, pp. 680-687.

LEENHEER, J. A., and others, 1974, *Occurrence of Dissolved Organic Carbon in Selected Ground-Water Samples in the United States*, U. S. Geol. Survey J. Research, v. 2, no. 3, pp. 361-369.

LeGRAND, H. E., 1958, *Chemical Character of Water in the Igneous and Metamorphic Rocks of North Carolina*, Econ. Geol., v. 53, pp. 178-189.

LEHR, J. H., and others, 1980, *Domestic Water Treatment*, National Water Well Assoc., 500 W. Wilson Bridge Rd., Worthington, Ohio 43085, 264 pp.

LLOYD, J. W., and HEATHCOTE, J. A., 1985. *Natural Inorganic Hydrochemistry in Relation to Groundwater- An Introduction*, Oxford Univ. Press, 200 Madison Ave., New York, New York 10016, 296 pp.

Water Quality

LONG, D. T., and SALEEM, Z. A., 1974, *Hydrogeochemistry of Carbonate Groundwaters of an Urban Area*, Water Resources Res., v. 10, no. 6, p 1229-1238.

MARSH, J. M., and LLOYD, J. W., 1980, *Details of Hydrochemical Variations in Flowing Wells*, Ground Water, v. 18, no. 4, Jul.-Aug., pp. 366-373.

MATISOFF, G., and others, 1982, *The Nature and Source of Arsenic in Northeastern Ohio Ground Water*, Ground Water, v. 20, no. 4, pp. 446-456.

MATTHESS, G., 1983, *The Properties of Groundwater,* John Wiley & Sons, Inc., One Wiley Dr., Somerset, New Jersey 08873, 406 pp.

MAZOR, E., and others, 1980, *Chemical Composition of Ground Waters in the Vast Kalahari Flatland*, J. Hydrology, v. 48, no. 1/2, Aug.

MEANS, J. L., 1982, *Organic Geochemistry of Deep Ground Waters*, DE83-005333, NTIS, Springfield, Virginia 22161.

MITCHELL, G. F., and HILL, D. O., 1982, *Methods for Treatment of Color in Groundwater*, PB82-157850, NTIS, Springfield, Virginia 22161, 99 pp.

MOORE, J. D., and HUGHES, T. H., 1980, *Effect of Quarry Blasting on Ground-Water Quality in a Limestone Terrane in Calhoun County, Alabama*, PB80-111735, NTIS, 5285 Port Royal Rd., Springfield, Virginia 22161.

MORGAN, C. O., and others, 1966, *Digital Computer Methods for Water Quality Data*, Ground Water, v. 4, no. 3.

MORSE, R. R., 1943, *The Nature and Significance of Certain Variations in Composition of Los Angeles Basin Ground Waters*, Econ. Geol., v. 38, pp. 475-511.

MUSSER, J. J., and WHETSTONE, G. W., 1964, *Geochemistry of Water*, in *Influences of Strip Mining in Beaver Creek Basin, Kentucky*, U. S. Geol. Survey Prof. Paper 427-B, pp. B25-48.

NEWCOMBE, R., Jr., 1975, *Formation Factors and Their Use in Estimating Water Quality in Mississippi Aquifers*, U. S. Geol. Surv. Water-Resources Inv., 2-75, 17 pp.

NIGHTINGALE, H. I., and BIANCHI, W. C., 1977, *Ground-Water Chemical Quality Management by Artificial Recharge*, Ground Water, v. 15, no. 1, pp. 15-21.

NIGHTINGALE, H. I., and BIANCHI, W. C., 1980, *Well-Water Quality Changes Correlated with Well Pumping Time and Aquifer Parameters — Fresno, California,* Ground Water, v. 18, no. 3, May-Jun., pp. 274-280.

NIGHTINGALE, H. I., and BIANCHI, W. C., 1980, *Correlation of Selected Well Water Quality Parameters With Soil and Aquifer Hydrologic Properties,* Water Resources Res., v. 16, no. 4, pp. 702-709.

O'CONNER, D. J., 1976, *The Concentration of Dissolved Solids and River Flow,* Water Resources Res., v. 12, no. 2, pp. 279-294.

OLSON, R. A., SEIM, E. C., and MUIR, J., 1973, *Influence of Agricultural Practices on Water Quality in Nebraska: a Survey of Streams, Groundwater, and Precipitation,* Water Resources Bull., v. 9, no. 2, pp. 301-311.

OSWALD, W. J., 1967, *Remote Sensing Data and Evaluation of Water Quality,* Proc. Third Ann. Conf. on Remote Sensing of Air and Water Pollution, North Am. Aviation, Inc., Autonetics Div., Anaheim, California, pp. 15-1 to 15-12.

PALMER, C., 1911, *The Geochemical Interpretation of Water Analyses,* U. S. Geol. Survey Bull. 479, 31 pp.

PETTYJOHN, W. A., 1976, *Monitoring Cyclic Fluctuations in Ground-Water Quality,* Ground Water, v. 14, no. 6, pp. 472-479.

PIMENTEL, K. D., 1980, *Chemical Fingerprints to Assess the Effects of Geothermal Development on Water Quality in Imperial Valley,* UCRL-81177, NTIS, 5285 Port Royal Rd., Springfield, Virginia 22161.

PIPER, A. M., and others, 1953, *Native and Contaminated Waters in the Long Beach-Santa Ana Area, California,* U. S. Geol. Survey Water Supply Paper 1136, 320 pp.

REARDON, E. J., 1981, *Kd's — Can They Be Used to Describe Reversible Ion Sorption Reactions in Contaminant Migration?* Ground Water, v. 19, no 3, pp. 279-286.

REEVES, C. C., Jr., and MILLER, W. D., 1978, *Nitrate, Chloride and Dissolved Solids, Ogallala Aquifer, West Texas,* Ground Water, v. 16, no. 3, pp. 167-173.

RENICK, B. C., 1925, *Base Exchange in Ground Water by Silicates as Illustrated in Montana,* U. S. Geol. Survey Water Supply Paper 520-D, pp. 53-72.

RHOADES, J. D., and BERNSTEIN, L., 1971, *Chemical, Physical and Biological Characteristics of Irrigation and Soil Water,* in *Water and Water Pollution Handbook,* L. L. Ciaccio (ed.), Marcel Dekker, Inc., New York, v. 1, pp. 142-222.

ROBERSON, C. E., and others, 1963, *Differences between Field and Laboratory Determinations of pH, Alkalinity, and Specific Conductivity of Natural Water,* U. S. Geol. Survey Prof. Paper 475-C, pp. 212-215.

ROBINSON, L. R., Jr., and DIXON, R. I., 1968, *Iron and Manganese Precipitation in Low Alkalinity Ground Waters,* Water and Sewage Works, v. 115, no. 11, Nov.

ROSSUM, J. R., 1948, *Chemical Quality of Underground Water Supplies,* Water and Sewage Works, v. 95, pp. 69-71.

RUNNELLS, D. D., 1976, *Wastewaters in the Vadose Zone of Arid Regions: Geochemical Interactions,* Ground Water, v. 14, no. 6, pp. 374-384.

SAWYER, C. N., and McCARTY, P. L., 1967, *Chemistry for Sanitary Engineers,* 2nd ed., McGraw-Hill, New York, 518 pp.

SCHAEFFER, D. J., and JANARDAN, K. G., 1977, *Communicating Environmental Information to the Public: A New Water Quality Index,* J. Environmental Education, v. 8, no. 4, pp. 18-26.

SCHMIDT, K. D., 1977, *Water Quality Variations for Pumping Wells,* Ground Water, v. 15, no. 2, Mar.-Apr., pp. 130-137.

SCHOELLER, H. J., 1977, *Geochemistry of Ground Water,* in *Ground-Water Studies — An International Guide for Research and Practice.* Supplement No. 3, UNIPUB, Box 433, Murray Hill Station, New York, New York 10016, 42 pp.

SCHOELLER, M., 1963, *Recherches sur l'Acquisition de la Composition Chimique des Eaux Souterraines,* Drouillard, Bordeaux, 231 pp.

SCHOEN, ROBERT, 1972, *Hydrochemical Study of the National Reactor Testing Station, Idaho,* Internat. Geol. Congress, 24th Session, Montreal, Section II-Hydrogeology, pp. 306-314.

SCHUSTER, E. T., and WHITE, W. B., 1971, *Seasonal Fluctuations in the Chemistry of Limestone Springs: A Possible Means for Characterizing Carbonate Aquifers,* J. Hydrology., v. 14, pp. 93-218.

SCHUSTER, E. T., and WHITE, W. B., 1972, *Source Areas and Climatic Effects in Carbonate Groundwaters Determined by Saturation Indices and Carbon Dioxide Pressures*, Water Resources Res., v. 8, no. 4, pp. 1067-1076.

SCHWARTZ, F. W., and DOMENICO, P. A., 1973, *Simulation of Hydrochemical Patterns in Regional Groundwater Flow*, Water Resources Res., v. 9, no. 3, pp. 707-720.

SCHWARTZ, F. W., MUEHLENBACHS, K., and CHORLEY, D. W., 1981, *Flow-System Controls of the Chemical Evolution of Ground Water*, J. Hydrology, v. 54, no. 1/3, Dec.

SEABER, P. R., 1962, *Cation Hydrochemical Facies of Ground Water in the Englishtown Formation, New Jersey*, U. S. Geol. Surv. Prof. Paper 450-B.

SEABER, P. R., 1965, *Variations in Chemical Character of Water in the Englishtown Formation, New Jersey*, U. S. Geol. Survey Prof. Paper 498-B, 35 pp.

SILVEY, W. D., 1967, *Occurrence of Selected Minor Elements in the Waters of California*, U. S. Geol. Survey Water-Supply Paper 1535-L, 25 pp.

SKOUGSTAD, M. W., and HORR, C. A., 1963, *Occurrence and Distribution of Strontium in Natural Water*, U. S. Geol. Survey Water-Supply Paper 1496-D, p. 55-97.

SPALDING, R. F., and others, 1978, *Carbon Contents and Sources in Ground Water of the Central Platte Region in Nebraska*, J. of Environmental Quality, v. 7, no. 3, pp. 428-434.

STALLMAN, R. W., 1965, *Steady One-Dimensional Fluid Flow in a Semi-Infinite Porous Medium with Sinusoidal Surface Temperature*, J. Geophys. Res., v. 70, no. 12, June.

STEELE, T. D., 1976, *A Bivariate-Regression Model for Estimating Chemical Composition of Streamflow or Groundwater*, Hydrological Sciences Bull., J. XXI, no. 1, pp. 149-161.

STUDLICK, J. R. J., and BAIN, R. C., 1980, *Bottled Waters — Expensive Ground Water*, Ground Water, v. 18, no. 4, Jul.-Aug., pp. 340-345.

SUFFET, I. H., and McGUIRE, M. J., (eds.), 1981, *Activated Carbon Adsorption of Organics from the Aqueous Phase*, 2 vols., Ann Arbor Science Publ., Inc., 10 Tower Office Park, Woburn, Massachusetts 01801.

SUGISAKI, R., 1962, *Geochemical Study of Ground Water*, Nagoya University, Japan, J. Earth Sci., v. 10, pp. 1-33.

SULAM, D. J., and KU, H. F. H., 1977, *Trends of Selected Ground-Water Constituents from Infiltration Galleries, Southeast Nassau County, New York*, Ground Water, v. 15, no. 6, pp. 439-445.

SUMMERS, W. K., 1972, *Factors Affecting the Validity of Chemical Analyses of Natural Water*, Ground Water, v. 10, no. 2, pp. 12-17, Mar.-Apr.

TAUSSIG, K., 1961, *Natural Groups of Ground Water and Their Origin*, Mekoroth, Tel Aviv.

TAYLOR, F. B., 1963, *Significance of Trace Elements in Public, Finished Water Suplies*, J. Am. Water Works Assoc., v. 55, pp. 619-623.

THORSTENSON, D. C., and FISHER, D. W., 1979, *The Geochemistry of the Fox Hills-Basal Hell Creek Aquifer in Southwestern North Dakota and Northwestern South Dakota*, Water Resources Res., v. 15, p. 1479. (coal basin geochemistry)

TIHANSKY, D. P., 1974, *Economic Damages from Residential Use of Mineralized Water Supply*, Water Resources Res., v. 10, no. 2, pp. 145-155.

TOFFLEMIRE, T. J., and CHEN, M., 1977, *Phosphate Removal by Sands and Soils*, Ground Water, v. 15, no. 5, pp. 377-387.

TRAINER, F. W., and HEATH, R. C., 1976, *Bicarbonate Content of Groundwater in Carbonate Rock in Eastern North America*, J. Hydrology, v. 31, (1/2), pp. 37-55.

UMANA, A., and others, 1978, *Inorganic Chemical Interactions During Groundwater Recharge*, 51st Annual Conf. Water Pollution Control Fed., Anaheim, California, Oct. 1978.

VALKENBURG, NICHOLAS, CHRISTIAN, ROBERT, and GREEN, MARGARET, 1979, *Occurrence of Iron Bacteria in Ground-Water Supplies of Alabama*, PB-289 575/3WP, NTIS, 5285 Port Royal Rd., Springfield, Virginia 22161.

VAN BEEK, C. G. E. M., and KOOPER, W. F., 1980, *The Clogging of Shallow Discharge Wells in the Netherlands River Region*, Ground Water, v. 18, no. 6, Nov-Dec, pp. 578-586.

VAN LIER, J. A., 1959, *The Solubility of Quartz*, Kemink en Zoon, Utrecht, The Netherlands, 54 pp.

WAGNER, G. H., and K. F., STEELE, 1980, *Chemistry of the Spring Waters of the Ouachita Mountains Excluding Hot Springs, Arkansas*, Arkansas Water Resources Res. Center, Fayetteville, NTIS, Springfield, Virginia 22161, PB80-203466, 182 pp. (ground water quality and its potential as mineral pathfinder)

WARD, C. H., GIGER, W. and McCARTY, P. L. (eds.), *Ground Water Quality*, John Wiley, New York, NY, 547 pp.

WARING, F. G., 1949, *Significance of Nitrate in Water Supplies*, J. Am. Water Works Assoc., v. 41, pp. 147.

WARNER, D. L., and DOTY, L. F., 1967, *Chemical Reaction Between Recharge Water and Aquifer Water*, Symp. Haifa, Internat. Assoc. Sci. Hydrology, pp. 278-288.

WHITE, D. E., 1957, *Magmatic, Connate, and Metamorphic Waters*, Bull. Geol. Soc. Am., v. 68, pp. 1659-1682.

WHITE, D. E., and others, 1963, *Chemical Composition of Sub-Surface Waters*, Chap. F, in *Data of Geochemistry*, 6th ed., U. S. Geol. Survey Prof. Paper 440-F, 67 pp.

WHITTEMORE, D. O., and LANGMUIR, DONALD, 1975, *The Solubility of Ferric Oxyhydroxides in Natural Waters*, Ground Water, v. 13, no. 4, pp. 360-365.

WOOD, W. W., 1976, *A Hypothesis of Ion Filtration in a Potable-Water Aquifer System*, Ground Water, v. 14, no. 4, pp. 233-244.

WOOD, W. W., 1978, *Use of Laboratory Data to Predict Sulfate Sorption During Artificial Ground-Water Recharge*, Ground Water, v. 16, no. 1, p. 22-31.

WOOD, W. W., and BASSETT, R. L., 1975, *Water Quality Changes Related to the Development of Anaerobic Conditions During Artificial Recharge*, Water Resources Res., v. 11, no. 4, pp. 553-558.

WOOD, P. R., and DeMARCO, JACK, 1979, *Treatment of Ground Water with Granular Activated Carbon*, J. Am. Water Works Assoc., v. 71, no. 11, Nov.

YOUNG, C. P., OAKES, D. B., and WILKINSON, W. B., 1976, *Prediction of Future Nitrate Concentrations in Ground Water*, Ground Water, v. 14, no. 6, pp. 426-438.

Water Quality

ZACK, A. L., 1980, *Geochemistry of Fluoride in the Black Creek Aquifer System of Horry and Georgetown Counties, S. C. — and its Physiological Implications.* Geol. Survey Water-Supply Paper 2067, GPO, Wash. D. C. 20402.

GROUNDWATER CONTAMINATION

(For sampling procedures see previous section)

BIBLIOGRAPHY

ANONYMOUS, 1978, *Landfill Disposal of Waste: A Bibliography* (1000 Abstracts) Waste Management Info. Bureau, Great Britain Dept. of the Environment, Hartwell Laboratory, Didcot, Oxfordshire, England.

ANONYMOUS, 1981, *Sanitary Landfills*, (87 Citations from the NTIS Data Base), NTIS, Springfield, Virginia 22161, 87 pp.

ANONYMOUS, 1981, *Sanitary Landfills*, (110 Citations from the Engineering Data Base), PB 81-802779 NTIS, Springfield, Virginia 22161, 154 pp.

ANONYMOUS, 1981, *Septic Tank and Household Sewage Systems Design and Use*, (Citations from Engineering Index Data Base), 1970-June, 1981, PB81-807315, NTIS, Springfield, Virginia 22161.

ANONYMOUS, 1982, *Ground Water Pollution: General Studies*, (Citations from the NTIS Data Base), 1977-Jan. 1980, PB82-800780, 272 pp; Period Feb. 1980-May, 1981, PB82-800798, 110 pp., NTIS, Springfield, Virginia 22161.

ANONYMOUS, 1982, *Sanitary Landfills, 1978-February, 1982* (Citations from the Engineering Index Data Base), PB82-807330, NTIS, Springfield, Virginia 22161.

ANONYMOUS, 1983, *Ground Water Pollution: Pollution from Irrigation and Fertilization, 1977-September, 1982 (206 Citations from the NTIS Data Base)*, PB83-800433 (See also 1964-1976, NTIS/PS-78/0140), NTIS, Springfield, Virginia 22161, 214 pp.

ANONYMOUS, 1984, *Ground-Water Pollution: General Studies, Feb. 1980 - Nov. 1984*, (Citations from the NTIS Data Base), PB85-850998/WNR, NTIS, Springfield, Virginia 22161.

BADER, J. S., and others, 1973, *Selected References — Ground-Water Contamination in the United States and Puerto Rico*, U. S. Geol. Survey Water Research Div., Washington, D. C., 103 pp.

171

BROWN, R. J., 1980, *Acid Mine Drainage* (A Bibliography with Abstracts), PB80-801475, NTIS, Springfield, Virginia 22161.

BROWN, R. J., 1980, *Ground-Water Pollution* (Citations from the American Petroleum Institute Data Base), PB80-806920, NTIS, Springfield, Virginia 22161.

BROWN, R. J., 1980, *Ground-Water Pollution: Saline Ground Water* (Citations from the NTIS Data Base), PB80-806417, NTIS, Springfield, Virginia 22161.

CALIFORNIA DEPT. WATER RESOURCES, 1969, *Sanitary Landfill Studies: Appendix A — Summary of Selected Previous Investigations*, Bull. 147-5, Sacramento, 115 pp.

HANDMAN, E. H., 1983, *Hydrologic and Geologic Aspects of Waste Mangement and Disposal: A Bibliography of Publications by U. S. Geological Survey Authors, 1950-81*, U. S. Geol. Survey, Circ. 907, 40 pp.

LEWIS, W. J., FOSTER, S. S. D., and DRASAR, B. S., 1982, *The Risk of Groundwater Pollution by On-Site Sanitation in Developing Countries — A Literature Review*, Internat. Ref. Centre for Wastes Disposal Rept. no. 01/82, c/o EAWAG Ueberlandstrasse CH-8600 Duebendorf, Switzerland, 79 pp.

NUS CORPORATION, 1984, *Bibliography: Salt Impacts on Vegetation and Soils*, DE85000384/WEP, NTIS, Springfield, Virginia 22161, 82 pp.

STEELMAN, B. L., and ECKER, R. M., 1984, *Organics Contamination of Ground-Water: An Open Literature Review*, DE85-003601/WNR, NTIS, Springfield, Virginia 22161.

SUMMERS, W. K., and SPIEGEL, ZANE, 1973, *Ground Water Pollution — A Bibliography*, Ann Arbor Science Publ. Inc., P.O. Box 1425, Ann Arbor, Michigan 48106, 83 pp.

TODD, D. K., and McNULTY, D. E., 1974, *Polluted Ground Water: A Review of the Significant Literature*, U. S. EPA-600/4-74-001, Water Information Center, Plainview, New York 11803, 179 pp.

U. S. ENVIRONMENTAL PROTECTION AGENCY, 1972, *Subsurface Water Pollution — A Selective Annotated Bibliography, Part I Subsurface Waste Injection, Part II Saline Water Intrusion, Part III Percolation from Surface Sources*, NTIS, Springfield, Virginia 22161, Accession Numbers: PB-211 340, PB-211 341, and PB-211 342.

WATER RESEARCH CENTRE (ENGLAND), 1979, *Utilization of Sewage Sludge on Land, A Bibliography of Selected Reference Covering the Period 1950-1977,* No. N79-12967/2WP, NTIS, Springfield, Virginia 22161.

GENERAL WORKS/CONTAMINANT TRANSPORT/ REGIONAL ASSESSMENT

(See also Radionuclides and Groundwater Models)

ALTHOFF, W. F., CLEARY, R. W., and ROUX, P. H., 1981, *Aquifer Decontamination for Volatile Organics: A Case History,* Ground Water, v. 19, no. 5, pp. 495-504.

AMERICAN SOCIETY FOR TESTING MATERIALS, 1981, *Permeability and Groundwater Contaminant Transport* (Based on June 1979 Philadelphia, Pennsylvania Symp.), ASTM Publ. STP 746, 245 pp.

AMERICAN WATER WORKS ASSOC., 1981, *Organic Chemical Contaminants in Groundwater: Transport and Removal,* Proc. AWWA Seminar June 7, 1981 St. Louis, Missouri, AWWA, Denver, Colorado 80235, 123 pp.

ANONYMOUS, 1970, *Ground Water Pollution,* Water Well J., v. 24, no. 7, pp. 31-61.

ANONYMOUS, 1973, *Ground Water Pollution From Subsurface Excavations.* Rept. EPA-430/9-73-012, U. S. Environmental Protection Agency, Washington, D. C., 224 pp.

ANONYMOUS, 1985, *Experiences with Groundwater Contamination,* Papers presented at AWWA's 1984 Annual Conference in Dallas, TX, AWWA, Denver, Colorado 80235, 156 pp.

ANONYMOUS, 1985, *1985 Suspect Chemicals Sourcebook,* Roytech Publications, 1499 Old Bayshore Hwy., Burlingame, California 94010 (guide to 6,000 chemicals).

BAARS, J. K., 1957, *Pollution of Ground Water,* Internat. Assoc. Sci. Hydrology, General Assembly Toronto, Pub. 44, v. 2, pp. 279-289.

BALLENTINE, R. K., and others, 1972, *Subsurface Pollution Problems in the United States,* U. S. Environmental Protection Agency Tech. Studies Rept. TS-00-72-02, Washington, D. C. 20460.

BARBASH, JACK, and ROBERTS, P. V., 1986, *Volatile Organic Chemical Contamination of Groundwater Resources in the U. S.*, J. Water Pollution Control Fed., v. 58, no. 5, pp. 343-348.

BARBER, A. J., and BRAIDS, O. C., 1982, *Application of a Portable Organic Vapor Analyzer in Ground-Water Contamination Investigations*, Proc. 2nd Natl. Symp. on Aquifer Restoration and Ground-Water Rehabilitation, May 1982, National Water Well Assoc., Dublin, Ohio.

BELL, L. A., and others, 1980, *Florida Surface Impoundment Assessment — Final Report*, Groundwater Section, Florida Dept. of Envir. Regulation, 2600 Blairstone Rd., Tallahassee, Florida 32301.

BERK, W. J., and YARE, B. S., 1977, *An Integrated Approach to Delineating Contaminated Ground Water*, Ground Water, v. 15, no. 2, pp. 138-145.

BOND, J. G., WILLIAMS, R. E., and SHADID, O., 1972, *Delineation of Areas for Terrestrial Disposal of Waste Water*, Water Resources Res., v. 8, no. 6, pp. 1560-1573.

BORN, S. M., and STEPHENSON, D. A., 1969, *Hydrogeologic Considerations in Liquid Waste Disposal*, J. Soil Water Conservation, v. 24, no. 2, pp. 52-55.

BROWN, R. H., 1964, *Hydrologic Factors Pertinent to Ground-Water Contamination*, Ground Water, v. 2, no. 1.

BUTLER, R. G., and others, 1954, *Underground Movement of Bacterial and Chemical Pollutants*, J. Am. Water Works Assoc., v. 46, pp. 97-111.

CANTER, L. W., and KNOX, R. C., 1985, *Ground Water Pollution Control*, Lewis Publishers, Inc., P.O. Drawer 519, Chelsea, Michigan 48118, 500 pp.

CARTWRIGHT, K. R., GILKESON, H., and JOHNSON, T. M., 1981, *Geological Considerations in Hazardous-Waste Disposal*, J. Hydrology, v. 54, no. 1/3, Dec.

CHERRY, J. A., and others, 1975, *Contaminant Hydrogeology — Part I, Physical Processes*, Geoscience Canada, v. 2, no. 2.

COLE, J. A., 1975, *Ground-Water Pollution in Europe*, Proc. Conference on Ground-Water Pollution, Reading Univ., Sept., 1972, Water Information Center, Plainview, New York 11803.

COLWILL, D. M., PETERS, C. J., and PERRY, R., 1984 *Water Quality of Motorway Runoff*, Transport and Road Research Lab., Crowthorne (England), TRRL/SR-823, NTIS, PB84-2011052, 28 pp.

COUNCIL ON ENVIRONMENTAL QUALITY, 1981, *Contamination of Ground Water by Toxic Chemicals*, Supt. Documents, U. S. Govt. Printing Office, Washington, D. C. 20402, 84 pp.

CRABTREE, K. T., 1970, *Nitrogen Cycle: Nitrification-Denitrification and Nitrate Pollution of Ground Water*, Univ. of Wisconsin Water Resources Center Rep. B-004-Wis.

CRABTREE, K. T., 1972, *Nitrate and Nitrite Variation in Ground Water*, Wisconsin Dept. of Natural Resources, 22 pp.

DEBUCHANNE, G. D., and LAMOREAUX, P. E., 1961, *Geologic Controls Related to Ground-Water Contamination*, in *Ground-Water Contamination*, Proc. 1961 Symp., U. S. Public Health Service Tech. Rept. W 61-5. pp. 3-7.

DEUTSCH, MORRIS, 1963, *Groundwater Contamination and Legal Controls in Michigan*, U. S. Geol. Survey Water-Supply Paper 1691.

DOMENICO, P. A., and PALCIAUSKAS, V. V., 1982, *Alternative Boundaries in Solid Waste Management*, Ground Water, v. 20, no. 3, pp. 303-311.

DOMENICO, P. A., and ROBBINS, G. A., 1985, *A New Method of Contaminant Plume Analysis*, Ground Water, v. 23, no. 4, pp. 476-486.

DVGW RESEARCH GROUP, 1983, *Verfahrentechnische Grundlagen für Anlagen zur Entfernung von Halogenkohlenwasserstoffen aus Gründwassern*, 1982 Water Technology Symp. Karlsruhe/Mulheim an Ruhr, DVGW Engler-Bunte Institute, University of Karlsruhe, W. Germany, Publ. no. 21, 267 pp.

EDWARDS, N. T., 1983, *Polycyclic Aromatic Hydrocarbons (PAH's) in the Terrestrial Environment - A Review*, J. Environ. Qual., v. 12, no. 4, pp. 427-441.

FRIED, J. J., 1975, *Groundwater Pollution, Theory, Methodology, Modelling and Practical Rules*, Elsevier, New York, 330 pp.

FRYBERGER, J. S., and BELLIS, W. H., 1977, *Quantifying the Natural Flushout of Alluvial Aquifers*, Ground Water, v. 15, no. 1, pp. 58-65.

FUHRIMAN, D. K., and BARTON, J. R., 1971, *Ground Water Pollution in Arizona, California, Nevada and Utah*, Environmental Protection Agency, Office of Research and Monitoring, U. S. Govt. Printing Office, Washington, D. C. 20402.

GERAGHTY, J. J., 1962, *Movements of Contaminants*, Water Well J., v. 16, Oct.

GERAGHTY, J. J., and MILLER, D. W., 1978, *Status of Ground-Water Pollution in the U. S.*, J. Am. Water Works Assoc., v. 70, no. 3, pp. 162-167.

GOTAAS, H. B., 1953, *Laboratory and Field Investigations of the Travel of Pollution from Direct Water Recharge into Underground Formations*, Annual Report Sanitary Engineering Res. Projects, Univ. of California, Berkeley.

GUEVEN, O., MOLZ, F. J., and MELVILLE, J. G., 1984, *Analysis of Dispersion in a Stratified Aquifer*, Auburn Univ., AL. Dept. of Civil Eng. Pub. no. 20, pp. 1337-1354.

HAJEK, B. F., 1969, *Chemical Interactions of Wastewater in a Soil Environment*, J. Water Poll. Control Fed., v. 41, no. 10, pp. 1775-1786.

HASSAN, A. A., 1974, *Water Quality Cycle — Reflection of Activities of Nature and Man*, Ground Water, v. 12, no. 1, pp. 16-21.

HUIBREGTSE, K. R., and KASTMAN, K. H., 1981, *Development of a System to Protect Ground Water Threatened by Hazardous Spills on Land*, PB81-209587, NTIS, Springfield, Virginia 22161.

INSTITUTE OF GEOLOGY & MINES and MASS. INSTITUTE OF TECHNOLOGY, 1979, *Ground-Water Pollution: Technology, Economics and Management*, FAO Irrigation and Drainage Paper No. 31, UNIPUB, 345 Park Ave. South, New York, New York 10010, 137 pp.

INTERNATIONAL ASSOC. OF HYDROLOGICAL SCIENCES, 1975, *Groundwater Pollution*, Proc. of the Moscow Symp. Aug. 1971, IAHS Publ. No. 103. 2000 Florida Ave., NW, Washington, D. C. 20009.

JACKSON, R. E. (ed.), 1983, *Aquifer Contamination and Protection*, UNESCO, 7 Place de Fontenoy, 75700 Paris, France, 440 pp.

JOHNSON, T. M., and CARTWRIGHT, KEROS, 1978, *Implications of Solid-Waste Disposal in the Unsaturated Zone*, Illinois State Geol. Survey Reprint.

KARUBIAN, J. F., 1974, *Polluted Groundwater: Estimating the Effects of Man's Activities*, Rept. EPA 600/4-74-002, U. S. Environmental Protection Agency, Washington, D. C., 99 pp.

KEELY, J. F., PIWONI, M. D., and WILSON, J. T., 1986, *Evolving Concepts of Subsurface Contaminant Transport*, J. Water Pollution Control Fed., v. 58, no. 5, pp. 349-357.

KELLY, W. E., 1982, *Ground-Water Dispersion Calculations with a Programmable Calculator*, Ground Water, v. 20, no. 6, p. 736-738.

KIRKHAM, DON, and AFFLECK, S. B., 1977, *Solute Travel Times to Wells*, Ground Water, v. 15, pp. 231-242.

KIRKHAM, DON, and VAN DER PLOEG, R. R., 1974, *Ground-Water Flow Patterns in Confined Aquifers and Pollution*, J. Am. Water Works Assoc., v. 66, no. 3, March.

KREITLER, C. W., 1975, *Determining the Source of Nitrate in Ground Water by Nitrogen Isotope Studies*, Bureau of Econ. Geol., Report of Invest. No. 83, Austin, Texas, 57 pp.

KREITLER, C. W., and JONES, D. C., 1975, *Natural Soil Nitrate: The Cause of Nitrate Contamination of Ground Water in Runnels County, Texas*, Ground Water, v. 13, no. 1, pp. 53-61.

KRUL, W. F., J. M., 1957, *Sanitary Engineering and Water Economy in Europe*, Bull. of the World Health Organization, v. 16, no. 4, Geneva.

LEGRAND, H. E., 1965, *Environmental Framework of Ground-Water Contamination*, Ground Water, v. 3, no. 2, pp. 11-15.

LEGRAND, H. E., 1965, *Patterns of Contaminated Zones of Water in the Ground*, Water Resources Res., v. 1, no. 1, pp. 83-95.

LEO, A., HANSCH, C., and ELKINS, D., 1971 *Partition Coefficients and Their Uses*, Chemical Reviews, v. 71, no. 6, pp. 525-553.

LOFGREN, B. E., 1973, *Hazards of Waste Disposal in Groundwater Basins*, in *Underground Waste Management and Artificial Recharge,* New Orleans, Louisiana, Am. Assoc. Petrol. Geol., v. 2, pp. 715-728.

LYMAN, W. J., and others, 1982, *Handbook of Chemical Property Estimation Methods: Environmental Behavior of Organic Compounds,* McGraw-Hill Book Co., New York, New York 10020, 960 pp.

MANAHAN, S. E., 1982, *A Simplified Scheme for the Analysis of Pollutants in Groundwater and Leachates Contaminated by Hazardous Chemicals*, Univ. Missouri Water Resources Res. Center, Columbia, Missouri, 56 pp.

MANN, J. F., Jr., 1976, *Wastewaters in the Vadose Zone of Arid Regions: Hydrologic Interactions*, Ground Water, v. 14, no. 6, pp. 367-373.

MARINO, M. A., 1974, *Distribution of Contaminants in Porous Media Flow*, Water Resources Res., v. 10, no. 5, pp. 1013-1018.

MATHER, J. R., and RODRIGUEZ, P. A., 1980, *The Use of the Water Budget in Evaluating Leaching Through Solid Waste Landfills*, PB80-180888, NTIS, Springfield, Virginia 22161.

MATTHESS, GEORG, 1972, *Hydrogeologic Criteria for the Self-Purification of Polluted Groundwater*, Internat. Geol. Congress, 24th Session, Montreal, Section II-Hydrogeology, pp. 296-304.

MAXEY, G. B., and FARVOLDEN, R. N., 1965, *Hydrogeologic Factors in Problems of Contamination in Arid Lands*, Ground Water, v. 3, no. 4, pp. 29-32.

McMILLION, L. G., and HOUSER, VICTOR, 1969, *Field Evaluation of Potential Pollution from Ground-Water Recharge*, Water Well J., v. 23, no. 8, Aug.

MERCADO, ABRAHAM, 1976, *Nitrate and Chloride Pollution of Aquifers: A Regional Study with the Aid of a Single-Cell Model*, Water Resources Res., v. 12, no. 4, pp. 731-747.

MEYER, C. F., (ed.), 1973, *Polluted Groundwater: Some Causes, Effects, Controls, and Monitoring*, Prepared by GE/TEMPO for Environmental Protection Agency, Office of Research and Development, EPA-600/4-73-0016, Washington, D. C. 282 pp.

MILLER, D. W., 1977, *Waste Disposal Effects on Ground Water;* Reprint of 1977 EPA Report to Congress *Waste Disposal Practices and Their Effects on Ground Water,* Premier Press, Berkeley, California, 512 pp. Also available from Water Information Center, Plainview, New York 11803.

MILLER, D. W., 1984, *Groundwater Contamination*, Water Information Center, Spec. Rept., Plainview, New York 11803, 23 pp.

MILLER, D. W., 1984, *Sources of Ground-Water Pollution*, EPA Journal, v. 10, pp. 17-19.

MILLER, J. C., and others, 1977, *Ground-Water Pollution Problems in the Southeastern United States*, Geraghty & Miller, Inc., NTIS, PB-268 234/2WP, Springfield, Virginia 22161.

MOORE, J. W., and RAMAMOORTHY, S., 1984, *Organic Chemicals in Natural Waters: Applied Monitoring and Impact Assessment*, Springer Verlag, P.O. Box 2485, Secaucus, New Jersey 07084, 289 pp.

MORRILL, L. G., MAHILUM, B. C., and MOHIUDDIN, S. H., 1983, *Organic Compounds in Soils: Sorption, Degradation and Persistence*, Ann Arbor Science Publ., 10 Tower Office Park, Woburn, Massachusetts 01801, 326 pp.

MORRISON, ROBERT, and others, 1980, *Effects of Upland Disposal of Dredged Material on Groundwater Quality: Technical Report EL-80-8*, U. S. Army Engineer Waterways Experiment Sta., Vicksburg, MS, 117 pp.

NATIONAL RESEARCH COUNCIL, 1984, *Groundwater Contamination*, Natl. Academy Press, Washington, DC 20418, 179 pp.

NATIONAL WATER WELL ASSOC., 1977, *Proceedings of the Third National Ground Water Quality Symposium*, Ground Water, v. 10, no. 4, pp. 21-23. 117 EPA-600/9-77-014, U. S. Environmental Protection Agency, Ada, Oklahoma, 232 pp.

NAYLOR, and others, 1981, *The Investigation of Landfill Sites*, Tech. Rept. 91, WRC Medmenham Laboratory, Medmenham, Marlow, Bucks SL7 2HD, England.

NELSON, R. W., 1978, *Evaluation the Environmental Consequences of Ground-Water Contamination. 1. An Overview of Contaminant Arrival Distributions as General Evaluation Requirements; 2. Obtaining Location/Arrival Time and Location/Outflow Quantity Distributions for Steady Flow Systems; 3. Obtaining Contaminant Arrival Distributions for Steady Flow in Heterogeneous Systems; 4. Obtaining and Utilizing Contaminant Arrival Distributions in Transient Flow Systems*, Water Resources Res., v. 14, no. 3, Jun., pp. 409-450.

NELSON, R. W., and ELIASON, J. R., 1966, *Prediction of Water Movement Through Soils — a First Step in Waste Transport Analysis*, Proc. 21st Industrial Waste Conf., Purdue Univ. Engr. Ext. Bull. no. 121, v. 1, pp. 516-526.

NEWSOM, J. M., 1985, *Transport of Organic Compounds Dissolved in Ground Water*, Ground Water Monitoring Rev., v. 5, no. 2, pp. 28-36.

NEW YORK STATE DEPT. HEALTH, 1972, *The Long Island Ground Water Pollution Study*, New York State Dept. Environmental Conservation, Albany, 396 pp.

NISHI, T. S., BRUCH, J. C., Jr., and LEWIS, R. W., 1976, *Movement of Pollutants in a Two-Dimensional Seepage Flow Field*, J. Hydrology, v. 31, no. 3/4, Dec.

O'CONNOR, J. T., GHOSH, M. M., and BENERJI, S. K., 1984, *Organic Groundwater Contamination Evaluation and Prediction*, Missouri Univ.-Rolla, PB85-230068/WEP, NTIS, Springfield, Virginia 22161, 74 pp.

OVERCASH, M. R., (ed.), 1981, *Decomposition of Toxic and Nontoxic Compounds in Soils*, Ann Arbor Science Publishers, The Butterworth Group, 10 Tower Office Park, Woburn, Massachusetts 01801, 445 pp.

PALMQUIST, R., and SENDLEIN, L. V. A., 1975, *The Configuration of Contamination Enclaves from Refuse Disposal Sites on Floodplains*, Ground Water, v. 13, no. 2, pp. 167-181.

PENNINGTON, D., 1974, *The Physical and Chemical Processes Controlling the Migration of Pollutants Resulting from Land Disposal of Liquid and Solid Wastes into Subsurface Waters*, Proc. 7th Mid-Atlantic Industrial Waste Conference, Drexel Univ., Philadelphia, Pennsylvania, pp. 414-424.

PERLMUTTER, N. M., and others, 1978, *Surface Impoundments and Their Effects on Ground-Water Quality in the United States — A Preliminary Survey*, Geraghty & Miller, Inc., EPA-570/9-78-004.

PERNER, E. R., and GIBSON, A. C., 1980, *Hydrologic Simulation of Solid Waste Disposal Sites*, SW-868 (draft), U. S. Environmental Protection Agency, Cincinnati, Ohio.

PETTYJOHN, W. A., 1982, *Cause and Effect of Cyclic Changes in Ground-Water Quality*, (contamination), Ground Water Monitoring Review, v. 2, no. 1, pp. 43-49.

PETTYJOHN, W. A., and HOUNSLOW, A. W., 1983, *Organic Compounds and Ground-Water Pollution*, Ground Water Monitoring Rev., v. 3, no. 4, pp. 41-47.

PETTYJOHN, W. A., and others, 1982, *Predicting Mixing of Leachate Plumes in Ground Water*, Oklahoma State Univ., Stillwater, 15 p.; Included in *Land Disposal of Hazardous Waste*, p. 122-136, NTIS, PB82-173105 (order as PB82-173022), Springfield, Virginia 22161.

PHILLIPS, K. J., and GELHAR, L. W., 1978, *Contaminant Transport to Deep Wells*, J. Hydraulics Div., Amer. Soc. Civil Engrs., v. 104, no. HY6, pp. 807-819.

PORTER, K. S., *An Evaluation of Sources of Nitrogen as Causes of Ground-Water Contamination in Nassau County, Long Island*, Ground-Water, v. 18, no. 6, Nov-Dec., pp. 617-625.

PRINCETON UNIVERSITY WATER RESOURCES PROGRAM, 1985, *Groundwater Contamination from Hazardous Wastes*, Prentice-Hall, Inc., Englewood Cliffs, New Jersey 07632, 161 pp.

PYE, V. I., PATRICK, RUTH, and QUARLES, JOHN, 1983, *Groundwater Contamination in the United States*, University of Pennsylvania Press, Philadelphia, Pennsylvania, 315 pp.

REILLY, T. E., 1978, *Convective Contaminant Transport to Pumping Well*, J. Hydraulics Div., Am. Soc. Civil Engrs., Dec., pp. 1565-1575.

RITCHIE, E. A., 1976, *Cathodic Protection Wells and Ground-Water Pollution*, Ground Water, v. 14, no. 3, pp. 146-149.

ROBERTS, P. V., REINHARD, M., and VALOCCHI, A. J., 1982, *Movement of Organic Contaminants in Ground-Water: Implications for Water Supply*, J. Am. Water Works Assoc., pp. 408-413.

RUNNELLS, D. D., 1976, *Wastewaters in the Vadose Zone of Arid Regions: Geochemical Reactions*, Ground Water, v. 14, pp. 374-385.

SCALF, M. R., and others, 1973, *Ground Water Pollution in the South Central States*, U. S. Environmental Protection Agency, Office of Research and Monitoring, Corvallis, Oregon 97330 and U. S. Govt. Printing Office, Washington, D. C. 20402.

SENDLEIN, L. V. A., and PALMQUIST, R. C., 1975, *A Topographic-Hydrogeologic Model for Solid Waste Landfill Siting*, Ground Water, v. 13, no. 3, pp. 260-268.

SHUCKROW, A. J., PAJAK, A. P. and TOUHILL, C. J., 1982, *Hazardous Waste Leachate Management Manual*, Noyes Data Corp., Park Ridge, New Jersey 07656, 379 pp.

SMITH, R., and others, 1983, *Toxic Hazards of Underground Excavation*, DE83-01442, NTIS, 5285 Port Royal Rd., Springfield, Virginia 22161, 130 pp, (compilation of case histories).

SOKOL, DANIEL, 1970, *Ground-Water Safety Evaluation — Project Gasbuggy*, NTIS No. PNE-1009, Springfield, Virginia 22161.

SPRUILL, T. B., 1983, *Relationship of Nitrate Concentrations to Distance of Well Screen Openings Below Casing Water Levels*, Water Resources Bull., v. 19, no. 6.

TENN, D. G., HANLEY, K. J., and DEGEARE, T. V., 1975, *Use of the Water Balance Method for Predicting Leachate Generation from Solid Water Disposal Sites*, EPA/630/SW-168, U. S. Environmental Protection Agency, Cincinnati, Ohio.

THOMAS, G. W., 1972, *The Relation Between Soil Characteristics, Water Movement and Nitrate Contamination of Ground Water*, Kentucky Water Resources Res. Institute, PB 220015.

TOLMAN, A. L., and others, 1980, *Guidance Manual for Minimizing Pollution from Waste Disposal Sites*, PB-286 905/5WP, NTIS, 5285 Port Royal Rd., Springfield, Virginia 22161.

TUCKER, R. K., 1981, *Ground Water Quality in New Jersey: An Investigation of Toxic Contaminants*, Office of Cancer and Toxic Substances Research, New Jersey Dept. Env. Protection, Trenton, New Jersey 08625, 60 pp.

URBISTONDO, R., and BAYS, L. R., 1985, *Contamination of Ground-Water and Ground-Water Treatment*, Proc. Symp. IWSA, Berlin, FRG, April 1985, IWSA, London, Pergamon Press, London, 243 pp.

U. S. COAST GUARD, 1985, *Hazardous Chemical Data Manual*, CHRIS Manual II, No. 050-012-00215-1, Supt. Doc., Govt. Printing Office, Washington, DC 20402, 2,200 pp.

U. S. PUBLIC HEALTH SERVICE, 1961, *Ground Water Contamination*, Proc. 1961 Symp., Tech. Rept. W61-5, U. S. Dept. of Health, Education, and Welfare, Cincinnati, 218 pp.

U. S. WATER RESOURCES COUNCIL, 1981, *A Summary of Groundwater Problems*, PB82-160904, NTIS, Springfield, Virginia 22161, 25 pp.

VALOCCHI, A. J., and ROBERTS, P. V., 1983, *Attenuation of Ground-Water Contaminant Pulses*, J. Hydraulic Engineering, v. 109, no. 12, pp. 1665-1682.

VAN DER LEEDEN, FRITS, CERRILLO, L. A., and MILLER, D. W., 1975, *Ground-Water Pollution Problems in the Northwestern United States*, Geraghty & Miller, Inc., EPA-660/3-75-018, U. S. Environmental Protection Agency, Corvallis Oregon 97330, 361 pp.

VAN DER WAARDEN, M., and others, 1971, *Transport of Mineral Oil Components to Groundwater — 1. Model Experiments on the Transfer of Hydrocarbons From a Residual Oil Zone to Trickling Water.* Water Research, v. 5, pp. 213-226.

VAN DUIJVENBOODEN, W., GLASBERGEN, P. and VAN LELYVELD, H., (eds.), *Quality of Groundwater*, Proc. Internat. Symp. Noordwijkerhout, The Netherlands 23-27 Mar. 1981, Studies in Environmental Science 17, Elsevier Sci. Publ. Co., New York, 1128 pp. (120 papers on groundwater contamination and groundwater quality management).

WALKER, W. H., 1969, *Illinois Ground Water Pollution*, J. Am. Water Works Assoc., v. 61, no. 1, pp. 31-40.

WALSH, J. J., and others, 1981, *Field Report — Waste Impoundment Assessment in the State of Indiana*, Ground Water, v. 19, no. 1., Jan.-Feb., pp. 81-87.

WATER WELL JOURNAL, 1970, *Primer on Ground Water Pollution*, Special Issue, Water Well J., v. 24, no. 7, Jul.

WENTSEL, R. S., and others, 1982, *Restoring Hazardous Spill-Damaged Areas: Technique Identification/Assessment,* PB82-103870, NTIS, Springfield, Virginia 22161, 374 pp.

WESTRICK, J. J., and others, 1983, *The Ground Water Supply Survey: Summary of Volatile Organic Contaminant Data*, USEPA Office of Drinking Water, Cincinnati, OH, 49 pp. (analyses from 945 ground water supplies)

WOLTERINK, T. J., and others, 1979, *Identifying Sources of Subsurface Nitrate Pollution with Stable Nitrogen Isotopes*, EPA-600/4-79-050, Ada, Oklahoma.

WOOD, E. F., 1984, *Groundwater Contamination From Hazardous Wastes*, Prentice-Hall, Inc., Route 59 at Brook Hill Dr., West Nyack, New York 10995, 192 pp.

WRIGHT, K. R., and ROVEY, C. K., 1979, *Land Application of Waste — State of the Art*, Ground Water, v. 17, no. 1, Jan.-Feb., pp. 47-62.

ZIMMIE, T. F. and RIGGS, C. O., (eds), 1979, *Permeability and Groundwater Contaminant Transport,* STP 746, ASTM, Sales Service Dept., 1916 Race St. Philadelphia, Pennsylvania 19103, 245 pp.

SITE EVALUATION/RANKING SYSTEMS/RESPONSE GUIDES

ALLER, L., and others, 1985, *DRASTIC: A Standardized System for Evaluating Ground Water Pollution Potential Using Hydrogeologic Settings,* National Water Well Assoc., Dublin, Ohio, PB85-228146/ WEP, NTIS, Springfield, Virginia 22161, 180 pp.

COCHRAN, R., and HODGE, V., 1985, *Guidance on Feasibility Studies Under CERCLA,* PB85-238590/WEP, NTIS, Springfield, Virginia 22161.

DONIGIAN, A. S., Jr., YO, R.T.Y., and SHANAHAN, E. W., 1983, *Rapid Assessment of Potential Ground-Water Contamination under Emergency Response Conditions,* Anderson-Nichols and Co., Inc., Palo Alto, CA, PB84-133123, NTIS, Springfield, Virginia 22161, 159 pp.

ENVIRONMENTAL PROTECTION AGENCY, 1980, *Guidance Manual for the Classification of Solid Waste Disposal Facilities,* EPA, Office of Solid Waste, WH 563, Washington D. C. 20460.

FORD, P. J., and TURINA, P. J., 1985, *Characterization of Hazardous Waste Sites - A Methods Manual, Volume 1, Site Investigations,* Lockheed Engineering and Management Services Co., Inc., PB85-215960/WEP, NTIS, Springfield, Virginia 22161, 252 pp.

GUSWA, J. H., and others, 1983, *Groundwater Contamination Response Guide,* Vol. 1-Methodology; Vol. 2-Desk Reference, A. D. Little Inc., Cambridge, MA, NTIS, Vol. 1, AD-A131 045/7, 46 pp.; Vol. 2, AD-A131 129/9, 325 pp.

GUSWA, J. H., and others, 1984, *Groundwater Contamination and Emergency Response Guide,* Pollution Technology Review No. 111, Noyes Publications, Mill Rd. at Grand Ave., Park Ridge, New Jersey 08656, 490 pp.

LEGRAND, H. E., 1964, *System for Evaluation of Contamination Potential of Some Waste Disposal Sites,* J. Amer. Water Works Assoc., v. 56, pp. 959-974.

LEGRAND, H. E., 1979, *A Standardized System for Evaluating Waste-Disposal Sites*, Water Well Journal Publ. Co., Dublin, Ohio 43017.

MELVOLD, R. W., and McCARTHY, L. T., Jr., 1983, *Emergency Response Procedures for Control of Hazardous Substance Releases*, Rockwell International Newbury Park, CA, PB84-128719, NTIS, Springfield, Virginia 22161, 28 pp.

NEW JERSEY GEOLOGICAL SURVEY, 1983, *A Ground Water Pollution Priority System*, NJGS Open File Rep. 83-4, N. J. Dept. of Environmental Protection, CN-402, Trenton, New Jersey 08625, 32 pp.

SILKA, L. R., and SWEARINGEN, T. L., 1978, *A Manual for Evaluating Contamination Potential of Surface Impoundments*, U. S. Environmental Protection Agency, Office of Drinking Water, Ground Water Protection Branch, Rep. No. EPA 570/9-78-003.

SISK, S. W., 1981, *NEIC Manual for Groundwater/Subsurface Investigations at Hazardous Waste Sites*, National Enforcement Investigations Center, Denver, Colorado, EPA-330/9-81-002, PB82-103755, NTIS, Springfield, Virginia 22161, 213 pp.

U. S. ENVIRONMENTAL PROTECTION AGENCY, 1980, *Landfill and Surface Impoundment Performance Evaluation*, EPA, Office of Water and Waste Management, Washington, D. C. SW-869.

U. S. ENVIRONMENTAL PROTECTION AGENCY, 1984, *The National Surface Impoundment Assessment Report*, USEPA, Press Office, 401 M St. S.W. Rm 311, West Tower, Washington, DC 20460.

AGRICULTURE

ADRIANO, D. C., and others, 1971, *Nitrate and Salt in Soils and Ground Waters from Land Disposal of Dairy Manure*, Soil Sci. Soc. Amer. Proc., v. 35, pp. 759-762.

ANONYMOUS, 1985, *Soil Salinity: Irrigation Practices and Effects on Crops and Groundwater, 1977-October 1985*, (Citations from the Selected Water Resources Abstracts Data Base), PB85-869667/WEP, NTIS, Springfield, Virginia 22161, 131 pp.

AYERS, R. S., and BRANSON, R. L., 1973, *Nitrates in the Upper Santa Ana River basin in Relation to Ground Water Pollution*, Agric. Experiment Station Bull. 861, Univ. of California, 59 pp.

BAIER, J. H., and RYKBOST, K. A., 1976, *The Contribution of Fertilizer to the Ground Water of Long Island*, Ground Water, v. 14, no. 6, pp. 439-447.

BAIER, J. H., 1982, *Report on the Occurrence and Movement of Agricultural Chemicals in Groundwater: North Fork of Suffolk Country*, Bureau of Water Resources, Suffolk County Dept. of Health Services, Hauppauge, New York 11788.

CALIFORNIA DEPT. WATER RESOURCES, 1968, *The Fate of Pesticides Applied to Irrigated Agricultural Land*, Bull. 174-1, Sacramento, 30 pp.

COHEN, S. Z., EIDEN, C., and LORBER, M. N., 1986, *Monitoring Ground Water for Pesticides in the USA*, ACS Symp. Series, Am. Chem. Society, (in preparation).

CONCANNON, T. J., and GENETELLI, E. J., 1971, *Groundwater Pollution due to High Organic Manure Loadings, Livestock Waste Management and Pollution Abatement*, Proc. Internat. Symp. on Livestock Wastes, Ohio State University, pp. 249-253.

COUNCIL FOR AGRICULTURAL SCIENCE AND TECHNOLOGY, 1986, *Agriculture and Ground Water Quality*, CAST, P.O. Box 1550, Iowa State Univ. Station, Ames, Iowa 50010, 62 pp.

CROSBY, J. W., III, and others, 1971, *Migration of Pollutants in a Glacial Outwash Environment*, Part 2, Water Resources Res., v. 7, no. 1, pp. 204-208. (feedlots).

DIERBERG, F. E., and GIVEN, C. J., 1986, *Aldicarb Studies in Ground Waters from Florida Citrus Groves and Their Relation to Ground-Water Protection*, Ground Water, v. 24, no. 1, pp. 16-22.

DONEEN, L. D., 1954, *Salination of Soil by Salts in the Irrigation Water*, Trans. Am. Geophys. Union, v. 35, pp. 943-950.

DREGNE, H. E., and others, 1969, *Movement of 2, 4-D in Soils*, Western Regional Research Project Progress Report, New Mexico Agric. Experiment Station, University Park, 35 pp.

DUKE, H. R., SMIKA, D. E., and HEERMANN, D. F., 1978, *Ground-Water Contamination by Fertilizer Nitrogen*, J. Irrigation and Drainage Div., Am. Soc. Civil Engrs., v. 104, no. IR3, Sep.

EXNER, M. E., and SPALDING, R. F., 1979, *Evolution of Contaminated Ground Water in Holt County, Nebraska*, Water Resources Res., v. 15, p. 139.

FITZSIMMONS, D. W., and others, 1972, *Nitrogen, Phosphorus, and Other Inorganic Materials in Waters in a Gravity-Irrigated Area,* Trans. Am. Soc. Agric. Engrs., v. 15, no. 2, pp. 292-295.

FLIPSE, W. J., Jr., and BONNER, F. T., 1985, *Nitrogen-Isotope Ratios of Nitrate in Groundwater Under Fertilized Fields, Long Island, New York,* Ground Water, v. 23, no. 1, 59-67.

GERWING, J. R., and others, 1979, *Fertilizer Nitrogen Distribution Under Irrigation Between Soil, Plant and Aquifer,* J. Environ. Qual., v. 8, p. 281.

GILLHAM, R. W., and WEBBER, L. R., 1969, *Nitrogen Contamination of Groundwater by Barnyard Leachates,* J. Water Poll. Control Fed., v. 41, no. 10, pp. 1752-1762.

GORMLY, J. R., and SPAULDING, R. F., 1979, *Sources and Concentrations of Nitrate-Nitrogen in Ground Water of the Central Platte Region, Nebraska,* Ground Water, v. 17, no. 3, May-June, pp. 291-301.

HARKIN, J. M., and others, 1984, *Pesticides in Groundwater Beneath the Central Sand Plain of Wisconsin,* Wisconsin Univ.-Madison, Water Res. Center, OWRT-A-094-WIS (1), NTIS, PB84-212372, 57 pp, (Aldicarb survey).

HEGG, R. O., KING, T. G., and WILSON, T. V., 1979, *The Effects on Ground Water from Seepage of Livestock Manure Lagoons,* PB81-115, NTIS, Springfield, Virginia 22161.

JENKE, A. L., 1974, *Evaluation of Salinity Created by Irrigation Return Flows,* Rept. EPA 430/9-74-006, U. S. Environmental Protection Agency, Washington, D. C., 128 pp.

JOHNSTON, W. R., and others, 1967, *Insecticides in Tile Drainage Effluent,* Water Resources Res. v. 3, no. 2, pp. 525-537.

LAW, J. P., Jr., and others, 1970, *Degradation of Water Quality in Irrigation Return Flows,* Bull. B-684, R. S. Kerr Water Research Center, Oklahoma Agric. Experiment Station, Ada; and Oklahoma State University, Stillwater, Dept. of Agronomy, 26 pp.

LEGRAND, H. E., 1970, *Movement of Agricultural Pollutants with Groundwater,* Chap. 22, pp. 303-313, in *Agricultural Practices and Water Quality,* ed. by T. L., Willrich and G. E. Smith: Iowa State Univ. Press, 415 pp.

LEHMANN, E. J., 1977, *Ground-Water Pollution*, Part 1, *General Studies*, PS-77/0007/3WP, Part 2, *Pollution from Irrigation and Fertilization*, PS-77/0-008/IWP, Part 3, *Saline Ground Water*, PS-77/0009/9WP, Part 4, *Am. Petroleum Inst. Citations*, PS-77/0010/7WP, NTIS, Springfield, Virginia 22161.

LIEBHARDT, W. C., and others, 1979, *Nitrate and Ammonium Concentrations of Ground Water Resulting from Poultry Manure Applications*, J. Environ. Qual., v. 8, p. 211.

LOEHR, R. C., 1967, *Effluent Quality from Anaerobic Lagoons Treating Feedlot Wastes*, J. Water Pollution Control Fed., v. 39, no. 3, pp. 384-391.

LORIMOR, J. C., and others, 1972, *Nitrate Concentrations in Groundwater Beneath a Beef Cattle Feedlot*, Water Resources Bull., no. 8, p. 999.

MILLER, W. D., 1971, *Subsurface Distribution of Nitrates Below Commercial Cattle Feedlots, Texas High Plains*, Water Resources Bull., v. 7, no. 5, p. 941-950.

MINK, L. L., and others, 1976, *The Selection and Management of Feedlot Sites and Land Disposal of Animal Waste in Boise Valley, Idaho*, Ground Water, v. 14, no. 6, pp. 411-424.

NATIONAL ACADEMY OF SCIENCE, 1986, *Pesticides and Groundwater Quality: Issues and Problems in Four States*, National Academy Press, 2101 Constitution Ave., NW, Washington, DC 20418, 136 pp

PANASEWICH, CAROL, 1985, IProtecting Groundwater from Pesticides, EPA Journal, v. 11, pp. 18-20.

PRATT, P. R., 1972, *Nitrate in the Unsaturated Zone Under Agricultural Lands*, U. S. Environmental Protection Agency, Water Pollution Control Research Ser., 16060 DOE 04/72, 45 pp.

PRATT, P. R., JONES, W. W., and HUNSAKER, V. E., 1972, *Nitrate in Deep Soil Profiles in Relation to Fertilizer Rates and Leaching Volume*, J. Environ. Qual., v. 1, pp. 97-102.

RAMLIT ASSOC., INC., 1984, *Groundwater Contamination by Pesticides: A California Assessment*, 2437 Durant Ave., Berkeley, California 94704, 209 pp.

RITTER, W. F., 1984, *Influence of Nitrogen Application and Irrigation on Ground-Water Quality in the Coastal Plain*, Delaware Univ., PB85-225852/WEP, NTIS, Springfield, Virginia 22161, 26 pp.

ROBBINS, J. W. D., and KRIZ, G. J., 1969, *Relation of Agriculture to Groundwater Pollution — A Review*, Trans. Am. Soc. Agr. Engrs., v. 12, no. 3, p. 397-403.

SAFFIGNA, P. G., and KEENEY, D. R., 1977, *Nitrate and Chloride in Ground Water Under Irrigated Agriculture in Central Wisconsin*, Ground Water, v. 15, no. 2, pp. 170-177.

SCALF, M. R., and others, 1968, *Fate of DDT and Nitrate in Ground Water*, U. S. Dept. Interior, Robert S. Kerr Water Res. Center, Ada, Oklahoma, p. 46.

SCALF, M. R., and others, 1969, *Movement of DDT and Nitrates During Ground-Water Recharge*, Water Resources Res., v. 5, no. 5, p. 1041.

SCHUMAN, G. E., and ELLIOT, L. F., 1978, *Cropping an Abandoned Feedlot to Prevent Deep Percolation of Nitrate-Nitrogen*, Soil Sci., v. 126, p. 237.

SHAW, E. J., 1968, *Western Fertilizer Handbook*, California Fertilizer Assoc., Sacramento, California, 200 pp.

SKOGERBOE, G. V., and LAW, J. V., 1971, *Research Needs for Irrigation Return Flow Quality Control*, U. S. Environmental Protection Agency, Water Pollution Control Research Ser. 13030-11/71, 98 pp.

SOREN, JULIAN, and STELZ, W. G., 1985, *Aldicarb-Pesticide Contamination of Ground Water in Eastern Suffolk County, Long Island, New York*, U. S. Geol. Survey Water-Resources Inv. Rept. 84-4251, 34 pp.

SPAULDING, R. F., 1975, *Effects of Land Use and River Seepage on Groundwater Quality in Hall County, Nebraska*, Univ. Nebraska Conservation and Surv. Div., Water-Surv. Paper 38, 95 pp.

STEWART, B. A., and others, 1967, *Distribution of Nitrates and Other Water Pollutants Under Fields and Corrals in the Middle South Platte Valley of Colorado*, U. S. Dept. of Agric., Agric. Research Service, Rept. ARS 41-134, 206 pp.

STEWART, B. A., VIETS, F. G., and HUTCHINSON, G. L., 1968, *Agriculture's Effect on Nitrate Pollution*, J. Soil Water Conservation, v. 23, no. 13, pp. 13-15.

UTAH STATE UNIVERSITY FOUNDATION, 1969, *Characteristics and Pollution Problems of Irrigation Return Flow*, Federal Water Pollution Control Adm., Ada, Oklahoma.

WALKER, W. H., 1973, *Ground-Water Nitrate Pollution in Rural Areas*, Ground Water, v. 11, no. 5, pp. 19-22.

WALKER, W. R., and KROEKER, B. E., 1982, *Nitrates in Groundwater Resulting from Manure Applications to Irrigated Croplands*, Colorado State Univ., Fort Collins, EPA-600/2-82-079, NTIS, PB82-255415, 94 pp.

WENDT, C. W., and others, 1977, *Effects of Irrigation Methods on Ground-Water Pollution by Nitrates and Other Solutes*, PB-268 322/ 5WP, NTIS, Springfield, Virginia 22161.

WENGEL, R. W., and GRIFFIN, G. F., 1980, *Potential Ground-Water Pollution from Sewage Sludge Application on Agricultural Land*, PB80-102957, NTIS, Springfield, Virginia 22161.

WILLRICH, T. L., and SMITH, G. E., 1970, *Agricultural Practices and Water Quality*, Iowa State Univ. Press, Ames, Iowa, 415 pp.

MUNICIPAL AND DOMESTIC

ALLEN, R. D., and RAYMOND, J. R., 1984, *Influence of Landfills on Ground-Water Quality*, Battelle Pacific NW Labs., Richland, WA, DE85004537/WEP, NTIS, Springfield, Virginia 22161, 21 pp.

AMERICAN CHEMICAL SOCIETY, 1969, *Cleaning Our Environment, The Chemical Basis for Action*, Report of the Committee on Chemistry and Public Affairs, Washington, D. C. (sewage effluent characteristics)

ANDERSEN, J. R., 1972, *Studies of the Influence of Lagoons and Landfills on Groundwater Quality*, Water Resources Institute, South Dakota State Univ., Brookings, NTIS, Springfield, Virginia 22161, PB-214 138, 47 pp.

ANDERSEN, J. R., and DORNBUSH, J. N., 1967, *Influence of Sanitary Landfill on Ground Water Quality*, J. Am. Water Works Assoc., v. 59, p. 457, Apr.

ANDREOLI, A., and others, 1979, *Nitrogen Removal in a Subsurface Disposal System*, J. Water Poll. Control Fed., v. 51, p. 841.

ANDREWS, F. L., and others, 1984, *Effects of Storm-Water Runoff on Water Quality of the Edwards Aquifer near Austin, Texas*, WRI 84-4124, U. S. Geol. Survey, Austin, Texas 78701, 50 pp.

ANONYMOUS, 1981, *Movement and Fate of Septic Tank Effluent in Soils of the North Carolina Coastal Plain*, North Carolina Div. of Health Services, Raleigh, North Carolina 27602.

APGAR, M. A., and LANGMUIR, D., 1971, *Ground-Water Pollution Potential of a Landfill Above the Water Table*, Ground Water, v. 9, no. 6, pp. 76-96.

BAARS, J. K., 1957, *Travel of Pollution and Purification en Route in Sandy Soils*, Bull. World Health Organization, v. 16, no. 4, pp. 727-747.

BAEDECKER, M. J., and BACK, WILLIAM, 1979, *Hydrogeological Processes and Chemical Reactions at a Landfill*, Ground Water, v. 17, no. 5, Sept.-Oct., pp. 429-437.

BEHNKE, J. J., and HASKELL, E. E., 1968, *Ground Water Nitrate Distribution Beneath Fresno, California*, J. Am. Water Works Assoc., v. 60, no. 4, pp. 477-480.

BOGAN, R. H., 1961, *Problems Arising from Ground Water Contamination by Sewage Lagoons at Tieton, Washington*, U. S. Public Health Service Tech. Rept. W61-5, pp. 83-87.

BOUMA, J., 1971, *Evaluation of the Field Percolation Test and Alternative Procedure to Test Soil Potential for Disposal of Septic Tank Effluent*, Proc. Soil Society of America, 35, pp. 871-875.

BOUMA, J., 1975, *Innovative On-Site Soil Disposal and Treatment Systems for Septic-Tank Effluent*, in *Home Sewage Disposal*, Proc. National Home Sewage Disposal Symp., Am. Soc. Agric. Engrs., St. Joseph, Michigan.

BOUMA, J., 1975, *Unsaturated Flow Phenomena During Subsurface Disposal of Septic Tank Effluent*, J. Am. Soc. Civil Engrs., Environ. Eng. Div., 101, pp. 967-983.

BRAIDS, O. C., and RAGONE, S. E., 1975, *Nitrogen in Long Island Ground Water*, Proc. Conference on *Nitrogen in Long Island Water Systems,* Am. Chemical Society, Environmental Improvement Committee, Nassau-Suffolk Subsection, pp. 64-88.

BUSH, A. F., 1954, *Studies of Waste Water Reclamation and Utilization*, Calif. State Water Pollution Control Bd., Pub. no. 9, Sacramento, California.

CALDWELL, E. L., 1937, *Pollution Flow From Pit Latrines When Impervious Stratum Closely Underlies the Flow*, J. Infectious Diseases, v. 61, no. 3, pp. 270-288.

Contamination

CALDWELL, E. L., 1937, *Study of an Envelope Pit Privy,* J. Infectious Diseases, v. 61, no. 3, pp. 264-269.

CALDWELL, E. L., 1938, *Pollution Flow From Pit Latrine When Permeable Soils of Considerable Depth Exist Below the Pit,* J. Infectious Diseases, v. 62, no. 3., pp 225-258.

CALDWELL, E. L., and PARR, L. W., 1937, *Ground Water Pollution and the Bored-Hole Latrine,* J. Infectious Diseases, v. 61, no. 2, pp. 148-183.

CALIFORNIA DEPARTMENT OF WATER RESOURCES, 1969, *Sanitary Landfill Studies,* Appendix A — Summary of Selected Previous Investigations, Bull. 147-5, 115 pp.

CALIFORNIA STATE WATER POLLUTION CONTROL BOARD, 1961, *Effects of Refuse Dumps on Ground Water Quality,* Calif. State Water Resources Control Board Publ. no. 24, 107 pp.

CAMERON, R. D., 1978, *The Effects of Solid Waste Landfill Leachates on Receiving Waters,* J. Am. Water Works Assoc., v. 70, p. 173-176.

CARRIERE, G. D., and CANTER, L. W., 1980, *Effects of Septic Tank Systems on Ground-Water Quality,* National Center for Ground-Water Research, Rept. 80-8, Norman, Oklahoma 73019.

CARTWRIGHT, KEROS, GRIFFIN, R. A., and GILKESON, R. H., 1977, *Migration of Landfill Leachate Through Glacial Tills,* Ground Water, v. 15, no. 4, pp. 294-305.

CARTWRIGHT, KEROS, and SHERMAN, F. B., 1971, *Ground-Water and Engineering Geology in Siting of Sanitary Landfills,* Illinois State Geol. Survey, Publ. 1971-E.

CHERRY, J. A., (ed.), 1984, *Migration of Contaminants in Groundwater at a Landfill: A Case Study,* Elsevier Science Publ. Co., Inc., P.O. Box 1663, Grand Central Sta., New York, New York 10163.

CHERRY, R. N., and BROWN, D. P., 1973, *Hydrogeologic Aspect of a Proposed Sanitary Landfill Near Old Tampa Bay, Florida,* Florida Dept. Natural Res. Rept. Inv. 68.

COE, J. J., 1970, *Effect of Solid Waste Disposal on Groundwater Quality,* J. Am. Water Works Assoc., v. 62, no. 12, pp. 776-783.

CONVERSE, J. C., and OTIS, R. J., 1973, *The Mound or Fill System for On-Site Wastewater Disposal for Rural Homes in Wisconsin,* Small-Scale Waste Management Project, Agriculture Hall, Univ. of Wisconsin, Madison, Wisconsin 53706.

CROSBY, J. W., III, and others, 1968, *Migration of Pollutants in a Glacial Outwash Environment*, Water Resources Res., v. 4, no. 5, pp. 1095-1115. (drain fields)

CROSBY, J. W., III, and others, 1971, *Investigation of Techniques to Provide Advance Warning of Ground-Water Pollution Hazards with Special Reference to Aquifers in Glacial Outwash*, (septic tanks) Wash. State. Univ., College of Engineering Research Div., Final Rept., Pullman, Washington 99163. PB-203 748, NTIS, Springfield, Virginia 22161.

DANIELS, R. B., and others, 1975, *Nitrogen Movement in a Shallow Aquifer System of the North Carolina Coastal Plain*, Water Resources Bull., v. 11, no. 6, pp. 1121-1130.

DeWALLE, F. B., and SCHAFF, R. M., 1980, *Ground-Water Pollution by Septic Tank Drainfields*, J. Environ. Eng. Div., Am. Soc. Civil Engrs., v. 106, no. EE3, Jun.

DUNLAP, W. J., and others, 1972, *Probable Impact of NTA on Ground Water*, Ground Water, v. 10, no. 1, pp. 107-116, Jan.-Feb. (detergents)

ECCLES, L. A., and BRADFORD, W. L., 1977, *Distribution of Nitrate in Ground Water, Redlands, California*, U. S. Geol. Survey Water-Resources Invest. 76-117, 44 pp.

ECCLES, L. A., and KLEIN, J. M., 1978. *Distribution of Dissolved Nitrate and Fluoride in Ground Water, Highland-East Highlands, San Bernardino County, California*, U. S. Geol. Survey Water Resources Invest. 78-14, 42 pp.

ECCLES, L. A., KLEIN, J. M., and HARDT, W. F., 1976, *Abatement of Nitrate Pollution in a Public-Supply Well by Analysis of Hydrologic Characteristics*, Ground Water, v. 14, no. 6, pp. 449-453.

ECCLES, L. A., and others, 1977, *Abatement of Nitrate Pollution in a Public-Supply Well by Analysis of Hydrologic Characteristics*, Proc. Third National Ground Water Quality Symp., EPA-600/9-77-014, Robert S. Kerr Environmental Research Laboratory, Ada, Oklahoma, pp. 93-98.

EISEN, CRAIG, and ANDERSON, M. P., 1979, *The Effects of Urbanization on Ground-Water Quality — A Case Study*, Ground Water, v. 17, no. 5, pp. 456-463.

ENGINEERING-SCIENCE, INC., 1961, *Effects of Refuse Dumps on Groundwater Quality*, Resources Agency, Calif. State Water Pollution Control Board, Pub. 211.

FIELD, R., and others, 1973, *Water Pollution and Associated Effects from Street Salting*, Rept. EPA-R2-73-257, U. S. Environmental Protection Agency, Cincinnati, Ohio, 48 pp.

FILIPPINI, M. G., and KROTHE, N. C., 1983, *Impact of Urbanization on a Flood-Plain Aquifer: Bloomington, Indiana*, Purdue Univ., Lafayette, IN., Water Resources Res. Center, TR-156, W83-03609, OWRT-A-062-IND (1), NTIS, PB83-235515, 114 pp.

FLYNN, J. M., and others, 1958, *Study of Synthetic Detergents in Ground Water*, J. Am. Water Works Assoc., v. 50, no. 12, pp. 1551-1562.

FRITTON, D. D., and others, 1983, *On-Site Sewage Disposal: Site Suitability, System Selection and Soil Absorption Area Sizing*, Pennsylvania State Univ., University Park, Inst. for Research on Land and Water Resources, W83-03268, OWRT-B-122-PA (3), NTIS, PB83-219725, 292 pp. (computer model).

FULLER, W. H., 1979, *Investigation of Landfill Leachate Pollutant Attenuation by Soils*, PB-286 995/6WP, NTIS, Springfield, Virginia 22161.

FUNGAROLI, A. A., and EMRICH, G. H., 1966, *Pollution of Subsurface Water by Sanitary Landfill*, Cooperative Study by Civil Eng. Dept., Drexel Inst. of Technology and Bur. Sanitary Eng. Penn. Dept. of Health.

FUNGAROLI, A. A., and STEINER, R. L., 1979, *Investigation of Sanitary Landfill Behavior*, v. 1, Final Report, EPA-600-2-79-053a, Municipal Envir. Res. Lab., Cincinnati, Ohio.

GIDDINGS, M. T., Jr., 1977, *The Lycoming County, Pennsylvania, Sanitary Landfill: State-of-the-Art in Ground-Water Protection*, Ground Water, v. 15, no. 1, pp. 5-12.

GOTAAS, H. B., and others, 1954, *Report on the Investigation of Travel of Pollution*, Calif. State Water Pollution Control Bd., Pub. no. 11, Sacramento, 218 pp.

HATHAWAY, S. W., 1980, *Sources of Toxic Compounds in Household Wastewater*, PB81-110942, NTIS, Springfield, Virginia 22161.

HUGHES, G., and others, 1971, *Pollution of Ground Water Due to Municipal Dumps*, Canada Dept. of Energy, Mines and Resources, Inland Waters Branch, Tech. Bull. no. 42, Ottawa.

HUGHES, G. M., 1972, *Hydrogeologic Considerations in the Siting and Design of Landfills*, Ill. Geol. Survey Environ. Geol. Note 51, 22 pp.

HUGHES, G. M., LANDON, R. A., and FARVOLDEN, R. N., 1971, *Summary of Findings on Solid Waste Disposal Sites in Northeastern Illinois*, Ill. Geol. Survey Environmental Geology Notes 45, 25 pp.

HUGHES, G. M., and others, 1971, *Hydrogeology of Solid Waste Disposal in Northeastern Illinois*, U. S. Environmental Protection Agency, Govt. Printing Office, Washington D. C., 154 pp; also available from U. S. Environmental Protection Agency, Cincinnati, Ohio 45268.

HUGHES, J. L., 1975, *Evaluation of Ground-Water Degradation Resulting from Water Disposal to Alluvium near Barstow, California*, U. S. Geol. Survey Prof. Paper 878, 33 pp.

HUGHES, J. L., and ROBSON, S. G., 1973, *Effects of Waste Percolation on Ground Water in Alluvium near Barstow, California*, in v. 1 of *Underground Waste Management and Artificial Recharge*, Am. Assoc. Petrol. Geol., pp. 91-134.

HUGHES, J. L., and others, 1974, *Dissolved Organic Carbon (DOC), an Index of Organic Contamination in Ground Water near Barstow, California*, Ground Water, v. 12, no. 5, pp. 283-290.

JEWELL, W. J., and SWAN, R., (eds.), 1975, *Water Pollution Control in Low Density Areas*, Proc. National Rural Engineering Conf., The University Press of New England, Hanover, New Hampshire.

KATZ, B. G., LINDNER, J. B., and RAGONE, S. E., 1980, *A Comparison of Nitrogen in Shallow Ground Water from Sewered and Unsewered Areas, Nassau County, New York from 1952 Through 1976*, Ground Water, v. 18, no. 6, Nov.-Dec., pp. 607-616.

KAUFMANN, R. F., 1977, *Land and Water Use Impacts on Ground-Water Quality in Las Vegas Valley*, Ground Water, v. 15, no. 1, pp. 81-89. (urban problems)

KELLY, W. E., 1976, *Ground-Water Pollution near a Landfill*, J. Environ. Eng. Div. Am. Soc. Civ. Eng., 102 (EE6), pp. 1189-1199.

KIMMEL, G. E., and BRAIDS, O. C., 1974, *Leachate Plumes in a Highly Permeable Aquifer*, Ground Water, v. 12, no. 6, pp. 388-393.

KIMMEL, G. E., and BRAIDS, O. C., 1980, *Leachate Plumes in Ground Water from Babylon and Islip Landfills, Long Island, New York*, U. S. Geol. Survey Prof. Paper 1085.

KLAER, F. H., Jr., 1963, *Bacteriological and Chemical Factors in Induced Infiltration*, Ground Water, v. 1, no. 1.

KLIGLER, I. J., 1921, *Investigation of Soil Pollution and Relation of Various Types of Privies to the Spread of Intestinal Infections*, Rockefeller Institute for Medical Res., Monograph 15.

KREITLER, C. W., RAGONE, S. E., and KATZ, B. G., 1980, N^{15}/N^{14} *Ratios of Ground-Water Nitrate, Long Island, New York*, Ground Water, v. 16, no. 6, pp. 404-409.

KU, H. F. H., and SULAM, D. J., 1976, *Distribution and Trend of Nitrate, Chloride, and Total Solids in Water in the Magothy Aquifer in Southeast Nassau County, New York*, U. S. Geol. Survey Water-Resources Inv. 76-44, 54 pp.

KUNKLE, G. R., and SHADE, J. W., 1976, *Monitoring Ground-Water Quality Near a Sanitary Landfill*, Ground Water, v. 14, no. 1, pp. 11-20.

LU, J. C. S., and others, 1984, *Production and Management of Leachate from Municipal Landfills: Summary and Assessment*, Calscience Research, Inc., Huntington Beach, CA, PB84-187913, NTIS, Springfield, Virginia 22161, 474 pp.

McDERMOTT, J. H., 1971, *The Home Ground-Water Supply Picture as We See It*, Ground Water, v. 9, no. 3, pp. 17-23. (rural ground water quality in Georgia)

McGAUGHEY, P. H., 1975, *Septic Tanks and Their Effect on the Environment*, in Jewell and Swan, (eds.), *Water Pollution Control in Low-Density Areas*, Proc. Rural Environmental Eng. Conf., The University Press of New England, Hanover, New Hampshire.

MILLER, J. C., 1975, *Nitrate Contamination of the Water-Table Aquifer by Septic Tank Systems in the Coastal Plain of Delaware*, in Jewell and Swan (eds.), *Water Pollution Control in Low-Density Areas*, Proc. Rural Environ. Eng. Conf., Univ. Press of New England, Hanover, New Hampshire, pp. 121-133.

MORRILL, G. B., and TOLER, L. G., 1973, *Effect of Septic-Tank Wastes on Quality of Water, Ipswich and Shawsheen River Basins, Massachusetts*, J.Research U. S. Geol. Survey, v. 1, no. 1, pp. 117-120.

NAYLOR, J. A., and others, 1978, *The Investigation of Landfill Sites*, Water Research Centre, Tech. Rept. 91, Medmenham Laboratory, P.O. Box 16, Marlow, Bucks SL7 2HD, England, 68 pp.

NEW YORK STATE DEPT. OF HEALTH, 1960, *Effect of Synthetic Detergents on the Ground Waters of Long Island, New York*, Research Rept. 6, Albany, New York, 17 p.

NIGHTENGALE, HARRY, HARRISON, DOUG, and SALO, JOHN, 1985, *An Evaluation Technique for Ground Water Quality Beneath Urban Runoff Retention and Percolation Basins*, Ground Water Monitoring Rev., v. 5, no. 1, pp. 43-50.

NIGHTINGDALE, H. I., 1970, *Statistical Evaluation of Salinity and Nitrate Content and Trends Beneath Urban and Agricultural Areas — Fresno, California*, Ground Water, v. 8, no. 1, pp. 22-29.

ORLOB, G. T., and KRONE, R. B., 1956, *Movement of Coliform Bacteria Through Porous Media*, Sanitary Engineering Res. Lab., Univ. of California, Berkeley.

OSTENDORF, D. W., NOSS, R. R., and LEDERER, D. O., 1984, *Landfill Leachate Migration Through Shallow Unconfined Aquifers*, Water Resources Res., v. 20, no. 2.

PAGE, H. G., and others, 1963, *Behavior of Detergents (ABS), Bacteria, and Dissolved Solids in Water-Saturated Soils*, U. S. Geol. Survey Prof. Paper 450-E, Art. 237, pp. 179-181.

PALMQUIST, R. C., and SENDLEIN, L. V. A., 1975, *The Configuration of Contamination Enclaves Resulting from Refuse Disposal Sites on Floodplains*, Ground Water, v. 13, no. 2, pp. 167-181.

PERLMUTTER, N. M., and GUERRERA, A. A., 1970, *Detergents and Associated Contaminants in Ground Water at Three Public-Supply Well Fields in Southwestern Suffolk County, Long Island, New York*, U. S. Geol. Survey Water Supply Paper 2001-B, 22 pp.

PERLMUTTER, N. M., and KOCH, ELLIS, 1972, *Preliminary Hydrogeologic Appraisal of Nitrate in Ground Water and Streams, Southern Nassau County, Long Island, N.Y.*, U. S. Geol. Survey Prof. Paper 800-B, Geol. Survey Research 1972, Chapter B, pp. 225-237.

PETTIJOHN, R. A., 1977, *Nature and Extent of Ground-Water Quality Changes Resulting From Solid-Waste Disposal, Marion County, Indiana*, U. S. Geol. Survey Water-Resources Inv., 77-40, 129 pp.

PITT, W. A. J., 1974, *Effects of Septic Tank Effluent on Ground-Water Quality, Dade County, Florida*, U. S. Geol. Survey Open-file Rept., 50 pp.

POLTA, R. C., 1969, *Septic Tank Effluents, Water Pollution by Nutrients — Sources Effects, and Controls*, Water Resources Res. Center, Univ. of Minnesota, Minneapolis, Minnesota, WRRC Bull. 13, pp. 53-57.

QASIM, S. R., and BURCHINAL, J. C., 1970, *Leaching of Pollutants from Refuse Beds*, J. Sanitary Eng. Div. Am. Soc. Civil Engrs., v. 96, no. SA1, Feb.

QUAN, E. L., SWEET, H. R., and ILLIAN, J. R., 1974, *Subsurface Sewage Disposal and Contamination of Ground Water in East Portland, Oregon*, Ground Water, v. 12, no. 6, pp. 356-368.

RAGONE, S. E., GUERRERA, A. A., and FLIPSE, W. J., Jr., 1976, *Changes in Methylene Blue Active Substances and Chloride Levels in Streams in Suffolk County, Long Island, New York, 1960-76*, U. S. Geol. Survey Open-file Rept. 76-600, 65 pp.

RAGONE, S. E., and others, 1981, *Nitrogen in Ground Water and Surface Water in Sewered and Unsewered Areas, Nassau County, Long Island, NY*, U. S. Geol. Survey, Water Res. Div., Albany, New York 12201, 72 pp.

REMSON, IRWIN, 1968, *Water Movement in an Unsaturated Sanitary Landfill*, J. Sanitary Eng., Proc. Am. Soc. Civil Engrs., v. 94, SA 2, pp. 307-317.

RICCIO, J. F., and HYDE, L. W., 1971, *Hydrogeology of Sanitary Landfill Sites in Alabama*, Geol. Survey of Alabama Circ. 71, University, Alabama 35486.

ROBECK, G. G., and others, 1963, *Degradation of ABS and Other Organics in Unsaturated Soils*, J. Water Poll. Control Fed., v. 35, pp. 1225-1237, Oct.

ROBERTSON, F. N., 1979, *Evaluation of Nitrate in the Ground Water in the Delaware Coastal Plain*, Ground Water, v. 17, no. 4, Jul.-Aug., pp. 328-337.

ROBERTSON, J. M., and others, 1974, *Organic Compounds Entering Ground-Water from a Landfill*, U. S. Environmental Protection Agency Rept. 660/2-74-077, Washington, D. C.

ROBSON, S. G., 1977, *Ground-Water Quality near a Sewage-Sludge Recycling Site and a Landfill near Denver, Colorado*, U. S. Geol. Survey, Water-Resources Invest. 76-132, 152 pp.

ROMERO, J. C., 1970, *The Movement of Bacteria and Viruses Through Porous Media*, Ground Water, v. 8, no. 2, pp. 37-49.

ROTH, D., and WALL, G., 1976, *Environmental Effects of Highway Deicing Salts*, Ground Water, v. 14, no. 5, pp. 286-289.

SANGREY, D. A., and PHILIPSON, W. R., 1980, *Detecting Landfill Leachate Contamination Using Remote Sensors*, PB80-174295, NTIS, Springfield, Virginia 22161, 78 pp.

SARTOR, J. D., and BOYD, G. B., 1972, *Water Pollution Aspects of Street Surface Contaminants*, U. S. Environmental Protection Agency, Technology Series, EPA-R2-72-081, 236 pp.

SCALF, M. R., DUNLAP, W. J., and KREISSL, J. F., 1977, *Environmental Effects of Septic Tank Systems*, Robert S. Kerr Environ. Research Lab., Office of Research and Development, U. S. Environmental Protection Agency, Ada, Oklahoma 74820; PB272 702/2 wp, NTIS, Springfield, Virginia 22161, 35 pp.

SCHMIDT, K. D., 1972, *Nitrate in Ground Water of the Fresno-Clovis Metropolitan Area, California*, Ground Water, v. 10, no. 1, pp. 50-64.

SCHMIDT, K. D., 1975, *Regional Sewering and Groundwater Quality in the Southern San Joaquin Valley*, Water Resources Bull., v. 11, no. 3, pp. 514-525.

SCHNEIDER, W. S., 1970, *Hydrologic Implications of Solid Waste Disposal*, Water in Urban Environment, U. S. Geol. Survey Circ. 601-F, 10 pp.

SEITZ, H. R., WALLACE, A. T., and WILLIAMS, R. E., 1972, *Investigation of a Landfill in Granite-Loess Terrane*, Ground Water, v. 10, no. 4, pp. 35-41.

SIKORA, L. J., and COREY, R. B., 1976, *Fate of Nitrogen and Phosphorus in Soils under Septic Tank Waste Disposal Fields*, Trans. Am. Soc. Agric. Engrs., v. 19, no. 5, pp. 866-870.

STILES, C. W., 1927, *Experimental Bacterial and Chemical Pollution of Wells via Ground Water and the Factors Involved*, U. S. Public Health Service Hygiene Lab., Bull. 147.

STUART, J. D., and others, 1984, *Detection and Effects of Aperiodic Leachate Discharges from Landfills*, Conn. Univ., Storrs, Inst. Water Resources, PB85-242949/WEP, NTIS, Springfield, Virginia 22161, 213 pp.

SUESS, M. J., 1964, *Retardation of ABS in Different Aquifers*, J. Am. Water Works Assoc., v. 56, p. 89, Jan.

Contamination

TERRY, R. C., Jr., 1974, *Road Salt, Drinking Water, and Safety*, Ballinger, Cambridge, Massachusetts, 161 pp.

TIMOTHY, J. H., 1984, *Septic-Tank Systems: A Consultant's Toolkit*, Butterworth Publishers, 10 Tower Office Park, Woburn, Massachusetts 01801, Vol. 1, 265 pp., Vol. 2, 144 pp.

VECCHIOLI, JOHN, and others, 1972, *Travel of Pollution-Indicator Bacteria Through the Magothy Aquifer, Long Island, N.Y.*, U. S. Geol. Survey Prof. Paper 800-B, Geol. Survey Research 1972- Chapter B, pp. 237-241.

WALDORF, E., and EVANS, J. L., (eds.), 1983, *Individual Onsite Wastewater Systems*, Natl. Sanitation Foundation, P.O. Box 1468, Ann Arbor, Michigan 48106, 350 pp.

WALKER, W. H., 1970, *Salt Piling — A Source of Water Supply Pollution*, Pollution Eng., Jul.-Aug., pp. 30-33.

WALTON, GRAHAM, 1960, *ABS Contamination*, J. Am. Water Works Assoc., v. 52, p. 1354, Nov.

WALTZ, J. P., 1971, *Methods of Geologic Evaluation of Pollution Potential at Mountain Home Sites*, Proc. National Ground Water Quality Symp., U. S. Environmental Protection Agency Water Pollution Control Research Series, 16060 GRB 08/71, p. 136-143.

WALTZ, J. P., 1972, *Methods of Geologic Evaluation of Pollution Potential at Mountain Homesites*, Ground Water, v. 10, no. 1, pp. 42-49, Jan.-Feb.

WEIST, W. G., and PETTIJOHN, R. A., 1975, *Investigation Ground-Water Pollution from Indianapolis Landfills — The Lessons Learned*, Ground Water, v. 13, no. 2, pp. 191-196.

WILSON, L. G., CLARK, W. L., III, and SMALL, G. C., 1973, *Subsurface Transformations During the Initiation of a New Stabilization Lagoon*, Water Resources Bull., v. 9, no. 9, pp. 243-257.

WILSON, L. G., 1974, *Quality Transformation in Recharged River Water During Possible Interactions with Landfill Deposits Along the Santa Cruz River*, Annual Rept. to Pima County Dept. of Sanitation, Water Resources Res. Center, Univ. of Arizona.

YATES, M. V., 1985, *Septic Tank Density and Ground-Water Contamination*, Ground Water, v. 23, no. 5, pp. 586-591.

ZANONI, A. E., 1972, *Ground Water Pollution and Sanitary Landfills —
A Critical Review*, Ground Water, v. 10, no. 1, pp. 3-16.

ZENONE, C., DONALDSON, D. E., and GRUNWALDT, J. J., 1975,
*Groundwater Quality Beneath Solid-Waste Disposal Sites at
Anchorage, Alaska*, Ground Water, v. 13, no. 2, pp. 182-190.

INDUSTRIAL

ABRIOLA, L. M., and PINDER, G. F., 1982, *Migration of Petroleum
Products Accidentally Introduced into the Subsurface Through Spills*,
Princeton Univ., NJ, DOE/EV/10257-1, NTIS, DE82008877, 54 pp.

AMERICAN PETROLEUM INSTITUTE — ENGINEERING AND
TECHNICAL RESEARCH COMMITTEE, 1972, *The Migration of
Petroleum Products in Soil and Ground Water*, Pub. 4149, API,
Washington, D. C., 36 pp.

ATWATER, J. W., 1984, *A Case Study of a Chemical Spill:
Polychlorinated Biphenyls (PCB's) Revisited*, Water Resources Res.,
v. 20, no. 2.

BARNES, I., and CLARKE, F. E., 1964, *Geochemistry of Ground Water
in Mine Drainage Problems*, U. S. Geol. Survey Prof. Paper 473-A,
6 pp.

BRAIDS, O. C., WILSON, G. R., and MILLER, D. W., 1977, *Effects of
Industrial Hazardous Waste Disposal on the Ground-Water Resource*,
in Pojasek, *Drinking Water Quality Enhancement Through Source
Protection*, Ann Arbor Science Publ. Inc., P. O. Box 1425, Ann
Arbor, Michigan 48106, pp. 179-207.

BROWN, MICHAEL, 1980, *Laying Waste: The Poisoning of America by
Toxic Chemicals*, Pantheon Books, 201 East 50th St., New York,
New York 10022, 335 pp.

BURROWS, W. D., 1979, *Development of Guidelines for Contaminated
Soil and Ground Water at U. S. Army Installations*, AD-A067 527/
2WP, NTIS, Springfield, Virginia 22161.

BURT, E. M., 1972, *The Use, Abuse, and Recovery of a Glacial Aquifer*,
Ground Water, v. 10, no. 1, pp. 65-72, (disposal of industrial
wastewater into disposal ponds)

CARTWRIGHT, KEROS, and HUNT, C. S., 1981, *Hydrogeologic Aspects
of Coal Mining in Illinois: An Overview*, Illinois State Geol. Survey
Environmental Geol. Note 90.

CASE, L. C., 1970, *Water Problems in Oil Production*, Petroleum Publ. Co., Tulsa, 133 pp.

COLLIER, C. R., and others, 1964, *Influences of Strip Mining on the Hydrologic Environment of Parts of Beaver Creek Basin, Kentucky, 1955-59*, U. S. Geol. Survey Prof. Paper 427-B, 83 pp.

COLLINS, A. G., 1977, *Possible Contamination of Ground Waters by Oil- and Gas-Well Drilling and Completion Fluids*, CONF-7505133, NTIS, Springfield, Virginia 22161.

COLORADO, UNIVERSITY, 1981, *Pollution of Ground Water Due to Inactive Uranium Mill Trailings, Summary of Progress, Oct. 1, 1979-Sept. 30, 1981*, DOE/ET/44206-1, NTIS, Springfield, Virginia 22161, 50 pp.

CONCAWE, 1974, *Oil Spill Clean-up Manual*, Concawe Report No. 7, van Hogenhoucklaan 60, The Hague 2018, The Netherlands.

CONCAWE, 1979, *Protection of Ground Water from Oil Pollution*, Concawe Report No. 3, van Hogenhoucklaan 60, The Hague 2018, The Netherlands.

DALLAIRE, G., 1978, *EPA's Hazardous-Waste Program: Will It Save Our Ground Water?* Civil Engineering, v. 48, no. 12, pp. 39-45.

DAVIDS, H. W., and LIEBER, MAXIM, 1951, *Underground Water Contamination by Chromium Wastes*, Water and Sewage Works, v. 98, no. 12, p. 528-534.

DEFFERDING, L. J., 1979, *State-of-the-Art of Liquid Waste Disposal for Geothermal Energy Systems: 1979*, Rept. PNL-2404, DOE/EV-0083, NTIS, Springfield, Virginia 22161.

DELL'ACQUA, R., BUSH, B., and EGAN, J., 1978, *Identification of Gasoline Contamination of Ground Water by Gas Chromatography*, J. Chromatography, v. 128, no. 2, pp. 271-280.

DEUTSCH, MORRIS, 1961, *Incidents of Chromium Contamination of Ground Water in Michigan*, Public Health Service Tech. Rept., W61-5, pp. 98-104.

EHRLICH, G. G., and others, 1979, *Chemical Changes in an Industrial Waste Liquid During Post-Injection Movement in a Limestone Aquifer, Pensacola, Florida*, Ground Water, v. 17, no. 6, Nov-Dec, pp. 562-573.

EMRICH, G. H., and MERRITT, G. L., 1969, *Effects of Mine Drainage on Ground Water*, Ground Water, v. 7, no. 3, p. 27.

EVERETT, L. G., 1979, *Ground-Water Quality Monitoring of Western Coal Strip Mining: Identification and Priority Ranking of Potential Pollution Sources*, Envir. Monitor. and Support Lab. Pub. EPA-600/ 7-79-024.

EVERETT, L. G., and HOYLMAN, E. W., 1980, *Ground-Water Quality Monitoring of Western Coal Strip Mining: Preliminary Designs for Reclaimed Mine Sources of Pollution*, PB80-203193, NTIS, Springfield, Virginia 22161.

FRIED, J. J., MUNTZER, P., and ZILLIOX, L., 1979, *Ground-Water Pollution by Transfer of Oil Hydrocarbons*, Ground Water, v. 17, no. 6, Nov-Dec, pp. 586-594.

FRYBERGER, J. S., 1975, *Investigation and Rehabilitation of a Brine-Contaminated Aquifer*, Ground Water, v. 13, no. 2, pp. 155-160.

GALBRAITH, J. H., and others, 1972, *Migration and Leaching of Metals from Old Mine Tailings Deposits*, Ground Water, v. 10, no. 3, pp. 33-44, May-Jun.

GERAGHTY & MILLER, INC., 1977, *The Prevalence of Subsurface Migration of Hazardous Chemical Substances at Selected Industrial Waste Land Disposal Sites*, U. S. Environmental Protection Agency, Solid Waste Management Series, Rept. EPA/530/SW-634, 166 pp.

GRUBB, H. F., 1970, *Effects of a Concentrated Acid on Water Chemistry and Water Use in a Pleistocene Outwash Aquifer*, Ground Water, v. 8, no. 5, pp. 4-8.

HALL, P. L., and QUAM, H., 1976, *Countermeasures to Control Oil Spills in Western Canada*, Ground Water, v. 14, no. 3, pp. 163-169.

HAM, R. K., and others, 1979, *Comparison of Three Leaching Tests* Ind. Envir. Research Lab., Cincinnati, Ohio, EPA-600/2-79-071. (industrial wastes)

HARRISON, S. S., 1985, *Contamination of Aquifers by Overpressuring the Annulus of Oil and Gas Wells*, Ground Water, v. 23, no. 3, pp. 317-324.

HOLZER, T. L., 1976, *Application of Ground-Water Flow Theory to a Subsurface Oil Spill*, Ground Water, v. 14, no. 3, pp. 138-145.

HUMENICK, M. J., TURK, L. J., and COLCHIN, M. P., 1980, *Methodology for Monitoring Ground Water at Uranium Solution Mines*, Ground Water, v. 18, no. 3, May-Jun., pp. 262-273.

HUMENICK, M. J., CHARBENEAU, R. J., and HASSLER, B., 1984, *Migration of Heavy Elements in Ground Water Following Uranium Solution Mining Operations*, Wyoming Univ., PB85-226314/WEP, NTIS, Springfield, Virginia 22161, 75 pp.

HUNTER, BLAIR, A., 1980, *Ground Water Pollution by Oil Products*, J. Institution of Water Engrs. and Scientists, v. 34, no. 6, pp. 557-570.

INTERNATIONAL TECHNICAL INFORMATION INSTITUTE, 1975, *Toxic and Hazardous Industrial Chemicals Safety Manual for Handling and Disposal with Toxicity and Hazard Data*, Toranomon-Tachikawa Bldg., 65, 1 Chome, Nishi-Shimbashi, Minato-ku, Tokyo, Japan.

JOHNSON, R. L., and others, 1985, *Migration of Chlorophenolic Compounds at the Chemical Waste Disposal Site at Alkali Lake, Oregon - 2. Contaminant Distribution, Transport, and Retardation*, Groundwater, v. 23, no. 5, pp. 652-666.

KAUFMAN, W. J., 1974, *Chemical Pollution of Ground Waters*, J. Amer. Water Works Assoc., v. 66, pp. 152-159.

KAUFMANN, R. F., and others, 1976, *Effects of Uranium Mining and Milling on Ground Water in the Grants Mineral Belt, New Mexico*, Ground Water, v. 14, no. 5, pp. 296-308.

KNOWLES, D. B., 1965, *Hydrologic Aspects of the Disposal of Oil-Field Brines in Alabama*, Ground Water, v. 3, no. 2, pp. 22-27.

KRIEGER, R. A., and HENDRICKSON, G. E., 1960, *Effects of Greensburg Oilfield Brines on the Streams, Wells, and Springs of the Upper Green River Bains, Kentucky*, Kentucky Geol. Survey Rept. of Inv. no. 2, Ser. X, 36 pp.

LIEBER, MAXIM, and others, 1964, *Cadium and Hexavalent Chromium in Nassau County Ground Water*, J. Am. Water Works Assoc., v. 56, no. 6, pp. 739-747.

MAST, V. A., 1985, *The Use of Ionic Mixing Curves in Differentiating Oil Field Brine from Natural Brine in a Fresh Water Aquifer*, Ground Water Monitoring Rev., v. 5, no. 3, pp. 65-70.

MATIS, J. R., 1971, *Petroleum Contaminaion of Ground Water in Maryland*, Ground Water, v. 9, no. 6, pp. 57-61.

McKEE, J. E., and others, 1972, *Gasoline in Groundwater*, J. Water Poll. Control Fed., v. 44, no. 2, pp. 293-302, Feb.

McMILLION, L. G., 1965, *Hydrologic Aspects of Disposal of Oil-Field Brines in Texas*, Ground Water, v. 3, no. 4, pp. 36-42.

McWHORTER, D. B., and NELSON, J. D., 1980, *Seepage in the Partially Saturated Zone Beneath Tailings Impoundments*, Mining Engineering, v. 32, no. 4, Apr.

MEAD, W. and RABER, E., 1980, *Environmental Controls for Underground Coal Gasification: Ground-Water Effects and Control Technologies*, UCRL-84075, NTIS, Springfield, Virginia 22161.

METRY, A. A., and others, 1980, *Handbook of Hazardous Waste Management*, Technomic Publishing Co., Inc., 265 Post Rd. West, Westport, Connecticut 06880, 400 pp.

MILLER, J. A., and others, 1978, *Impact of Potential Phosphate Mining on the Hydrology of Osceola National Forest*, Florida, U. S. Geol. Survey Water Resources Invest. 78-6, 169 pp.

MILLIGAN, J. D., and RUANE, R. J., 1980, *Effects of Coal-Ash Leachate on Ground Water Quality*, PB81-178535, NTIS, Springfield, Virginia 22161, 128 pp.

MINK, L. L., and others, 1972, *Effect of Early Day Mining Operations on Present Day Water Quality*, Ground Water, v. 10, no. 1, pp. 17-26, Jan.-Feb.

MURPHY, E. C., and KEHEW, A. E., 1984, *Effect of Oil and Gas Well Drilling Fluids on Shallow Groundwater in Western North Dakota*, North Dakota Geol. Survey, 165 pp.; DE85900622/WEP, NTIS, Springfield, Virginia 22161.

NATIONAL FIRE PROTECTION ASSOC., 1972, *Hazardous Chemicals Data 1972*, NFPA No. 49, 60 Batterymarch St., Boston, Massachusetts 02110, 261 pp.

NATIONAL RESEARCH COUNCIL, 1981, *Coal Mining and Ground-Water Resources in the United States*, National Academy Press, Washington, D. C. 20418, 197 pp.

NATIONAL WATER WELL ASSOCIATION, 1985, *Petroleum Hydrocarbons and Organic Chemicals in Ground Water*, Proc. NWWA/API Conf., Nov., 1984, Houston, Texas; NWWA, Dublin, Ohio 43017, 519 pp.

OSGOOD, J. O., 1974, *Hydrocarbon Dispersion in Ground Water: Significance and Characteristics*, Ground Water, v. 12, no. 6, pp. 427-438.

Contamination

PARIZEK, R. R., 1971, *Prevention of Coal Mine Drainage Formation by Well Dewatering*, Special Research Rept. No. SR-82, Dept. of Environmental Resources, Commonwealth of Pennsylvania.

PERLMUTTER, N. M., and LIEBER, MAXIM, 1970, *Dispersal of Plating Wastes and Sewage Contaminants in Ground Water and Surface Water South Farmingale-Massapequa Area, Nassau County, New York*, U. S. Geol. Survey Water-Supply Paper 1879-G.

PERLMUTTER, N.M., and others, 1963, *Movement of Waterborne Cadmium and Hexavalent Chromium Wastes in South Farmingdale, Nassau County, Long Island, New York*, U. S. Geol. Survey Prof. Paper 475-C, pp. 179-184.

PETTYJOHN, W. A., 1972, *Water Pollution by Oil-Field Brines and Related Industrial Wastes in Ohio*, in *Water Quality in a Stressed Environment*, Burgess Publ. Co. Minneapolis, Minnesota.

PETTYJOHN, W. A., 1975, *Pickling Liquors, Strip Mines and Ground-Water Pollution*, Ground Water, v. 13, no. 1, pp. 4-10.

PFANNKUCH, H. O., 1983, *Hydrocarbon Spills, Their Retention in the Subsurface and Propagation into Shallow Aquifers*, PB83-196477, NTIS, 5285 Port Royal Rd., Springfield, Virginia 22161, 55 pp.

PITT, W. A. J., Jr., MEYER, F. W., and HULL, J. E., 1977, *Disposal of Salt Water During Well Construction: Problems and Solutions*, Ground Water, v. 15, no. 4, pp. 276-283.

PURTYMUN, W. D., 1977, *Hydrologic Characteristics of the Los Alamos Well Field with Reference to the Occurrence of Arsenic in Well LA-6*, Los Alamos Science Lab. Rept. LA-7212-MS, 63 pp.

RABER, E., and STONE, R., 1980, *Ground-Water Hydrologic Effects Resulting from Underground Coal Gasification Experiments at the Hoe Creek Site Near Gillette, Wyoming*, Interim Report October 1979-March 1980, UCID-18627, NTIS, Springfield, Virginia 22161, 63 pp.

RABER, E., and others, 1982, *Cleanup of Groundwater Contaminated by Underground Coal Gasification*, DE82-005824 NTIS, Springfield, Virginia 22161, 22 pp.

ROBERTSON, F. N., 1975, *Hexavalent Chromium in the Ground Water in Paradise Valley, Arizona*, Ground Water, v. 13, no. 6, pp. 516-527.

ROBINSON, J. S., 1983, (ed.), *Hazardous Chemical Spill Cleanup*, Pollution Technology Review No. 59, Noyes Data Corp., Park Ridge, New Jersey 07656, 406 pp.

ROUX, P. H., and ALTHOFF, W. F., 1980, *Investigation of Organic Contamination of Ground Water in South Brunswick Township, New Jersey*, Ground Water, v. 18, no. 5, Sept.-Oct., pp. 464-471.

SCHMIDT, K. D., and others, 1981, *Brine Pollution at Fresno — Twenty-Six Years Later*, Ground Water, v. 19, no. 1, Jan.-Feb., pp. 12-19.

SCHUBERT, J. P., 1979, *Groundwater Contamination Problems Resulting from Coal Refuse Disposal*, CONF-7905106-1, NTIS, Springfield, Virginia 22161.

SGAMBAT, J. P., LaBELLA, E. A., and ROEBUCK, SHEILA, 1980, *Effects of Underground Coal Mining on Ground Water in the Eastern United States*, Geraghty & Miller, Inc., Annapolis, Maryland, PB80-216757, NTIS, Springfield, Virginia 22161, 201 pp.

SITTIG, M., 1980, *Landfill Disposal of Hazardous Wastes and Sludges*, Noyes Data Corp., Mill Rd. at Grand Ave., Park Ridge, New Jersey 07656.

SLAWSON, G. C., Jr., (ed.), 1979, *Ground-Water Quality Monitoring of Western Oil Shale Development: Identification and Priority Ranking of Potential Pollution Sources*, USEPA Interagency Energy-Environment Res. and Dev. Program Rept. EPA-600/7-79-023 NTIS, Springfield, Virginia 22161.

SLAWSON, G. C., Jr., 1980, *Ground-Water Quality Monitoring of Western Oil Shale Development: Monitoring Program Development*, PB80-203219, NTIS, Springfield, Virginia 22161.

STEIN, R. B., NOYES, J. A., 1981, *Field Report — Ground-Water Contamination Potential at 21 Industrial Waste-Water Impoundments in Ohio*, Ground Water, v. 19, no. 1, Jan.-Feb., pp. 70-80.

SUMMERS, K., GHERINI, S., and CHEN, C., 1981, *Methodology to Evaluate the Potential for Ground Water Contamination from Geothermal Fluid Releases*, PB81-111114, NTIS, Springfield, Virginia 22161, 178 pp.

SWEET, H. R., and FETROW, R. H., 1975, *Ground Water Pollution by Wood Waste Disposal*, Ground Water, v. 13, no. 2, pp. 227-231.

Contamination

THEIS, T. L., and MARLEY, J. J., 1975, *Contamination of Ground Water by Heavy Metals from the Land Disposal of Fly Ash*, TID-26973, NTIS, Springfield, Virginia 22161.

THOMPSON, W. E., and others, 1978, *Ground-Water Elements of In Situ Leach Mining of Uranium*, NUREG/CR-0311, NTIS, Springfield, Virginia 22161.

TRACY, J. V., and DION, N. P., 1975, *Evaluation of Ground-Water Contamination From Cleaning Explosive-Projectile Casings at the Bangor Annex, Kitsap County, Wash., Phase II*, U. S. Geol. Survey Water-Resources Inv. 62-75.

VAN DAM, J., 1967, *The Migration of Hydrocarbons in a Water-Bearing Stratum*, P. Hepple, Institute of Petroleum, London.

VAN DER LEEDEN, FRITS, BRAIDS, O. C., and FLEISHELL, J. L., 1980, *The Brooklyn Oil Spill*, Proc. May 1980 Nat. Conf. on Control of Hazardous Material Spills, Louisville, Kentucky, pp. 245-249.

VILLAUME, J. E., 1985, *Investigations of Sites Contaminated with Dense, Non-Aqueous Phase Liquids (NAPLS)*, Ground Water Monitoring Rev., v. 5, no. 2, pp. 28-36.

WALKER, T. R., 1961, *Ground Water Contamination in the Rocky Mountain Arsenal Area, Denver, Colorado*, Geol. Soc. Am. Bull., v. 72, no. 3, pp. 489-494.

WALKER, W. H., 1974, *Monitoring Toxic Chemical Pollution From Land Disposal Sites in Humid Regions*, Ground Water, v. 12, no. 4, pp. 213-218.

WALTERS, E. A., and NIEMCZYK, T. M., 1984, *Effect of Underground Coal Gassification on Groundwater*, New Mexico Univ., Albuquerque, EPA-600/2-84-123, 118 pp.

WALTON, GRAHAM, 1961, *Public Health Aspects of the Contamination of Ground Water in the Vicinity of Derby, Colorado*, Proc. Symp. Ground Water Contamination, Robert A. Taft San. Eng. Center, Tech. Rept. W61-5, Cincinnati, Ohio.

WALZ, D. H., and CHESTNUT, K. T., 1977, *Land Disposal of Hazardous Wastes: an Example from Hopewell, Virginia*, Ground Water, v. 15, no. 1, pp. 75-79.

WILLIAMS, D. E., and WILDER, D. G., 1971, *Gasoline Pollution of a Ground-Water Reservoir — A Case History*, Ground Water, v. 9, no. 6, pp. 50-56, Nov.-Dec.

WILLIAMS, R. E., 1975, *Waste Production in Mining, Milling, and Metallurgical Industries*, Miller Freeman Publications, Inc., San Francisco, California.

WILSON, J. L., and HAMILTON, D. A., 1978, *Influence of Strip Mines on Regional Ground-Water Flow*, J. Hydraul. Div. Am. Soc. Civ. Eng., v. 104 (HY9), pp. 1213-1223.

YANG, J. T. and BYE, W. E., 1982, *Protection of Ground Water Resources from the Effects of Accidental Spills of Hydrocarbons and Other Hazardous Substances (Guidance Document)*, PB82-204900, NTIS, Springfield, Virginia 22161, 166 pp.

MONITORING

ANONYMOUS, 1985, *Groundwater Pollution Monitoring*, 1976-January 1985, (218 Citations from the Energy Data Base), PB85-853752/WEP, NTIS, Springfield, Virginia 22161, 217 pp.

BRANNAKA, LARRY, and KEEFE, LAWRENCE, 1986, *Micro Computers Applied to Ground Water Monitoring and Testing*, Ground Water Monitoring Rev., v. 6, no. 2, pp. 135-140.

BRYDEN, G.W., MABEY, W.R., and ROBINE, K.M., 1986, *Sampling for Toxic Contaminants in Ground Water*, Ground Water Monitoring Rev., v. 6, no. 2, pp. 67-73.

CLARK, T. P., 1975, *Survey of Ground-Water Protection Methods for Illinois Landfills*, Ground Water, v. 13, no. 4, pp. 321-331.

CLARK, T. P., and SABEL, G. V., 1980, *Requirements of State Regulatory Agencies for Monitoring Ground-Water Quality at Waste Disposal Sites*, Ground Water, v. 18, no. 2, Mar.-Apr., pp. 168-174.

CROUCH, R. L., ECKERT, R. D., and RUGG, D. D., 1976, *Monitoring Groundwater Quality: Economic Framework and Principles*, U. S. Environmental Protection Agency, Las Vegas, Nevada.

DUNLAP, W. J., and others, 1977, *Sampling for Organic Chemicals and Microorganisms in the Subsurface*, EPA-600/2-77-176, U. S. Environmental Protection Agency, Ada, Oklahoma, 27 pp.

EVERETT, L. G., 1981, *Monitoring in the Vadose Zone*, Ground Water Monitoring Review, v. 1, no. 2, pp. 44-51.

EVERETT, L. G., 1981, *Monitoring in the Zone of Saturation*, Ground Water Monitoring Review, v. 1, no. 1, pp. 38-41.

EVERETT, L.G., 1985, *Groundwater Monitoring Handbook for Coal and Oil Shale Development*, Kaman Tempo, 816 State St., Santa Barbara, California 93102, 310 pp.

EVERETT, L. G., and HOYLMAN, E. W., (eds.), 1980, *Groundwater Quality Monitoring of Western Coal Strip Mining: Preliminary Designs for Active Mine Sources of Pollution*, EPA 600/7-80-110, NTIS, Springfield, Virginia, 22161, 104 pp.

EVERETT, L. G., SCHMIDT, K. D., and TINLIN, R. M., *Monitoring Groundwater Quality: Methods and Costs*, Rept. EOA-600/4-76-023, U. S. Environmental Protection Agency, Las Vegas, Nevada, 140 pp.

EVERETT, L.G., WILSON, L.G., and HOYLMAN, E.W., 1984, *Vadose Zone Monitoring for Hazardous Waste Sites*, PB84-212752, NTIS, 5285 Port Royal Rd., Springfield, Virginia 22161, 379 pp.

EVERETT, L. G., WILSON, L. G., and McMILLION, L. G., 1982, *Vadose Zone Monitoring Concepts for Hazardous Waste Sites*, Ground Water, v. 20, no. 3, pp. 312-324.

FENN, DENNIS, and others, 1977, *Procedures Manual for Ground-Water Monitoring at Solid Waste Disposal Facilities*, EPA/530/SW-611, U. S. Environmental Protection Agency, Cincinnati, Ohio 45268.

FISHBAUGH, TIMOTHY, 1985, *Monitoring in the Vadose Zone and Saturated Zones Utilizing Fluoroplastic*, Ground Water Monitoring Rev., v. 4, no. 4, pp. 183-187.

GIBB, J. P., SCHULLER, R. M., and GRIFFIN, R. A., 1981, *Procedures for the Collection of Representative Water Quality Data from Monitoring Wells*, Coop. Groundwater Rept. 7, Illinois State Water Survey, Champaign, Illinois 61820, 61 pp.

GRAVES, L. S., 1981, *Ground-Water Monitoring Requirements of RCRA*, Ground Water Monitoring Review, v. 1, no. 1, pp. 34-36.

HIRSCHFELD, T., and others, 1984, *Feasibility of Using Fiber Optics for Monitoring Groundwater Contaminants*, PB84-201607, NTIS, 5285 Port Royal Rd., Springfield, Virginia 22161, 92 pp.

HUMENICK, M. J., TURK, L. J., and COLCHIN, M. P., 1978, *Sampling of Ground Water-Baseline and Monitoring Data In-Situ Processes*, Center for Research in Water Resources Tech. Rept. CRWR-157, 10100 Burnet Rd., Austin, Texas 78758.

JOHNSON, T. M., and CARTWRIGHT, KEROS, 1980, *Monitoring of Leachate Migration in the Unsaturated Zone in the Vicinity of Sanitary Landfills*, Circ. 514, Illinois State Geol. Survey, Champaign, Illinois 61820, 82 pp.

KAZMANN, R. G., 1981, *An Introduction to Ground-Water Monitoring*, Ground Water Monitoring Review, v. 1, no. 1, pp. 28-29.

KETELLE, R. H., and TRIEGEL, E. K., 1981, *Interpreting the Factors Related to Groundwater Impact Assessment of Coal Conversion Solid Wastes*, CONF-800957-1, NTIS, Springfield, Virginia 22161, 15 pp.

KLEIN, W. L., DUNSMORE, D. A., and HORTON, R. K., 1968, *An Integrated Monitoring System for Water Quality Management in the Ohio Valley*, Env. Science and Technology, v. 2, Am. Chemical Society, pp. 764-771.

LOBASSO, THOMAS., and BARBER, A.J., 1982, *A Monitoring and Removal Plan for Leaked Propane in the Vadose Zone*, Proc. Symp. on Characterization and Monitoring of Vadose Zone, Dec. 1982, National Water Well Assoc., Dublin, Ohio.

McBEAN, E.A., and ROVERS, F.A., 1984, *Alternatives for Assessing Significance of Changes in Concentration Levels*, Ground Water Monitoring Rev., v. 4, no. 3, pp. 39-41.

McBEAN, E.A., and ROVERS, F.A., 1985, *Analysis of Variances as Determined from Replicates vs. Successive Sampling*, Ground Water Monitoring Rev., v. 5, no. 3, pp. 61-65.

MILLER, D. W., 1981, *Guidelines for Developing a Statewide Ground-Water Monitoring Program*, Ground Water Monitoring Review, v. 1, no. 1, pp. 32-33.

MORRISON, R.D., 1983 *Ground Water Monitoring Technology: Procedures, Equipment and Applications*, Timco Mfg., Inc.

NELSON, R.W., 1982, *Subsurface Hydrologic Monitoring to Evaluate Contaminant Migration: Requirements and Solutions*, DE82003849, NTIS, 5285 Port Royal Rd., Springfield, Virginia 22161, 18 pp.

NIELSEN, D.M., and YEATES, G.L., 1985, *Comparison of Sampling Mechanisms Available for Small-Diameter Ground Water Monitoring Wells*, Ground Water Monitoring Rev., v. 5, no. 2, pp. 83-99.

NORMAN, W.R., 1986, *An Effective and Inexpensive Gas-Drive Ground Water Sampler*, Ground Water Monitoring Rev., v. 6, no. 2, pp. 56-60.

PFANNKUCH, H. O., 1982, *Problems of Monitoring Network Design to Detect Unanticipated Contamination*, Ground Water Monitoring Review, v. 2, no. 1, p. 67-75.

PFANNKUCH, H. O., and LABNO, B. A., 1976, *Design and Optimization of Ground-Water Monitoring Networks for Pollution Studies*, Ground Water, v. 14, no. 6, pp. 455-462.

POWELL, W. J., and others, 1973, *Water Resources Monitoring in Alabama*, Geol. Survey of Alabama, Info. Ser. 44.

QUINLAN, J.F., and EWERS, R.O., 1985, *Ground Water Flow in Limestone Terranes: Strategy, Rationale, and Procedure for Reliable, Efficient Monitoring of Ground Water Quality in Karst Areas*, Proc. 5th Nat. Symp. on Aquifer Restoration, Columbus, OH, pp. 197-234.

ROVERS, F. A., and McBEAN, E. A., 1981, *Significance Testing for Impact Evaluation* (student t-test), Ground Water Monitoring Review, v. 1, no. 2, pp. 39-43.

SCHMIDT, K. D., 1975, *Monitoring Groundwater Pollution*, Article 9-4 in Proc. Internat. Conf. on Environmental Sensing and Assessment, v. 2, Las Vegas, Nevada, sponsored by U. S. Environmental Protection Agency, Las Vegas, Nevada.

SGAMBAT, J. P., PORTER, K. S., and MILLER, D. W., 1978, *Regional Ground-Water Quality Monitoring*, Proc. Symp. Am. Water Resources Assoc., June, 1978, San Francisco, California, pp. 181-195.

SGAMBAT, J. P., and STEDINGER, J. R., 1981, *Confidence in Ground-Water Monitoring*, Ground Water Monitoring Review, v. 1, no. 1, pp. 62-69.

SLAWSON, G. C., 1979, *Groundwater Quality of Monitoring of Western Oil Shale Development: Identification and Priority Ranking of Potential Pollution Sources*, U. S. Environmental Protection Agency, EPA-600/7-79-023.

THEIS, T. L., 1979, *Contamination of Groundwater by Heavy Metals from the Land Disposal of Fly Ash, Final Report*, Notre Dame Univ., IN., Dept. of Civil Engrg., NTIS, C00-2727-7, 82 pp.

TINLIN, R. M., and SCHMIDT, K. D., 1976, *Monitoring Ground Water Quality: Illustrative Examples*, U. S. Environmental Protection Agency, Environmental Monitoring Series, EPA 600/4-76-036, Las Vegas, Nevada, 92 pp.

TODD, D. K., and others, 1976, *Monitoring Ground Water Quality: Monitoring Methodology*, U. S. Environmental Protection Agency, Environmental Monitoring Series, EPA-600/4-76-026, Las Vegas, also J. Am. Water Works Assoc., v. 68, Nov., pp. 586-593.

U. S. ENVIRONMENTAL PROTECTION AGENCY, 1974, *Water Quality and Pollutant Source Monitoring*, Federal Register, v. 39, no. 168, Part III.

U. S. ENVIRONMENTAL PROTECTION AGENCY, 1975, *Model State Water Monitoring Program*, EPA-440/9-74-002, Office of Water and Hazardous Materials Monitoring and Data Support Division, Washington, D. C., 36 pp.

WARNER, D. L., 1974, *Rationale and Methodology for Monitoring Ground-Water Polluted by Mining Activities*, Rept. EPA-600/4-74-003, U. S. Environmental Protection Agency, Washington, D. C., 76 pp.

WAY, SHAO-CHIN, MCKEE, C.R., and WAINWRIGHT, H.K., 1985, *A Computerized Ground Water Monitoring System*, Ground Water Monitoring Rev., v. 4, no. 1, pp. 21-25.

WILSON, L. G., 1980, *Monitoring in the Vadose Zone: A Review of Technical Elements and Methods*, General Electric Co., Santa Barbara, CA, Center for Advanced Studies, EPA-600/7-80-134, PB81-125817, NTIS, Springfield, Virginia 22161, 186 pp.

WILSON, L.G., 1983, *Monitoring in the Vadose Zone: Part III*, Ground Water Monitoring Rev., v. 3, no. 1, pp. 155-166.

WONG, J., 1977, *The Design of a System for Collecting Leachate from a Lined Landfill Site*, Water Resources Res., v. 13, no. 2, pp. 404-410.

YARE, B. S., 1975, *The Use of a Specialized Drilling and Ground-Water Sampling Technique for Delineation of Hexavalent Chromium Contamination in an Unconfined Aquifer, Southern New Jersey Coastal Plain*, Ground Water, v. 13, no. 2, pp. 151-154.

REMEDIAL MEASURES/TREATMENT/LINERS/ CLOSURE

ALLEN, C.C., and BLANEY, B.L., 1985, *Techniques for Treating Hazardous Wastes to Remove Volatile Organic Constituents*, Research Triangle Inst., Research Triangle Park, NC, 24 pp,; EPA/ 600/D-85/127, PB85-218782/WEP NTIS, Springfield, Virginia 22161.

AMERICAN PETROLEUM INSTITUTE, 1985, *Protecting Groundwater: What We've Learned Through Research*, API, Washington, DC, 15 pp.

ANONYMOUS, 1984, *Remedial Response at Hazardous Waste Sites*, Environmental Law Inst., Washington, D.C., PB85-12721/WEP, NTIS, Springfield, Virginia 22161, 99 pp. (nationwide survey of 395 uncontrolled hazardous sites).

ANONYMOUS, 1985, *Hydraulic Barriers in Soil and Rock*, Am. Soc. Testing and Materials, Spec. Tech. Publ. 874 Philadelphia, Pennsylvania 19103, 329 pp. (21 papers).

ANONYMOUS, 1985, *Liners for Waste Disposal and Waste Storage Facilities, 1976-April 1985*, (84 Citations from the Energy Data Base), PB85-858132/WEP, NTIS, Springfield, Virginia 22161, 86 pp.

ATWOOD, D.F. and GORELICK, S.M., 1985, *Hydraulic Gradient Control for Groundwater Contaminant Removal*, J. Hydrology, v. 76, no. 1/2.

BAETSLE, L. H., and SOUFFRIAU, J., 1967, *Installation of Chemical Barriers in Aquifers and Their Significance in Accidental Contamination*, in *Disposal of Radioactive Wastes into the Ground*, Internat. Atomic Energy Agency Conf., Vienna, May 29-June 2.

BAETSLE, L. H., and others, 1968, *Remedial Actions in Case of Groundwater Contamination of Sandy Aquifers* — Final Report, Centre d'Etude de l'Energie Nucleaire, Rept. EUR-4095, Mol, Belgium; also publ. in Nuclear Sci. Abs., v. 23, 1969.

BOWDERS, J.J., Jr., 1985, *The Influence of Various Concentrations of Organic Liquids on the Hydraulic Conductivity of Compacted Clay*, Geotech. Eng. Center, Univ. of Texas at Austin, GT85-2.

BROWN, K.W., and ANDERSON, D.C., 1983, *Effects of Organic Solvents on the Permeability of Clay Soils*, USEPA, Rept. EPA-600/ S283-016 MERL, Cincinnati, OH.

CLARK, R.M., and EILERS, R.G., 1983, *Treatment for the Control of Organic Chemical Contamination of Drinking Water: Cost and Performance*, Municipal Environmental Research Lab., Cincinnati, OH, EPA-600/D-83-060, NTIS, PB83-219592, 38 pp.

CLARK, R.M., EILERS, R.G., and GOODRICH, J.A., 1984, *VOC'S (Volatile Organic Chemicals) in Drinking Water: Cost of Removal*, EPA, Cincinnati, OH, J. Environ. Engrg. Div., ASCE, v. 110, no. 6, pp. 1146-1162; also PB85-166429/WEP, NTIS, Springfield, Virginia 22161, 18 pp.

COCHRAN, R., and HODGE, V., 1985, *Guidance on Remedial Investigations Under CERCLA*, PB85-238616/WEP, NTIS, Springfield, Virginia 22161, 172 pp.

DANIEL, D.E., and LILJESTRAND, H.M., 1984, *Effects of Landfill Leachates on Natural Liner Systems*, Geotech. Eng, Center, Univ. of Texas at Austin, GR83-6.

DAY, S.R., 1985, *A Field Permeability Test for Compacted Clay Liners*, Geotech. Eng. Center, Univ. of Texas at Austin, 6T84-1.

DAKESSIAN, S., FONG, M., and WHITE, R., 1981, *Lining of Waste Impoundment and Disposal Facilities*, PB81-166365, NTIS, Springfield, Virginia 22161, 411 pp.

ENGINEERING and RESEARCH FOUNDATION, 1981, *Reference Handbook for Hazardous Waste Management*, Engineering and Science Research Foundation, 600 Bancroft Way, Berkeley, California 94710, 290 pp.

FOREMAN, D.E., 1984, *The Effects of Hydraulic Gradient and Concentrated Organic Chemicals on the Hydraulic Conductivity of Compacted Clay*, Geotech. Eng. Center, Univ. of Texas at Austin, 6T84-2.

FULLER, W. H., 1981, *Liners of Natural Porous Materials to Minimize Pollutant Migration*, PB81-221863, NTIS, Springfield, Virginia 22161.

FUNG, R., (ed.), 1980, *Protective Barriers for Containment of Toxic Materials*, Noyes Data Corp., Park Ridge, New Jersey 07656, 288 pp.

Contamination

GALEGAR, J., 1984, *Annotated Literature References on Land Treatment of Hazardous Waste*, East Central Oklahoma State Univ., Ada, OK, PB84-195270, NTIS, Springfield, Virginia 22161, 475 pp.

GESWEIN, A. J., 1975, *Liners for Land Disposal Sites*, U. S. Environmental Protection Agency, Publ. SW-137, Washington, D. C.

GIDDINGS, TODD, 1982, *The Utilization of a Ground Water Dam*, Ground Water Monitoring Rev., v. 2, no. 4, pp. 26-28.

GORELICK, S. M. and WAGNER, B. J., 1986, *Evaluating Strategies for Ground-Water Contaminant Plume Stabilization and Removal*, U. S. Geol. Survey Water-Supply Paper 2290, pp. 81-89.

GREEN, W. J., and others, 1983, *Interaction of Clay Soils With Water and Organic Solvents: Implications for the Disposal of Hazardous Wastes*, Env. Science and Technology, v. 17, pp. 278.

GREEN, W. J., LEE, G. F., and JONES, R. A., 1981, *Impact of Organic Solvents on the Integrity of Clay Liners for Industrial Waste Disposal Pits: Implications for Groundwater Contamination*, PB81-213423, NTIS, Springfield, Virginia 22161, 149 pp.

HAXO, H. E., Jr., 1982, *Effects on Liner Materials of Long-Term Exposure in Waste Environments*, Matrecon, Inc., Oakland, California, 21 p., included in *Land Disposal of Hazardous Waste*, pp. 191-211, NTIS, PB82-173162 (order as PB82-173022).

HUIBREGTSE, K. R., and KASTMAN, K. H., 1981, *Development of a System to Protect Groundwater Threatened by Hazardous Spills on Land*, PB81-209587, NTIS, Springfield, Virginia 22161, 143 pp. (grout injection)

JENNINGS WALLER, MURIEL, and DAVIS, J. L., 1982, *Assessment of Techniques to Detect Landfill Liner Failings*, EarthTech Research Corp., Baltimore, Maryland, 11 p.; included in *Land Disposal of Hazardous Waste*, pp. 239-249, NTIS, PB 82-173196 (order as PB 82-173022).

KAYS, W. B., 1977, *Construction of Linings for Reservoirs, Tanks, and Pollution Control Facilities*, John Wiley & Sons, New York, New York 10016.

KEELY, J.F., 1984, *Optimizing Pumping Strategies for Contaminant Studies and Remedial Actions*, Ground Water Monitoring Rev., v. 4, no. 3, pp. 63-74.

KNOX, R.C., and others, 1984, *State-of-the Art of Aquifer Restoration*, Volume 1, Sections 1 through 8, PB85-181071/WEP, 399 pp; and Volume 2, Appendices A-G, PB85-181089/WEP, NTIS, Springfield, Virginia 22161.

LANE, L.J., and NYHAN, J.W., 1984, *Water and Contaminant Movement: Migration Barriers*, Los Alamos Natl. Lab., NM, DE85005331/WEP, NTIS, Springfield, Virginia 22161, 21 pp.

LINDORFF, D. E., and CARTWRIGHT, KEROS, 1978, *Ground Water Contamination: Problems and Remedial Actions*, (116 Case Histories), Bull. EGN 81, Information Clearing House, Commission on Rural Water, Suite 2026, 221 North LaSalle St., Chicago, Illinois 60601 also Illinois Geol. Survey, Urbana, Illinois 61801, 58 pp.

LUTTON, R. J., 1980, *Evaluating Cover System for Solid and Hazardous Waste*, U. S. Army Engineer Waterways Experiment Station, Vicksburg, Mississippi, No. EPA-1AG-D7-01097.

LUTTON, R. J., 1981, *Evaluating Cover Systems for Solid and Hazardous Waste*, PB81-166340, NTIS, Springfield, Virginia 22161, 68 pp.

LUTTON, R. J., REGAN, G. L., and JONES, L. W., 1980, *Design and Construction of Covers for Solid Waste Landfills*, PB80-100381, NTIS, Springfield, Virginia 22161.

MATRECON, INC., 1980, *Lining of Waste Impoundment and Disposal Facilities*, Oakland, California.

McKOWN, G.L., 1984,*Location of Volatile Buried Wastes by Field Portable Instrumentation*, Battelle Pacific NW Labs., Richland, WA, AD-P004 147/5/WEP, NTIS, Springfield, Virginia 22161, 7 pp.

MOORE, C. A., 1980, *Landfill and Surface Impoundment Performance Evaluation Manual*, SW-869 (draft) U. S. Environmental Protection Agency, Cincinnati, Ohio. McKOWN, G.L., 1984, *Location of Volatile Buried Wastes by Field Portable Instrumentation*, Battelle Pacific NW Labs., Richland, WA, AD-P004 147/5/WEP, NTIS, Springfield, Virginia 22161, 7 pp.

MOORE, C. A., and ROULIER, MICHAEL, 1982, *Evaluating Landfill Containment Capability*, Geotechnics, Inc., Columbus, Ohio, 14 pp., included in *Land Disposal of Hazardous Waste*, pp. 53-66, NTIS, PB82-173063 (order as PB82-173022).

NATIONAL WATER WELL ASSOCIATION, 1985, *Proceedings of the 5th National Symposium and Exposition on Aquifer Restoration and Ground Water Monitoring*, May, 1985, Columbus, Ohio; NWWA, Dublin, Ohio 43017, 734 pp.

OHNECK, R.J., and GARDNER, G.L., 1982, *Restoration of an Aquifer Contaminated by a Accidental Spill of Organic Chemicals*, Ground Water Monitoring Rev., v. 2, no. 4, pp. 50-53.

QUINCE, J.R., and GARDNER, G.L., 1982, *Recovery and Treatment of Contaminated Ground Water: Part II*, Ground Water Monitoring Rev., v. 2, no. 4, pp. 18-25.

ROGOSHEWSKI, P., BRYSON, H., and WAGNER, K., 1984, *Remedial Action Technology for Waste Disposal Sites*, Pollution Tech. Review No. 101, Noyes Data Corp., Mill Rd. at Grand Ave., Park Ridge, New Jersey 07656, 497 pp.

SALVATO, J. A., and others, 1971, *Sanitary Landfill — Leaching Prevention and Control*, J. Water Poll. Control Fed., v. 43, no. 10, pp. 2084-2100.

SHAFER, R., and others, 1984, *Landfill Liners and Covers: Properties and Application to Army Landfills*, Construction Engineering Lab., (Army), Champaign, IL, AD-A144003/1, NTIS, Springfield, Virginia 22161, 66 pp.

SHAREFKIN, MARK, SHECHTER, MORDECHAI, and KNEESE, ALLEN, 1984, *Impacts, Costs, and Techniques for Mitigation of Contaminated Groundwater: A Review*, Water Resources Res., v. 16, no. 12, pp. 1771-1784.

SHECHTER, MORDECHAI, 1985, *Economic Aspects in the Investigation of Ground-Water Contamination Episodes*, Ground Water, v. 23, no. 2, pp. 190-197.

SIMS, R. and BASS, J., 1984, *Review of In-Place Treatment Techniques for Contaminated Surface Soils*, Vol. 1. Technical Evaluation, PB85-124881/WEP, 176 pp; Vol. 2. Background Information, PB85-124899/WEP, 389 pp.; JRB Associates, Inc., McLean, VA, NTIS, Springfield, Virginia 22161.

SPOONER, PHILIP, and others, 1984, *Slurry Trench Construction For Pollution Migration Control*, Noyes Publications, Mill Rd. at Grand Ave., Park Ridge, New Jersey 07636, 237 pp.

STOVER, E.L., 1982, *Removal of Volatile Organics from Contaminated Ground Water*, Ground Water Monitoring Rev., v. 2, no. 4, pp. 57-62.

SULLIVAN, J.M., Jr., LYNCH, D.R., and ISKANDAR, L.K., 1984, *Economics of Ground Freezing for Management of Uncontrolled Hazardous Waste Sites*, Thayer School of Engrng., Hanover, NH, PB85-121127/WEP, NTIS, Springfield, Virginia 22161, 39 pp.

TIMMERMAN, C.L., 1984, *Stabilization of Contaminated Soils by In Situ Vitrification*, Battelle Pacific NW Labs., Richland, WA, AD-P004 144/2/WEP, NTIS, Springfield, Virginia 22161, 13 pp.

U. S. ENVIRONMENTAL PROTECTION AGENCY, 1980, *Evaluating Cover Systems for Solid and Hazardous Waste*, EPA, Office of Water and Waste Management, Washington D. C., SW-867.

U. S. ENVIRONMENTAL PROTECTION AGENCY, 1980, *Lining of Waste Impoundment and Disposal Facilities*, EPA, Office of Water and Waste Management, Washington D. C., SW-870.

WALSH, J., LIPPITT, J., and SCOTT, M., 1983, *Costs of Remedial Actions at Uncontrolled Hazardous Waste Sites - Impacts of Worker Health and Safety Considerations*, SCS Engineers, Inc., Covington, KY, PB84-128701, NTIS, Springfield, Virginia 22161, 27 pp.

WARE, S. A., and JACKSON, G. S., 1979, *Liners for Sanitary Landfills and Chemical and Hazardous Waste Disposal Sites*, PB-293 335/6WP, NTIS, Springfield, Virginia 22161.

WYSS, A. W., and others, 1981, *Closure of Hazardous Waste Surface Impoundments*, PB81-166894, NTIS, Springfield, Virginia 22161, 103 pp.

MICROBIOLOGY, BIODEGRADATION

ANONYMOUS, 1982, *Biodeterioration of Oil Spills. 1964-March 1982* (Citations from the NTIS Data Base), 148 pp. PB82-807371, NTIS, Springfield, Virginia 22161.

ANONYMOUS, 1984, *Microbiology of Groundwater, 1977-May 1984*, (Citations from the Selected Water Resources Abstracts Data Base), NTIS, Springfield, VA, PB84-867365, 120 pp.

ANONYMOUS, 1984, *Microbiology of Groundwater, 1978-May 1984*, (Citations from the Life Sciences Collection Data Base), NTIS, Springfield, VA, PB84-867357 52 pp.

Contamination

ATLAS, R. M., 1981, *Microbial Degradation of Petroleum Hydrocarbons: An Environmental Perspective*, Microbiol. Rev., v. 45, 180-209 pp.

BITTON, GABRIEL, and GERBA, C. P. (eds.), 1984, *Groundwater Pollution Microbiology*, John Wiley & Sons, P. O. Box 092, Somerset, New Jersey 08873, 377 pp.

BOETHLING, R. S., and ALEXANDER, M., 1979, *Effect of Concentration of Organic Chemicals on Their Biodegradation by Natural Microbial Communities*, Appl. Environ. Microbiol, v. 37, pp. 1211-1216.

DIETZ, D. N., 1980, *The Intrusion of Polluted Water into a Groundwater Body and the Biodegradation of a Pollutant*, Proc. 1980 National Conf. on Control of Hazardous Materials, May 13-15, Louisville, Kentucky, pp. 236-244. (hydrocarbons in ground water)

DUNLAP, W. J., and McNABB, J. F., 1973, *Subsurface Biological Activity in Relation to Ground Water Pollution*, Rept. EPA-660/ 2-73-014, U. S. Environmental Protection Agency, Corvallis, Oregon, 60 pp.

GERBA, C. P. and MCNABB, J. F., 1982, *Microbial Aspects of Groundwater Pollution*, PB82-249343, NTIS, 5285 Port Royal Rd., Springfield, Virginia 22161, 6 pp, (state of knowledge).

HERBES, S. E., and SCHWALL, L. R., 1978, *Microbial Transformation of Polycyclic Aromatic Hydrocarbons in Pristine and Petroleum-Contaminated Sediments*, Appl. Environ. Microbiol., v. 35, pp. 306-316.

JAMISON, V. W., RAYMOND, R. L., and HUDSON, J. O., 1975, *Biodegradation of High Octane Gasoline in Ground Water*, J. Dev. Int. Microbiology, v. 16, pp. 305-312.

LITCHFIELD, J. H., and CLARK, L. E., 1973, *Bacterial Activity in Groundwaters Containing Petroleum Products*, API Publ. No. 4211.

McNABB, J. F., and DUNLAP, W. J., 1975, *Subsurface Biologial Acitivity in Relation to Ground-Water Pollution*, Ground Water, v. 13, no. 1, pp. 33-44.

McNABB, J. F., and others, 1977, *Nutrient, Bacterial, and Virus Control as Related to Ground-Water Contamination*, EPA/600/8-77-010, Robert S. Kerr Environmental Research Lab., Office of Research and Development, U. S. Environmental Protection Agency, Ada, Oklahoma 74820.

RAYMOND, R. L., HUDSON, J. O., and JAMISON, V. W., 1976, *Oil Degradation in Soil*, Applied and Environmental Microbiology, v. 31, no. 4, p. 522-535, Apr.

RAYMOND, R. L., JAMISON, V. W., and HUDSON, J. O., 1976, *Beneficial Stimulation of Bacterial Activity in Groundwaters Containing Petroleum Products*, Am. Inst. Chem. Eng., v. 73, no. 166, pp. 390-404.

RITTMANN, B. E., McCARTY, P. L., and ROBERTS, P. V., 1980, *Trace-Organics Biodegradation in Aquifer Recharge*, Ground Water, v. 18, no. 3, May-Jun., pp. 236-243.

SMITH, M. S., THOMAS, G. W., and WHITE, R. E., 1983, *Movement of Bacteria Through Macropores to Ground Water*, PB83-246546, NTIS, 5285 Port Royal Rd., Springfield, Virginia 22161 42 pp.

STETZENBACH, L. D. and others, 1985, *Decreases In Hydrocarbons by Soil Bacteria*, Proc. Symp. Groundwater Contamination and Reclamation, A. W. R. A., Tucson, Arizona August 1985, pp. 55-60.

TABAK, H. H., and others, 1981, *Biodegradability Studies with Organic Priority Pollutant Compounds*, J. Water Poll. Control Fed., v. 53, pp. 1503-1518.

VAUGHN, J. M., and LANDRY, E. F., 1981, *Viruses in Soil and Groundwater*, DE84001267, NTIS, 5285 Port Royal Rd., Springfield, Virginia 22161, 136 pp.

WENTSEL, R. S., and others, 1982, *Restoring Hazardous Spill-Damaged Areas: Technique Identification/Assessment*, PB82-103870, NTIS, Springfield, Virginia 22161, 374 pp, (biological techniques).

WESTLAKE, D. W., and others, 1974, *Biodegradability and Crude Oil Composition*, Canadian J. Microbiol., v. 20, pp. 915-928.

WILSON, J. T., and others, 1983, *Biotransformation of Selected Organic Pollutants in Ground Water*, Robert S. Kerr Environmental Research Lab., Ada, OK., EPA-600/J-83-042, NTIS, PB84-101526, 11 pp.

WILSON, J. T., and others, 1983, *Enumeration and Characterization of Bacteria Indigenous to a Shallow Water-Table Aquifer*, GroundWater, v. 21, no. 2, pp. 134-142.

WILSON, J. T., COSBY, R. L., and SMITH, G. B., 1984, *Potential for Biodegradation of Organo-Chlorine Compounds in Groundwater*, PB84-194612, NTIS, 5285 Port Royal Rd., Springfield, Virginia 22161, 17 pp.

ZAJIC, J. E., SUPPLISSON, B., and VOLESKY, B., 1974, *Bacterial Degradation and Emulsification of No. 6 Fuel Oil*, J. Environ. Science and Technology, v. 8, pp. 664-668.

RADIONUCLIDES IN GROUNDWATER/ NUCLEAR WASTE DISPOSAL

BIBLIOGRAPHY

NATIONAL TECHNICAL INFORMATION SERVICE, 1981, *Radioactive Waste Disposal in Salt Deposits (1964-78)*. Bibliography contains 190 citations, PB 81-802290, 197 pp.; *Period 1979-December 1980*, (173 Citations), PB81-802308, 180 pp., NTIS, Springfield, Virginia 22161.

ROBINSON, B. P., 1962, *Ion-Exchange Minerals and Disposal of Radioactive Wastes, A Survey of Literature*, U. S. Geol. Survey Water-Supply Paper 1616, 132 pp.

GENERAL WORKS

ANONYMOUS, 1980, *Spatial Patterns of Radiological Dose from Wells Drilled Near Nuclear Waste Repositories*, UCRL-15235, NTIS, Springfield, Virginia 22161, 63 pp.

ARNDT, R. H., and KURODA, P. K., 1953, *Radioactivity of Rivers and Lakes in Parts of Garland and Hot Springs Counties, Arkansas*, Econ. Geol., v. 48, pp. 551-567.

AZIZ, A., and MUBARAK, M. A., 1968 *Ion Exchange Properties of Pinstech Soil for the Disposal of Liquid Radioactive Waste Directly into the Ground*, Rept. Pinstech/HP-7, Pakistan Inst. Nuclear Sci. and Technology, Islamabad, Pakistan, also in Nuclear Sci. Abs., 23, 21927 (1969).

BARKER, F. B., and SCOTT, R. C., 1958, *Uranium and Radium in the Ground Water of the Llano Estacado, Texas and New Mexico*, Trans. Am. Geophys. Union, v. 39, pp. 459-466.

BARKER, F. B., and SCOTT, R. C., 1961, *Uranium and Radium in Ground Water from Igneous Terranes of the Pacific Northwest*, U. S. Geol. Survey Prof. Paper 424-B, pp. 298-299.

BATZEL, R. E., 1960, *Radioactivity Associated with Underground Nuclear Explosions*, J. Geophys. Res., v. 65, pp. 2897-2902.

Radionuclides

BELIN, R. E., 1959, *Radon in the New Zealand Geothermal Regions*, Geochim. et Cosmochim. Acta, v. 16, pp. 181-191.

BELTER, W. G., 1963, *Waste Management Activities of the U. S. Atomic Energy Commission*, Ground Water, v. 1, pp. 17-24.

BIERSCHENK, W. H., 1961, *Observational and Field Aspects of Ground-Water Flow at Hanford (Washington)*, in *Ground Disposal of Radioactive Wastes*, Sanitary Engineering Res. Lab., Univ. of California, Berkeley, Conf. Proc., pp. 147-156.

BOWEN, B. M., and others, 1960, *Geological Factors Affecting Ground Disposal of Liquid Radioactive Wastes into Crystalline Rocks at the Georgia Nuclear Laboratory Site*, Twenty-First Internat. Geol. Congress Section 20, pp. 32-48.

BREDEHOEFT, J. D., and others, 1978, *Geologic Disposal of High-Level Radioactive Wastes — Earth-Science Perspectives*, U. S. Geol. Survey, Arlington, Virginia 22202, Circ. 779, 15 pp.

BROWN, D. J., and RAYMOND, J. R., 1962, *Radiologic Monitoring of Ground Water at the Hanford Project*, J. Am. Water Works Assoc., v. 54, pp. 1201-1212.

BRUTSAERT, W. F., and others, 1981, *Geologic and Hydraulic Factors Controlling Radon-222 in Ground Water in Maine*, Ground Water, v. 19, no. 4, pp. 407-417.

CLEBSCH, ALFRED, Jr., 1961, *Tritium-Age of Ground Water at the Nevada Test Site, Nye County, Nevada*, U. S. Geol. Survey Prof. Paper 424-C, pp. 122-125.

COHEN, B. L., 1977, *The Disposal of Radioactive Wastes from Fission Reactors*, Sci. American, v. 236, no. 6, pp. 21-31. (waste isolation in salt beds)

COMMITTEE ON RADIOACTIVE WASTE MANAGEMENT, 1972, *An Evaluation of the Concept of Storing Radioactive Wastes in Bedrock below the Savannah River Plant Site*, National Academy of Sciences, Washington, D. C., 86 pp.

COUDRAIN, A., and others, 1982, *Underground Disposal for Radioactive Wastes: Study of the Thermal Impact in a Fractured Medium*, Commission of the European Communities, Luxembourg, in French, EUR-8186, DE84700984, NTIS, Springfield, Virginia 22161, 181 pp.

DELAGUNA, WALLACE, 1962, *Engineering Geology of Radioactive Waste Disposal,* Geol. Soc. America, Reviews in Engineering Geology, v. 1, pp. 129-160.

DeMARSILY, GHISLAIN, and MERRIAM, D. F., (eds), *Predictive Geology: Emphasis on Nuclear-Waste Disposal,* Pergamon Press, Elmsford, New York 10523, 222 pp.

DRURY, J. S., and others, 1982, *Uranium in U. S. Surface, Ground and Domestic Waters. Vol. 1,* PB82-258740, NTIS, Springfield, Virginia 22161, 344 pp. (concentration in 89,994 waters by state)

FEDERAL CIVIL DEFENSE DIVISION, 1957, *Water Contaminated by Radioactive Fallout from Atomic Weapons,* Municipal Utilities, v. 99, no. 3, Toronto, Canada.

FRUCHTER, J.S., and others, 1985, *Radionuclide Migration in Ground Water-Final Report,* Battelle Pacific NW Labs., Richland, WA, NUREG/CR-4030/WEP, NTIS, Springfield, Virginia 22161, 53 pp.

GILCREAS, F. W., 1961, *Radioactive Pollution of Water Supplies,* Water Works Engineering, v. 114, no. 3, New York.

HIGGINS, G. H., 1959, *Evaluation of the Ground-Water Contamination Hazard from Underground Nuclear Explosions,* J. Geophys. Res., v. 64, pp. 1509-1519.

HORNER, J.K., 1985, *Natural Radioactivity in Water Supplies,* Westview Press, Inc., 5500 Central Ave., Boulder, Colorado 80301, 325 pp.

INTERNATIONAL ATOMIC ENERGY AGENCY, 1984, *Isotope Hydrology,* Proc. International Symposium September 1983, Vienna, 874 pp.

JUDSON, SHELDON, and OSMOND, J. K., 1955, *Radioactivity in Ground and Surface Water,* Am. J. Sci., v. 253, pp. 104-116.

KOCHER, D.C., SJOREEN, A.L., and BARD, C.S., 1983, *Uncertainties in Geologic Disposal of High-Level Wastes - Groundwater Transport of Radionuclides and Radiological Consequences,* Oak Ridge National Lab., TN., ORNL-5838, NTIS, NUREG/CR-2506, 223 pp.

LANDIS, E.R., 1960, *Uranium Content of Ground and Surface Waters in a Part of the Central Great Plains,* U. S. Geol. Survey Bull. 1087-G, pp. 223-258.

LEONARD, R. B., and JANZER, V. J., 1978, *National Radioactivity in Geothermal Waters, Alahamba Hot Springs and Nearby Areas, Jefferson County, Montana*, U. S. Geol. Survey J. Research, v. 6, no. 4, pp. 529-539.

LEONHART, L. S., and others, 1982, *Devising a Ground-Water Monitoring Strategy for a Geologic Repository for Radioactive Waste*, Ground Water Monitoring Review, v. 2, no. 1, pp. 50-55.

LEWIS, B. D. and GOLDSTEIN, F. J., 1982, *Evaluation of a Predictive Ground-Water Solute-Transport Model at the Idaho National Engineering Laboratory, Idaho*, PB82-204066, NTIS, Springfield, Virginia 22161, 80 pp.

LINDEROTH, C. E., and PEARCE, D. W., 1961, *Operating Practices and Experiences at Hanford (Washington)* in *Ground Disposal of Radioactive Wastes*, Sanitary Engineering Res. Lab., Univ. of California, Berkeley, Conf. Proc., pp. 7-16.

MAUL, P.R., 1984, *Releases to Groundwater Following a Core-Melt Accident at the Sizewell PWR (England)*, NTIS, 5285 Port Royal Rd., Springfield, Virginia 22161, Publ. No. DE84703138/WEP (risk-assessment study).

MAZOR, E., 1962, *Radon and Radium Content of some Israeli Water Sources and a Hypothesis on Underground Reservoirs of Brines, Oils, and Gases in the Rift Valley*, Geochim. et Cosmochim. Acta, v. 26, pp. 765-786.

MERCER, J.W., THOMAS, S.D., and ROSS, B., 1983, *Parameters and Variables Appearing in Repository Siting Models*, NUREG/CR-3066, NRC/GPO Sales Program, U.S. Nuclear Regulatory Comm., Washington, DC 20555.

MUKHOPADHYAY, N.C., *Theory of Transport in Fractured Media for the Safety Analysis of a Nuclear Waste Repository*, Eidgenoessisches Inst. fuer Reaktorforschung, Wuerenlingen (Switzerland), EIR-467, DE84701341, NTIS, Springfield, Virginia 22161, 36 pp.

MUNNICH, K. O., and VOGEL, J. C., 1960, C^{14} *Determination of Deep Ground-Waters*, Internat. Assoc. Sci. Hydrology, General Assembly of Helsinki, Pub. 52, pp. 537-541.

PAPADOPULOS, S. S., and WINOGRAD, I. J., 1974, *Storage of Low-Level Radioactive Wastes in the Ground: Hydrogeologic and Hydrochemical Factors*, Rept. EPA-520/3-74-009, U. S. Environmental Protection Agency, Washington, D. C., 49 pp.

PARSONS, BRINCKERHOFF, QUADE and DOUGLAS, INC., 1979, *Technical Support for GEIS: Radioactive Waste Isolation in Geologic Formations, v. 8. Repository Preconceptual Design Studies: Salt*, No. Y/OWI/TM-36/8, NTIS, Springfield, Virginia 22161.

PRUDIC, D. E., 1978, *Installation of Water- and Gas-Sampling Wells in Low-Level Radioactive-Waste Burial Trenches, West Valley, New York*, U. S. Geol. Survey Open-file Rept. 78-718, 70 p.

PRUDIC, D. E., and RANDALL, A. D., 1979, *Ground-Water Hydrology and Subsurface Migration of Radioisotopes at a Low-Level Solid Radioactive-Waste Disposal Site, West Valley, New York*, in Carter, M. W., Kahn, B., and Moghissi, A. A. (eds.), *Management of Low-Level Radioactive Waste*, Pergamon Press., v. 1, p. 853-882.

REICHERT, S. O., 1962, *Radionuclides in Ground Water at the Savannah River Plant Waste Disposal Facilities*, J. Geophys. Res., v. 67, pp. 4363-4374.

RICKARD, W.H., and KIRBY, L.J., 1984, *Trees as Indicators of Subterranean Migration of Tritium at a Commercial Shallow Land Radioactive Waste Disposal Site*, Battelle Pacific NW Labs., Richland, WA, DE85003640/WEP, NTIS, Springfield, Virginia 22161, 12 pp.

ROBERTSON, J. B., and BARRACLOUGH, J. T., 1973, *Radioactive-and-Chemical-Waste Transport in Groundwater at National Reactor Testing Station, Idaho: 20-year Case History and Digital Model*, in *Underground Waste Management and Artificial Recharge*, Braunstein, Jules (ed.), Am. Assoc. Petrol. Geologists, Tulsa, Oklahoma, pp. 291-322.

ROBERTSON, J. B., and others, 1974, *The Influence of Liquid Waste Disposal on the Geochemistry of Water at the National Reactor Testing Station, Idaho: 1952-70*, U. S. Geol. Survey Open-file Rept. IDO-22053, 231 pp.

ROEDDER, EDWIN, 1959, *Problems in the Disposal of Acid Aluminum Nitrate High-Level Radioactive Waste Solutions by Injection into Deep-Lying Permeable Formations*, U. S. Geol. Survey Bull. 1088, 65 pp.

RYDBERG, J., 1981, *Groundwater Chemistry of a Nuclear Waste Repository in Granite Bedrock*, Lawrence Livermore National Lab., California (UCRL-53155), DE82-016316, NTIS, Springfield, Virginia 22161, 97 pp., (dissolution rates).

SCHMALZ, B. L., 1961, *Operating Practices, Experiences, and Problems at the National Reactor Testing Station, Idaho,* in *Ground Disposal of Radioactive Wastes,* Sanitary Eng. Res. Lab., Univ. of California, Berkeley, Conf. Proc., pp. 17-33.

SCHROEDER, M. C., and JENNINGS, A. R., 1963, *Laboratory Studies of the Radioactive Contamination of Aquifers,* Univ. of California, Lawrence Radiation Lab. Pub. UCRL-13074, 51 pp. plus 66 pp. Appendices.

SCHWARTZ, F. W., 1975, *On Radioactive Waste Management: An Analysis of the Parameters Controlling Subsurface Contaminant Transfer,* J. Hydrology, v. 27, (1/2).

SCOTT, R. C., and BARKER, F. B., 1958, *Radium and Uranium in Ground Water in the United States,* Second United Nations Internat. Conf. on Peaceful Uses of Atomic Energy, 10 pp.

SMITH, B. M., and others, 1961, *Natural Radioactivity in Ground Water Supplies in Maine and New Hampshire,* J. Am. Water Works Assoc., Jan.

STEAD, F. W., 1963, *Tritium in Ground Water Around Large Underground Fusion Explosions,* Sci., v. 142, pp. 1163-1165.

SUN, R. J., 1977, *Possibility of Triggering Earthquakes by Injection of Radioactive Wastes in Shale at Oak Ridge National Laboratory, Tennessee,* U. S. Geol. Survey J. Res., v. 5, no. 2, pp. 253-262.

TANNER, A. B., 1978, *Radon Migration in the Ground: A Supplementary Review,* U. S. Geol. Survey Open-File Report 78-1050, Open-File Service Section, Box 25425, Federal Center, Denver, Colorado 80225.

THATCHER, L., and others, 1961, *Dating Desert Ground Water,* Sci., v. 134, pp. 105-106.

TOKAREV, A. N., and SHCHERBAKOV, A. V., 1956, *Radiohydrogeology,* Moscow, State Publ. of Sci.-Tech. Literature on Geol. and Conservation of Natl. Resources (English trans. by U. S. Atomic Energy Comm. AEC-tr-4100), 346 pp.

TREVORROW, L. E., WARNER, D. L., and STEINDLER, M. J., 1977, *Considerations Affecting Deep-Well Disposal of Tritium-Bearing Low-Level Aqueous Waste from Nuclear Fuel Processing Plants,* Argonne National Lab., Argonne, Illinois ANL-76-76, NTIS, Springfield, Virginia 22161, 190 pp.

WILLIAMS, C. C., 1948, *Contamination of Deep Water Wells in Southeastern Kansas*, Kansas Geol. Survey Bull. 76, pt. 2, pp. 13-28.

WINOGRAD, I. J., 1974, *Radioactive Waste Storage in the Arid Zone*, EOS, Trans. Am. Geophys. Union, v. 55, no. 10, pp. 884-894.

SUBSURFACE DISPOSAL/DEEP WELL INJECTION

BIBLIOGRAPHY

ANONYMOUS, 1984, *Deep-Well Disposal, 1976-October, 1984 (211 Citations from the Energy Data Base)*, PB84-877042, NTIS, Springfield, Virginia 22161, 193 pp.

ANONYMOUS, 1982, *Deep Well Disposal, 1977-July, 1982 (Citations from the Selected Water Resources Abstracts Data Base)*, PB82-871880, NTIS, Springfield, Virginia, 22161, 253 pp.

FREY, J. H., 1980, *Deep-Well Disposal, June, 1963-May, 1980*, 34 citations from the Energy Data Base, PB80-857428 NTIS, Springfield, Virginia 22161, 51 pp.

RIMA, D. R., and others, 1971, *Subsurface Waste Disposal by Means of Wells — A Selective Annotated Bibliography*, U. S. Geol. Survey Water-Supply Paper 2020, 305 pp.

GENERAL WORKS

ABEGGLEN, D. E., and others, 1970, *The Effects of Drain Wells on the Ground-Water Quality of the Snake River Plain*, Pamphlet 148, Idaho Bur. Mines and Geol., Moscow, Idaho, 51 pp.

AMERICAN PETROLEUM INSTITUTE, 1960, *Subsurface Saltwater Disposal*, API Washington, D. C., 102 pp.

AMERICAN SOCIETY FOR TESTING AND MATERIALS, 1981, *Water for Subsurface Injection*, (STP 735), ASTM, 1916 Race St., Philadelphia, Pennsylvania 19103, 150 pp.

AMY, V. P. 1980, *Disposal Wells Really Can Work*, Water & Wastes Engineering, July, pp. 20-23. (West Palm Beach, Florida disposal wells)

ANONYMOUS, 1966, *Injection Well Earthquake Relationship — Rocky Mountain Arsenal, Denver, Colorado*, Rept. of Inv., U. S. Army Eng. Dist., Omaha Corps of Engrs., Omaha, Nebraska.

ARNOLD, S.C., 1984, *Near-Surface Groundwater Responses to Injection of Geothermal Wastes*, Idaho Water and Energy Resources Research Inst., Moscow, DE84015139, NTIS, Springfield, Virginia 22161, 148 pp.

BARRACLOUGH, J. T., 1966, *Waste Injection into a Deep Limestone in Northwestern Florida*, Ground Water, v. 4, no. 1., pp. 22-25.

BEAR, JACOB, and JACOBS, M., 1964, *The Movement of Injected Water Bodies in Confined Aquifers*, Underground Water Storage Study Rept. no. 13, Technion, Haifa, Israel.

BERGSTROM, R. E., 1968, *Feasibility of Subsurface Disposal of Industrial Wastes in Illinois*, Ill. Geol. Survey Circ.

BOEGLY, W. J., Jr., and others, 1969, *The Feasibility of Deep-Well Injection of Waste Brine from Inland Desalting Plants*, Office of Saline Water Res. and Dev. Progress Rept. no. 432, U. S. Dept. of the Interior.

BRADLEY, J.S., 1985, *Safe Disposal of Toxic and Radioactive Liquid Wastes*, (in underpressured reservoir rocks), Geology, v. 13, pp. 328-329.

BREEDEN, C. H., 1980, *Measuring the Risks of Waste Disposal by Deep-Well Injection*, PB80-139629, NTIS, Springfield, Virginia 22161, 46 pp.; also Bull. 122, Virginia Water Resources Res. Center, Blacksburg, Virginia 24060.

BROWN, P. M., and REID, M. S., 1977, *Geologic Evaluation of Waste-Storage Potential in Selected Segments of the Mesozoic Aquifer System Below the Zone of Freshwater, Atlantic Coastal Plain, North Carolina through New Jersey*, U. S. Geol. Survey Prof. Paper 881, 47 pp.

CLEARY, E. J., and WARNER, D. L., 1969, *Perspective on the Regulation of Underground Injection of Wastewaters*, Ohio River Valley Water Sanitary Comm., Cincinnati, Ohio.

CLEARY, E. J., and WARNER, D. L., 1970, *Some Considerations in Underground Wastewater Disposal*, J. Am. Water Works Assoc., v. 62, no. 8, p. 489.

CONRAD, E. T., and HOPSON, N. E., 1975, *Outlooks for the Future of Deep Well Disposal*, Water Resources Bull., v. 11, no. 2, pp. 370-378.

Subsurface Disposal

COOK, T. D., (ed.), 1972, *Underground Waste Management and Environmental Implications*, Proc. Symp. Dec. 6-9, 1971, Houston, Texas, sponsored by U. S. Geol. Survey and Am. Assoc. of Petrol. Geol., AAPG, Tulsa, Oklahoma, Mem. 18, 412 pp.

DEAN, B. T., 1965, *The Design and Operation of a Deep-Well Disposal System*, J. Water Poll. Control Fed., v. 37, no. 2, pp. 245-254.

DE LAGUNA, WALLACE, and others, 1968, *Engineering Development of Hydraulic Fracturing as a Method for Permanent Disposal of Radioactive Wastes*, Oak Ridge Natl. Lab., Rept. no. 4259.

DiTOMASO, A., and ELKAN, G. H., 1973, *Role of Bacteria in Decomposition of Injected Liquid Waste at Wilmington, North Carolina*, in *Underground Waste Management and Artificial Recharge*, New Orleans, v. 1, Am. Assoc. Petrol. Geol., pp. 585-599.

DONALDSON, E. C., 1964, *Subsurface Disposal of Industrial Wastes in the United States*, U. S. Bur. Mines Inf. Circ. 8212, 34 pp.

DRESCHER, W. J., 1965, *Hydrology of Deep-Well Disposal of Radioactive Liquid Wastes*, in *Fluids in Subsurface Environments*, Am. Assoc. Petrol. Geol., Mem. 4, pp. 399-407.

ELKAN, G. H., and HORVATH, EDWARD, 1977, *The Role of Microorganisms in the Decomposition of Deep-Well Injection Industrial Wastes*, PB-268 646/7WP, NTIS, Springfield, Virginia 22161.

EVERNDINGEN, A. F., 1968, *Fluid Mechanics of Deep-Well Disposals*, Am. Assoc. Petrol. Geol. Mem. no. 10, Aug.

FAULKNER, G. L., and PASCALE, C. A., 1975, *Monitoring Regional Effects of High Pressure Injection of Industrial Waste Water in a Limestone Aquifer*, Ground Water, v. 13, no. 2, pp. 197-208.

FINK, B. E., 1969, *State Regulates Subsurface Waste Disposal in Texas*, Water and Sewage Works, 116, IW-20.

FOSTER, J. B., and GOOLSBY, D. A., 1972, *Construction of Waste-Injection Monitor Wells near Pensacola, Florida*, Florida Dept. Natl. Res. Inf. Circ. 74.

GALLEY, J. E., (ed.), 1968, *Subsurface Disposal in Geologic Basins — A Study of Reservoir Strata*, Am. Assoc. Petrol. Geol., Tulsa, Oklahoma.

GOOLSBY, D. A., 1971, *Hydrogeological Effects of Injecting Wastes into a Limestone Aquifer near Pensacola, Florida*, Ground Water, v. 9, no. 1, pp. 13-19.

GOOLSBY, D. A., 1972, *Geochemical Effects and Movement of Injected Industrial Waste in a Limestone Aquifer*, Am. Assoc. Petrol. Geol. Mem. No. 18.

GRAHAM, W. G., and others, 1977, *Irrigation Wastewater Disposal Well Studies — Snake Plain Aquifer*, Rept. EPA-600/3-77-071, U. S. Environmental Protection Agency, Ada, Oklahoma, 51 pp.

GROVE, D. B., 1976, *A Model for Calculating Effects of Liquid Waste Disposal in Deep Saline Aquifers*, Water Resources Inv. 76-61, PB 256903/AS, NTIS, Springfield, Virginia 22161.

HALL, C. W., and BALLENTINE, R. K., 1973, *U. S. Environmental Protection Agency Policy on Subsurface Emplacement of Fluids by Well Injection*, in Braunstein, Jules, (ed.), *Underground Waste Management and Artificial Recharge*, Am. Assoc. Petrol. Geol., Tulsa, Oklahoma, pp. 783-793.

HANBY, K. P., KIDD, R. E., and LaMOREAUX, P. E., 1973, *Subsurface Disposal of Liquid Industrial Wastes in Alabama — A Current Status Report*, Geol. Survey of Alabama, Reprint Series 27.

HANSHAW, B. B., 1972, *Natural Membrane Phenomena and Subsurface Waste Emplacement*, in Cook, T. D., (ed). *Underground Waste Management and Environmental Implications,* Am. Assoc. Petrol. Geol. Mem. 18, Tulsa, Oklahoma, pp. 308-315.

HEALY, J. H., and others, 1968, *The Denver Earthquakes*, Sci., v. 161, no. 3484, p. 1301.

HEIDARI, MANOUTCHEHR, and others, 1974, *Analysis of Liquid Waste Injection Wells in Illinois by Mathematical Models*, Univ. Illinois, Water Resources Center Res. Rept. no. 77, Urbana, Illinois 61801.

HENRY, H. R., and KOHOUT, F. A., 1972, *Circulation Patterns and Movement of Injected Industrial Waste in a Limestone Aquifer*, Am. Assoc. Petrol. Geol. Mem. 18, pp. 202-221.

HICKEY, J.J., 1984, *Subsurface Injection of Treated Sewage into a Saline-Water Aquifer at St. Petersburg, Florida - Aquifer Pressure Buildup*, Ground Water, v. 22, no. 1, pp. 48-55.

HITE, R. J., and LOHMAN, S. W., 1973, *Geologic Appraisals of Paradox Basin Salt Deposits for Waste Emplacement*, U. S. Geol. Survey Open-file Rept.

HOLLISTER, J. C., and WEIMER, R. J., 1968, *Geophysical and Geological Studies of the Relationships Between the Denver Earthquake and the Rocky Mountain Arsenal Well*, Pt. A., Colorado School of Mines Quarterly, v. 63, no. 1., Jan.

HORVATH, E., and ELKAN, G. H., 1978, *A Model Aquifer System for Biological Compatibility Studies of Proposed Deep Well Disposal Systems*, Ground Water, v. 16, no. 3, pp. 174-185.

HOWARD, G. C., and FAST, C. R., 1970, *Hydraulic Fracturing*, Monograph Soc. Petrol. Eng., AIME, v. 2, 210 pp.

IVES, R. E., and EDDY, G. E., 1968, *Subsurface Disposal of Industrial Wastes*, Interstate Oil Compact Comm., Oklahoma City, Oklahoma, with supplement publ. Jan. 1970.

JOHNSON, J. L., (ed.), 1981, *Water for Subsurface Injection*, 2nd Symp. on Subsurface Injection ASTM, Ft. Lauderdale, Jan. 1980, Am. Soc. of Testing and Materials, 1916 Race St., Philadelphia, Pennsylvania 19103, 150 pp.

KAUFMAN, M. I., 1973, *Subsurface Wastewater Injection, Florida*, Am. Soc. Civil Engrs. J. Irrigation and Drainage Div., v. 99, (IR1), pp. 53-70.

KAUFMAN, M. I., GOOLSBY, D. A., and FAULKNER, G. L., 1974, *Injection of Acidic Industrial Waste into a Saline Carbonate Aquifer, Geochemical Aspects*, in *Underground Waste Management and Artificial Recharge*, v. 1, Am. Assoc. Petrol. Geol.

KAUFMAN, M. I., and McKENZIE, D. J., 1975, *Upward Migration of Deep-Well Waste Injection Fluids in Floridan Aquifer, South Florida*, U. S. Geol. Survey, J. Res., v. 3, no. 3, pp. 261-271.

KEHLE, R. O., 1964, *The Determination of Tectonic Stresses Through Analysis of Hydraulic Well Fracturing*, J. Geophys. Res., v. 69, no. 2, pp. 259-273.

KOELZER, V. A., and others, 1969, *The Chicago Area Deep Tunnel Project — A Use of the Underground Storage Resource*, J. Water Poll. Control Fed., v. 41, no. 4, Apr.

KREIDLER, W. L., 1975, *Underground Disposal of Liquid Waste in New York*, New York State Museum and Science Service, Map and Chart Ser. 26, Albany, New York 12224.

LAW, A. G., 1979, *Ground-Water Hydrology and Radioactive Waste Disposal at the Hanford Site*, RHO-SA-26, NTIS, Springfield, Virginia 22161.

LEENHEER, J. A., and MALCOLM, R. L., 1973, *Case History of Subsurface Waste Injection of an Industrial Organic Waste*, in *Underground Waste Management and Artificial Recharge*, v. 1, Am. Assoc. Petrol. Geol., pp. 565-584.

LEENHEER, J. A., and MALCOLM, R. L., 1973, *Chemical and Microbial Transformations of an Industrial Organic Waste During Subsurface Infection*, Inst. Environmental Sci. Proc., Anaheim, California, pp. 351-360.

LEENHEER, J. A., and others, 1977, *Physical, Chemical and Biological Aspects of Subsurface Organic Waste Injection near Wilmington, N. C.*, U. S. Geol. Survey Prof. Paper 987, Washington, D. C. 20402.

LeGROS, P. G., and others, 1969, *A Study of Deep Well Disposal of Desalination Brine Waste*, Office of Saline Water Res. and Dev., Progress Rept. no. 456, U. S. Dept. of the Interior.

MARSH, J. H., 1968, *Design of Waste Disposal Wells*, Ground Water, v. 6, no. 2, pp. 4-9.

McCLAIN, W. C., 1969, *Disposal of Radioactive Wastes by Hydraulic Fracturing*, Nuclear Eng. Des., v. 9, no. 3, p. 315, also Eng. Index Abs., Jan., 1970.

MOSELEY, W., and MALINA, J. F., 1969, *Relationships Between Selected Physical Parameters and Cost Responses for the Deep Well Disposal of Aqueous Industrial Wastes*, Water and Wastes Eng., v. 6, no. 9, p. 99.

NACE, R. L., 1973, *Problems of Underground Storage of Wastes*, U. S. Geol. Survey J. Res., v. 1, no. 6, pp. 719-723.

NIELSEN, D.M., and ALLER, LINDA, 1984, *Methods for Determining the Mechanical Integrity of Class II Injection Wells*, National Water Well Assoc., Dublin, Ohio 43017, 263 pp.

ORSANCO ADVISORY COMMITTEE ON UNDERGROUND INJECTION OF WASTEWATERS, 1973, *Underground Injection of Wastewaters in the Ohio Valley Region,* Ohio Valley Water Sanitation Comm., Cincinnati, Ohio 45202, 66 pp.

PARKER, F. L., 1969, *Status of Radioactive Waste Disposal in U. S. A.*, J. San. Eng. Div., Proc. Am. Soc. Civil Engrs., v. 95, p. 439.

PASCALE, C. A., 1976, *Construction and Testing of Two Waste-Injection Monitor Wells in Northwest Florida*, U. S. Geol. Survey Water-Resoures Inv., 76-1, 50 pp.

PEEK, H. M. and HEATH, R. C., 1973, *Feasibility Study of Liquid-Waste Injection into Aquifers Containing Salt Water, Wilmington, North Carolina*, in *Underground Waste Management and Artificial Recharge*, Braunstein, Jules (ed.), Am. Assoc. Petrol. Geol., Tulsa, Oklahoma, pp. 851-878.

PIPER, A. M., 1969, *Disposal of Liquid Wastes by Injection Underground – Neither Myth nor Millennium*, Chem. Eng. Progr., v. 65, p. 97.

PURI, H. S., FAULKNER, G. L., and WINSTON, G. O., 1973, *Hydrogeology of Subsurface Liquid-Waste Storage in Florida*, in *Underground Waste Management and Artificial Recharge*, Am. Assoc. Petrol. Geol., Tulsa, Oklahoma v. 2.

RAGONE, S. E., and VECCHIOLI, JOHN, 1975, *Chemical Interaction During Deep Well Recharge, Bay Park, New York*, Ground Water, v. 13, no. 1, pp. 17-23.

RALEIGH, C. B., 1972, *Earthquakes and Fluid Injection*, in Cook, T.D. (ed.), *Underground Waste Management and Environmental Implications*, Am. Assoc. Petrol. Geol., Me. no. 18, pp. 273-279.

REEDER, L. R., and others, 1977, *Review and Assessment of Deep-Well Injection of Hazardous Waste*, 4 vol. set, PB 269 000, NTIS, Springfield, Virginia 22161.

RICE, R. C., and RAATS, PETER, 1980, *Underground Travel of Renovated Wastewater*, J. Environ. Engineering Div., Am. Soc. Civil Engrs., v. 106, no. EE6, Dec.

SCEVA, J. E., 1968, *Liquid Waste Disposal in the Lava Terranes of Central Oregon*, U. S. Fed. Water Poll. Control Admin., Corvallis, Oregon, 2 vols., 162 pp.

SCHICHT, R. J., 1971, *Feasibility of Recharging Treated Sewage Effluent into a Deep Sandstone Aquifer*, Ground Water, v.9, no. 6, pp. 29-35, Nov.-Dec.

SELM, R. P., and HULSE, B. T., 1959, *Deep-Well Disposal of Industrial Wastes*, Proc. Purdue Univ. Indus. Waste Conf., pp. 566-586.

SHELDRICK, G. M., 1969, *Deep Well Disposal: are Safeguards Being Ignored?*, Chem Eng., v. 76., no. 7, p. 74.

SINGH, U. P., GARAIA-BENGOCHEA, J. I., and SPROUL, C. R., 1980, *Deep Well Disposal of Secondary Effluent in Dade County, Florida,* Water Resources Bull., v. 16, no. 5, pp. 812-817.

SLAGEL, K. A., and STROGNER, J. M., 1969, *Oil Fields Yield New Deep Well Disposal Technique,* Water and Sewage Works, v. 116, p. 238.

SMITH, H. F., 1971, *Subsurface Storage and Disposal in Illinois,* U. S. Environmental Protection Agency, Water Poll. Control Research Series, 16060 GRB08-71, pp. 20-28.

STEWART, R. S., 1968, *Techniques of Deep Well Disposal — A Safe and Efficient Method of Pollution Control,* Proc. 15th Ontario Ind. Waste Conf., Ontario Water Res. Comm., Toronto, Ontario.

TALBOT, J. S., 1968, *Some Basic Factors in Consideration and Installation of Deep Well Disposal Systems,* Water and Sewage Works, v. 115, p. 213.

TALBOT, J. S., 1972, *Requirements for Monitoring of Industrial Deep Well Disposal Systems,* in Cook, T.D., (ed.), *Underground Waste Management and Environmental Implications,* Am. Assoc. Petrol. Geol. Mem. 18, pp. 85-92.

TAMURA, T., and WEEREN, H., 1984, *Disposal of Waste by Hydraulic Fracturing,* Oak Ridge National Lab., TN, DE84014052, NTIS, Springfield, Virginia 22161, 16 pp.

TUCKER, W. E., 1971, *Subsurface Disposal of Liquid Industrial Wastes in Alabama — A Current Status Report,* Proc. National Ground Water Quality Symp., U. S. Environmental Protection Agency, Water Poll. Control Research Series, 16060 GRB 08/71, pp. 10-19.

U. S. ENVIRONMENTAL PROTECTION AGENCY, 1974, *Compilation of Industrial and Municipal Injection Wells in the United States,* 2 vols., EPA-520/9-74-020, Washington, D. C.

VECCHIOLI, JOHN, 1979, *Monitoring of Subsurface Injection of Wastes, Florida,* Ground Water, v. 17, no. 3, May-June, pp. 244-249

VEIR, B. B., 1969, *Deep Well Disposal Pays off at Celanese Chemical Plant,* Water and Sewage Works, v. 116, no. 5, pp. IW-21.

WALLER, R. M., TURK, J. T., and DINGMAN, R. J., 1978, *Potential Effects of Deep-Well Waste Disposal in Western New York,* U. S. Geol. Surv. Prof. Paper 1053, 39 pp.

WARNER, D. L., 1965, *Deep-Well Injection of Liquid Waste*, U. S. Public Health Service Environ. Health Ser. Pub. 999-WP-21, Washington, D. C., 55 pp.

WARNER, D. L., 1967, *Deep Wells for Industrial Waste Injection in the United States — Summary of Data*, Fed. Water Poll. Control Adm. publ. WP-20-10, Washington, D. C., 45 pp.

WARNER, D. L., 1968, *Subsurface Disposal of Liquid Industrial Wastes by Deep Well Injection*, in *Subsurface Disposal in Geologic Basins — a Study of Reservoir Strata*, Am. Assoc. Petrol. Geol., Mem. 10, pp. 11-20.

WARNER, D. L., 1975, *Monitoring Disposal-Well Systems*, U. S. Environmental Protection Agency, Office of Research and Development, Environmental Monitoring and Support Lab., Las Vegas, Nevada, EPA 608/7-74-008.

WARNER, D. L., 1975, *Underground Liquid Waste Disposal*, J. Hydraul. Div., Am. Soc. Civ. Eng., v. 101 (HY3), pp. 421-435.

WARNER, D. L., and LEHR, J. H., 1977, *An Introduction to the Technology of Subsurface Wastewater Injection*, Rept. EPA-600/2-77-240 U.S. Environmental Protection Agency, Ada, Oklahoma, 344 pp.; reprinted by Premier Press (1981), P. O. Box 4428, Berkeley, California.

WARNER, D. L. and ORCUTT, D. H., 1973, *Industrial Wastewater-Injection Wells in the United States — Status of Use and Regulations, 1973*, in Braunstein, Jules, (ed.), *Underground Waste Management and Artificial Recharge*, Am. Assoc. Petrol. Geol., Tulsa, Oklahoma, pp. 697-697.

WARNER, D. L., and others, 1979, *Radius of Pressure Influence of Injection Wells*, Robert S. Kerr Environmental Res. Lab. Rep. 71, Ada, Oklahoma 74820, also PB80-100498, NTIS, Springfield, Virginia 22161.

WESNER, G. M., and BAIER, D. C., 1970, *Injection of Reclaimed Wastewater into Confined Aquifers*, J. Am. Water Works Assoc., v. 62, no. 3, pp. 203-210, Mar.

WILLIS, R., 1976, *Optimal Groundwater Quality Management: Well Injection of Wastes*, Water Resoures Res., v. 12, no. 1, pp. 47-53.

WILSON, W. E., and others, 1973, *Hydrologic Evaluation of Industrial-Waste Injection at Mulberry, Florida*, in Braunstein, Jules, (ed.), *Underground Waste Management and Artificial Recharge,* Am. Assoc. Petrol. Geol., Tulsa, Oklahoma, pp. 552-564.

WITHERSPOON, P. A. and NEUMAN, S. P., 1972, *Hydrodynamics of Fluid Injection*, in Cook, T.D. (ed.), *Underground Waste Management and Environmental Implications,* Am. Assoc. Petrol. Geol. Mem. 18, Tulsa, Oklahoma.

WOOD, L. A., 1974, *Use of Underground Space for Waste Storage Through Injection Wells*, in Deju, P. A., (ed.), *Extraction of Minerals and Energy, Today's Dilemmas*, Ann Arbor Science Publishers, Inc., Ann Arbor, Michigan, pp. 193-202.

WRIGHT, J. L., 1969, *Underground Waste Disposal*, Ind. Waste Eng., v. 6, no. 5, p. 24.

SALT-WATER INTRUSION

BIBLIOGRAPHY

TODD, D. K., 1953, *An Abstract of Literature Pertaining to Sea Water Intrusion and its Control*, Tech. Bull. 10, Sanitary Eng. Res. Project, Univ. California, Berkeley, 74 pp.

SZELL, G. P., 1980, *Salt-Water Intrusion in Coastal Aquifers: A Bibliography*, St. Johns River Water Management Dist. Info. Circ. No. 4, P. O. Box 1429, Palatka, Florida 32077.

GENERAL WORKS

ACKERMANN, N. L., and CHANG, Y. Y., 1971, *Salt-Water Interface During Ground-Water Pumping*, J. Hydraulics Div., Amer. Soc. Civil Engrs., v. 97, no. HY2, pp. 223-232.

ANDERSON, M. P., and BERKEBILE, C. A., 1976, *Evidence of Salt-Water Intrusion in Southeastern Long Island*, Ground Water, v. 14, no. 5, pp. 315-319.

ANONYMOUS, 1958, *Sea Water Intrusion in California*, State Dept. of Water Resources, Div. of Resources Planning, Bull. 63.

ATKINSON, S.F., and others, 1985, *Salt Water Intrusion--Status and Potential in the Contiguous United States*, Lewis Publishers, Inc., Chelsea, Michigan 48118, 400 pp., (includes annotated bibliography of 125 references).

BANKS, H. O., and RICHTER, R. C., 1953, *Sea-Water Intrusion into Ground-Water Basins Bordering the California Coast and Inland Bays*, Trans. Am. Geophys. Union, v. 34, pp. 575-582.

BARKSDALE, H. C., 1940, *The Contamination of Ground-Water by Salt Water near Parlin, New Jersey*, Trans. Am. Geophys. Union, v. 21, pp. 471-474.

BAUMANN, P., 1953, *Experiments with Fresh-Water Barrier to Prevent Sea Water Intrusion*, J. Am. Water Works Assoc., v. 45, pp. 521-534.

BEAR, JACOB, 1961, *Some Experiments in Dispersion*, J. Geophys. Res., v. 66, pp. 1563-1572.

BEAR, JACOB, and DAGAN, G., 1963, *Some Exact Solutions of Interface Problems by Means of the Hodograph Method*, J. Geophys. Res., v. 69.

BEAR, JACOB and DAGAN, G., 1964, *Intercepting Freshwater Above the Interface in a Coastal Aquifer*, Int. Assoc. Sci. Hydrology Publ. 64, pp. 154-181.

BEAR, JACOB and DAGAN, G., 1964, *Moving Interface in Coastal Aquifers*, J. Hydraul. Div., Am. Soc. Civil Engrs., v. 90, no. HY4, pp. 193-216.

BEAR, JACOB and TODD, D. K., 1960 *Transition Zone Between Fresh and Salt Waters in Coastal Aquifers'*, Hydraulic Lab., Univ. of California, Berkeley.

BENNETT, G. D., and others, 1968, *Electric-Analog Studies of Brine Coning Beneath Freshwater Wells in the Punjab Region, West Pakistan*, U. S. Geol. Survey Water-Supply Paper 1608-J, 31. pp.

BIEMOND, C., 1957, *Dune Water Flow and Replenishment in the Catchment Area of the Amsterdam Water Supply*, J. Inst. Water Engrs., v. 11, pp. 195-213.

BLACK, A. P., and others, 1953, *Salt Water Intrusion in Florida*, Florida Water Survey and Res. Paper no. 9, 38 pp.

BOGGESS, D. H., MISSMER, T. M., and O'DONNELL, T. H., 1977, *Saline-Water Intrusion Related to Well Construction in Lee County, Florida*, U. S. Geol. Survey Water-Resoures Inv. 77-33, 34 pp.

BRAITHWAITE, F., 1855, *On the Infiltration of Salt Water into the Springs of Wells under London and Liverpool*, Proc. Inst. Civil Engrs., v. 14, pp. 507-523.

BRENNEKE, A. M., 1945, *Control of Salt-Water Intrusion in Texas*, J. Am. Water Works Assoc., v. 37, pp. 579-584.

BROWN, J. S., 1925, *A Study of Coastal Ground Water with Special Reference to Connecticut*, U. S. Geol. Survey Water-Supply Paper 537, 101 pp.

BROWN, R. H., and PARKER, G. G., 1945, *Salt-Water Encroachment in Limestone of Silver Bluff, Miami, Florida*, Econ. Geol., v. 40, pp. 235-262.

BRUINGTON, A. E., 1969, *Control of Sea-Water Intrusion in a Ground Water Aquifer*, Ground Water, v. 7, no. 3, pp. 9-15.

BRUINGTON, A. E., 1972, *Salt-Water Intrusion into Aquifers*, Water Resources Bull., v. 8, no. 1, pp. 150-160.

BRUINGTON, A. E., and SEARES, F. D., 1965, *Operating A Sea Water Barrier Project*, J. Irrigation and Drainage Div., Am. Soc. of Civil Engrs., v. 91, no. IR2, pp. 117-140.

CAHILL, J. M., 1967, *Hydraulic Sand-Model Study of the Cyclic Flow of Salt-Water in a Coastal Aquifer*, U. S. Geol. Survey Prof. Paper 575-B, pp. 240-244.

CALIFORNIA DEPT. OF WATER RESOURCES, 1958, *Sea-Water Intrusion in California*, Bull. 63, Sacramento, 91 pp. plus apps.

CALIFORNIA DEPT. OF WATER RESOURCES, 1965, *Sea-Water Intrusion — Oxnard Plain of Ventura County*, Bull. 63-1. 100 pp.

CALIFORNIA DEPT. OF WATER RESOURCES, 1966, *Santa Ana Gap Salinity Barrier, Orange County*, Bull. 147-1, Sacramento. 178 pp.

CALIFORNIA DEPT. OF WATER RESOURCES, 1968, *Sea-Water Intrusion — Bolsa-Sunset Area*, Bull. 63-2., 167 pp.

CALIFORNIA DEPT. OF WATER RESOURCES, 1970, *Oxnard Basin Experimental Extraction-Type Barrier*, Bull. 147-6, Sacramento, 157 pp.

CALIFORNIA DEPT. OF WATER RESOURCES, 1970, *Sea-Water Intrusion, Pismo-Guadelupe Area*, Bull. 63-3, 82 pp.

CARLSON, E. J., 1968, *Removal of Saline Water from Aquifers*, U. S. Bureau of Reclamation Res. Rept. no. 13.

CHANDLER, R. A. and McWHORTER, D. B., 1975, *Upconing of the Salt-Water — Fresh-Water Interface Beneath a Pumping Well*, Ground Water, v. 13, pp. 354-359.

CHARMONMAN, S., 1965, *A Solution of the Pattern of Fresh-Water Flow in an Unconfined Coastal Aquifer*, J. Geophys. Res., v. 70, pp. 2813-2819.

CHILDS, E. C., 1950, *The Equilibrium of Rain-Fed Groundwater Resting on Deeper Saline Water — The Ghyben-Herzberg Lens*, J. Soil Sciences, v. 1, no. 2, pp. 173-181.

COLLINS, M. A., and GELHAR, L. W., 1971, *Seawater Intrusion in Layered Aquifers*, Water Resources Res., v. 7, no.4, pp. 971-979.

CONTRACTOR, D. N., 1980, *A Review of Techniques for Studying Freshwater/Seawater Relationships in Coastal and Island Ground-Water Flow Systems*, PB80-193857, NTIS, Springfield, Virginia 22161.

COOPER, H. H., Jr., 1959, *A Hypothesis Concerning the Dynamic Balance of Fresh Water and Salt Water in a Coastal Aquifer*, J. Geophys. Res., v. 64, no. 4.

COOPER, H. H., Jr. and others, 1964, *Sea Water in Coastal Aquifers, Relation of Salt Water to Fresh Ground Water*, U. S. Geol. Survey, Water-Supply Paper 1613C, 84 pp.

DAGAN, G. and BEAR, JACOB, 1968, *Solving the Problem of Local Interface Upconing in a Coastal Aquifer by the Method of Small Perturbations*, J. Hydr. Research, v. 6, pp. 15-44, 1968.

DAS GUPTA, A., 1985, *Approximation of Salt-Water Interface Fluctuation in an Unconfined Coastal Aquifer*, Ground Water, v. 23, no. 6, pp. 783-794.

DAY, P. R., 1956, *Dispersion of a Moving Salt-Water Boundary Advancing Through Saturated Sand*, Trans. Am. Geophys. Union, v. 37, pp. 595-601.

DEJOSSELIN DEJONG, G., and VAN DUYN, C.J., 1984, *Transverse Dispersion from an Originally Sharp Fresh-Salt Water Interface Caused by Shear Flow*, PB85-209229/WNR, NTIS, Springfield, Virginia 22161.

ERNEST, L. F., 1969, *Groundwater Flow in the Netherlands Delta Area and its Influence on the Salt Balance of the Future Lake Zeeland*, J. Hydrology, v. 8, pp. 137-172.

FETTER, C. W., Jr., 1972, *Position of the Saline Water Interface Beneath Oceanic Islands*, Water Resources Res., v. 8, no. 5, pp. 1307-1315.

FITTERMAN, D.V., and HOEKSTRA, PIETER, 1984, *Mapping of Salt Water Instrusions with Transient Electromagnetic Soundings*, in *Surface and Borehole Geophysical Methods in Ground Water Investigations (D.M. Nielsen, ed.)*, pp.429-454, National Water Well Assoc., Dublin, OH.

FRIND, E.O., 1982, *Seawater Intrusion in Continous Coastal Aquifer-Aquitard Systems*, Advances in Water Resources, v. 5, no. 2, June.

GHYBEN, W., (BADON), 1889, *Nota in Verband met de Voorgenomen Putboring nabij Amsterdam*, Koninkl. Inst. Ing. Tijdschr., The Hague.

GLOVER, R. E., 1959, *The Pattern of Fresh-Water Flow in a Coastal Aquifer*, J. Geophys. Res., v. 64, no. 4.

GREGG, D. O., 1971, *Protective Pumping to Reduce Aquifer Pollution, Glynn County, Georgia*, Ground Water, v. 9, no. 5, pp. 21-29, Sept.-Oct. (upconing of saline water)

GREGG, D. O., and ZIMMERMAN, E. A., 1974, *Geologic and Hydrologic Control of Chloride Contamination in Aquifers at Brunswick, Glynn County, Georgia*, U. S. Geol. Survey Water-Supply Paper 2029-D.

HANSHAW, B. B., and other, 1965, *Relation of Carbon 14 Concentrations to Saline Water Contamination of Coastal Aquifers*, Water Resources Res., v. 1, no. 1, pp. 109-114.

HANTUSH, M. S., 1968, *Unsteady Movement of Fresh Water in Thick Unconfined Saline Aquifers*, Bull. Internat. Assoc. of Sci. Hydrology, v. 13, no. 2, pp. 40-60, Jun.

HARDER, J. A., and others, 1953, *Laboratory Research on Sea Water Intrusion into Fresh Ground-Water Sources and Methods of its Prevention — Final Report*, Sanitary Eng. Res. Lab., Univ. California, Berkeley, 68 pp.

HARRIS, W. H., 1967, *Stratification of Fresh and Salt Water on Barrier Islands as a Result of Differences in Sediment Permeability*, Water Resources Res., v. 3, no. 1, pp. 89-97.

HAUBOLD, R. G., 1975, *Approximation for Steady Interface Beneath a Well Pumping Fresh Water Overlying Salt Water*, Ground Water, v. 13, no. 3, pp. 254-259.

HAYAMI, S., 1951, *On the Saline Disaster and Variation of Coastal Underground Water Caused By Land Subsidence Accompanying the Great Earthquake of December 21, 1947*, General Assembly, Brussels, Internat. Assoc. Sci. Hydrology v. 2, pp. 249-251.

HENRY, H. R., 1959, *Salt Intrusion into Fresh-Water Aquifers*, J. Geophys. Res., v. 64, no. 11. (interface equations)

HENRY, H. R., 1964, *Interfaces Between Salt Water and Fresh Water in Coastal Aquifers*, U. S. Geol. Survey Water-Supply Paper, 1613-3.

HERZBERG, B., 1901, *Die Wasserversorgung Einiger Nordseebader*, J. Gasbeleuchtung und Wasserversorgung, v. 44, pp. 815-819, 842-844.

HUGHES, J. L., 1979, *Saltwater-Barrier Line in Florida: Concepts, Considerations and Site Examples*, PB-101 306, NTIS Springfield, Virginia 22161.

HUNT, BRUCE, 1985, *Some Analytical Solutions for Seawater Intrusion Control With Recharge Wells*, J. Hydrology, v. 80, no. 1/2, Sep. 15.

ISAACS, L.T., 1985, *Estimating Interface Advance Due to Abrupt Changes in Replenishment in Unconfined Coastal Aquifers*, J. Hydrology, v. 78, no. 6, June 20.

JACOBS, M. and SCHMORAK, S., 1960, *Sea Water Intrusion and Interface Determination Along the Coastal Plain of Israel*, Jerusalem Hydrological Service, v. 1, Hydrological Paper no. 6.

KASHEF, A. I., 1971, *On the Management of Ground Water in Coastal Aquifers*, Ground Water, v. 9., no. 2, pp. 12-20.

KASHEF, A. I., 1972, *What do We Know About Salt-Water Intrusion?*, Water Resources Bull., v. 8 no. 2, pp. 282-293.

KASHEF, A. I., 1976, *Control of Salt-Water Intrusion by Recharge Wells*, J. Irrig. Drainage Div., Am. Soc. Civ. Eng., v. 102 (IR4), pp. 445-457.

KASHEF, A.I., 1983, *Harmonizing Ghyben-Herzberg Interface with Rigorous Solutions*, Ground Water, v. 21, no. 2, pp. 153-159.

KASHEF, A. I., and SAFAR, M. M., 1975, *Comparative Study of Fresh-Salt Water Interfaces Using Finite Element and Simple Approaches*, Water Resoures Bull., v. 11, no. 4, pp. 651-665.

KASHEF, A. I., and SMITH, J. C., 1975, *Expansion of Salt-Water Zone due to Well Discharge*, Water Resoures Bull., v. 11, no. 6, pp. 1107-1120.

KEMBLOWSKI, MARION, 1985, *Salt-Water--Freshwater Transient Upconing-An Implicit Boundary-Element Solution*, J. Hydrology, v. 78, no. 1/2, May 30.

KOHOUT, F. A., 1960, *Cyclic Flow of Salt Water in the Biscayne Aquifer of Southeastern Florida*, J. Geophys. Res., v. 65, no. 7.

KOHOUT, F. A., 1961, *Case History of Salt Water Encroachment Caused by a Storm Sewer in Miami*, J. Am. Water Works Assoc., v. 53, pp. 1406-1416.

KOHOUT, F. A. and KLEIN, H., 1967, *Effect of Pulse Recharge on the Zone of Diffusion in the Biscayne Aquifer*, Internat. Assoc. Sci. Hydrology Publ. 72, pp. 252-270.

KWABATA, I., 1965, *Coning up of Confined Two-Layers' Liquids Through Porous Media by Pumping up*, Bull. Kyoto Gakugei Univ., Ser. B., no. 27, pp. 19-29.

LAU, L. S., 1967, *Seawater Encroachment in Hawaiian Ghyben-Herzberg Systems*, Am. Water Resources Assoc., Proc. Symp. on Ground-Water Hydrology, San Francisco, California, pp. 259-271.

LAU, L. S., and MINK, J. F., 1967, *A Step in Optimizing the Development of the Basal Water Lens of Southern Oahu, Hawaii*, Internat. Assoc. Sci. Hydrology Publ. 72, pp. 500-508.

LAVERTY, F. B., and VAN DER GOOT, H. A., 1955, *Development of a Fresh-Water Barrier in Southern California for the Prevention of Sea Water Intrusion*, J. Am. Water Works Assoc., v. 47, pp. 886-908.

LEE, CHUN-HIAN, and CHENG, R, T-S, 1974, *On Seawater Encroachment in Coastal Aquifers*, Water Resources Res., v. 10, no. 5, pp. 1039-1044.

LEGGETTE, R. M., 1947, *Salt Water Encroachment in the Lloyd Sand on Long Island, New York*, Water Works Eng., v. 100, pp. 1076-1079, 1107-1109.

LIEFRINCK, F. A., 1930, *Water Supply Problems in Holland*, Public Works, v. 61, no. 9, pp. 19-20, 65-66, 69.

LONG, R. A., 1965, *Feasibility of a Scavenger-Well System as a Solution to the Problem of Vertical Salt-Water Encroachment*, Water Resources Pamphlet 15, Louisiana Geol. Survey, Baton Rouge, 27 pp.

LOUISIANA WATER RESOURCES RESEARCH INSTITUTE, 1968, *Salt Water Encroachment into Aquifers*, Bull. 3, Louisiana State Univ., Baton Rouge, Louisiana, 192 pp.

LOVE, S. K., 1944, *Cation-Exchange in Ground Water Contaminated with Sea Water near Miami, Florida*, Trans. Am. Geophys. Union, v. 25, pp. 951-955.

LUSCZYNSKI, N. J., 1961, *Head and Flow of Ground Water of Variable Density*, J. Geophys. Res., v. 66, no. 12, pp. 4247-4256.

LUSCZYNSKI, N. J., and SWARZENSKI, W. V., 1962, *Fresh and Salty Ground Water in Long Island, New York*, Proc. Am. Soc. Civil Engrs., v. 88, no. 3207, pp. 173-194.

LUSCZYNSKI, N. J., and SWARZENSKI, W. V., 1966, *Salt-Water Encroachment in Southern Nassau and Southeastern Queens Counties, Long Island, New York*, U. S. Geol. Survey Water-Supply Paper 1613-F, 76 pp.

MATHER, J. D., 1975, *Development of the Groundwater Resources of Small Limestone Islands*, Quarterly J. Engrng. Geol., pp. 141-150.

MAURIN, V. and ZOETL, J., 1967, *Salt-Water Encroachment in the Low Altitude Karst Water Horizons of the Island of Kephallinia (Ionic Islands)*, in *Hydrology of Fractured Rocks*, Proc. Dubrovnik Symp. Oct. 1965, Internat. Assoc. Sci. Hydrology, Pub. 74, pp. 423-438.

McWHORTER, D.B., 1980, *Summary of Skimming Well Investigation*, Colorado State Univ., Fort Collins, Engineering Research Center, Water Management-TR-63, AID-PN-AAJ-014, (fresh-saline aquifers in Pakistan), NTIS, PB82-241811, 92 pp.

MEHNERT, E. and JENNINGS, A.A., 1985, *The Effect of Salinity-Dependent Hydraulic Conductivity on Saltwater Intrusion Episodes*, J. Hydrology, v. 80, no. 3/4, Oct. 15.

MERCER, J. W., LARSON, S. P. and FAUST, C. R., 1980, *Simulation of Salt-Water Interface Motion*, Ground Water, v. 18, no. 4, Jul.-Aug., pp. 374-385.

NEUMAN, S. P., and WITHERSPOON, P. A., 1971, *Flow in Multiple-Aquifer Systems in Seawater Intrusion: Aquitards in the Coastal Ground Water Basin of Oxnard Plain, Ventura County, California*, Dept. of Water Resources, Bull. 63-4, pp. 21-62.

NEWPORT, B. D., 1977, *Salt Water Intrusion in the United States*, Rept. EPA-600/8-77-011, U. S. Environmental Protection Agency, Ada, Oklahoma., 30 pp.

NOMITSU, T., and others, 1927, *On the Contact Surface of Fresh and Salt-Water near a Sandy Sea-Shore*, Mem. College Sci., Kyoto Imp. Univ. Ser. A, v. 10, no. 7, pp. 279-302.

OHRT, F., 1947, *Water Development and Salt Water Intrusion on Pacific Islands*, J. Am. Water Works Assoc., v. 39, pp. 979-988, Oct.

PARKER, G. G., 1945, *Salt-Water Encroachment in Southern Florida*, J. Am. Water Works Assoc., v. 37, pp. 526-542.

PENNINK, J. M. K., 1905, *Investigations for Ground-Water Supplies*, Trans. Am. Soc. Civil Engrs., v. 54-D, pp. 169-181.

PERLMUTTER, N. M., and GERAGHTY, J. J., 1963, *Geology and Ground-Water Conditions in Southern Nassau and Southeastern Queens Counties, Long Island, New York*, U. S. Geol. Survey Water-Supply Paper 1613-A, 205 pp.

PERLMUTTER, N. M., and others, 1959, *The Relation Between Fresh and Salty Ground Water in Southern Nassau and Southeastern Queens Counties, Long Island, New York*, Econ. Geol. v. 54, no. 3, pp. 416-435.

PINDER, G. F., and COOPER, H. H., Jr., 1970, *A Numerical Technique for Calculating the Transient Position of the Saltwater Front*, Water Resources Res., v. 6, no. 3, pp. 875-882.

POLAND, J. F., 1943, *Saline Contamination of Coastal Ground Water in South California*, Western City, v. 19, pp. 46, 48, 50, Oct.

RADER, E. M., 1955, *Salt Water Encroachment into Well Water in the Miami Area*, Proc. Am. Soc. Civil Engrs., v. 81, sep. 669, 11 pp.

REDDELL, D.L., 1984, *Single Location Doublet Well to Reduce Salt-Water Encroachment: Phase I, Numerical Simulation*, PB84-110873, NTIS, 5285 Port Royal Rd., Springfield, Virginia 22161, 51 pp.

REILLY, T.E., and GOODMAN, A.S., 1985, *Quantitative Analysis of Saltwater-Freshwater Relationships in Groundwater Systems-A Historical Perspective*, J. Hydrology, v. 80, no. 1/2, Sept. 15.

REVELLE, R. 1941, *Criteria for Recognition of Sea Water in Ground-Waters*, Trans. Am. Geophys. Union, v. 22, pp. 593-597.

RHODES, A. D., 1951, *Puddled-Clay Cutoff Walls Stop Sea-Water Infiltration*, Civil Eng., v. 21, no. 2, pp. 21-23.

RIDDEL, J. O., 1933, *Excluding Salt Water from Island Wells — A Theory of the Occurrence of Ground Water Based on Experience at Nassau, Bahama Islands*, Civil Eng., v. 3, pp. 383-385.

ROCHESTER, E. W., Jr., and KRIZ, G. J., 1970, *Potable Water Availability on Long Oceanic Islands*, J. Sanitary Engrng. Div., Amer. Soc. Civil Engrs., v. 96, no. SA5, pp. 1235-1248.

RUMER, R. R., Jr., and HARLEMAN, D. F., 1963, *Intruded Salt-Water Wedge in Porous Media*, J. Hydraul. Div., Am. Soc. Civil Engrs., v. 89, no. HY6, pp. 193-220.

RUMER, R. R., Jr., and SHIAU, J. C., 1968, *Salt Water Interface in a Layered Coastal Aquifer*, Water Resources Res., v. 4, no. 6, pp. 1235-1249.

SA DA COSTA, ANTONIO and WILSON, J. L., 1979, *A Numerical Model of Sea-Water Intrusion in Aquifers*, PB80-163, NTIS, Springfield, Virginia 22161.

SAHNI, B. M., 1973, *Physics of Brine Coning Beneath Skimming Wells*, Ground Water, v. 11, no. 1, pp. 19-24, Jan.-Feb.

SCHMORAK, S., 1967, *Salt Water Encroachment in the Coastal Plain of Israel*, Internat. Assoc. Sci. Hydrology, Haifa, Pub. 72, pp. 305-318.

SCHMORAK, S., AND MERCADO, A., 1969, *Upconing of Fresh Water − Sea Water Interface Below Pumping Wells*, Water Resources Res., v. 5, no. 6, pp. 1290-1311.

SEGOL, GENEVIEVE and PINDER, G. F., 1976, *Transient Simulation of Saltwater Intrusion in Southeastern Florida*, Water Resources Res., v. 12, no. 1, pp. 65-70.

SEGOL, GENEVIEVE, PINDER, G. F. and GRAY, W. G., 1975, *A Galerkin-Finite Element Technique for Calculating the Transient Position of the Saltwater Front*, Water Resources Res., v. 11, no. 2, pp. 343-347.

SENIO, K., 1951, *On the Ground Water Near the Seashore*, General Assembly Brussels, Internat. Assoc. Sci. Hydrology, v. 2, pp. 175-177.

SHAMIR, U., and DAGAN, G., 1971, *Motion of the Seawater Interface in Coastal Aquifers: A Numerical Solution*, Water Resources Res., v. 7, no. 3, pp. 644-657.

SHEAHAN, N. T., 1977, *Injection/Extraction Well System − A Unique Seawater Intrusion Barrier*, Ground Water, v. 15, no. 1, pp. 32-50.

SHECHTER, M., and SCHWARTZ, J., 1970, *Optimal Planning of a Coastal Collector*, Water Resources Res., v. 6, no. 4, pp. 1017-1025.

SIMPSON, T. R., 1946, *Salinas Basin Investigation*, Bull. 52, Calif. Div. Water Resources, Sacramento, 230 pp.

SPROUL, C. R., BOGGES, D. H., and WOODARD, H. J., 1972, *Saline-Water Intrusion from Deep Artesian Sources in the McGregor Isles Area of Lee County, Florida*, Florida Dept. Nat. Res. Inf. Circ. 75.

STRINGFIELD, V. T. and LeGRAND, H. E., 1971, *Effects of Karst Features on Circulation of Water in Carbonate Rocks in Coastal Areas*, J. Hydrology, v. 14, pp. 139-157.

TASK COMMITTEE ON SALTWATER INTRUSION, 1969, *Saltwater Intrusion in the United States*, J. Hydraulics Div., Am. Soc. Civil Engrs., v. 95, no. HY5, pp. 1651-1669, Sept.

THOMPSON, D. G., 1933, *Some Relations Between Ground-Water Hydrology and Oceanography*, Trans. Am. Geophys. Union, v. 14, pp. 30-33.

TODD, D. K., 1953, *Sea-Water Intrusion in Coastal Aquifers*, Trans. Am. Geophys. Union, v. 34, pp. 749-754.

TODD, D. K., 1960, *Salt Water Intrusion of Coastal Aquifers in the United States*, Internat. Assoc. Sci. Hydrology Publ. 52, pp. 452-461.

TODD, D. K., 1974, *Salt Water Intrusion and its Control*, J. Am., Water Works Assoc., v. 66, pp. 180-187.

TODD, D. K., and HUISMAN, L., 1959, *Ground Water Flow in the Netherlands Coastal Dunes*, J. Hydraulics Div., Am., Soc. Civil Engrs., v. 85, no. HY7, pp. 63-81.

TODD, D. K., and MEYER, C. F., 1971, *Hydrology and Geology of the Honolulu Aquifer*, J. Hydraulics Div., Am., Soc. Civil Engrs., v. 97, no. HY2, pp. 233-256.

TOLMAN, C. F., and POLAND, J. F., 1940, *Ground-Water, Salt-Water Infiltration and Ground-Surface Recession in Santa Clara Valley, Santa Clara County, California*, Trans. Am. Geophys. Union, v. 21, pp. 23-35.

TOYOHARA, Y., 1935, *A Study on the Coastal Ground Water at Yumigahama, Tottori*, Mem. College Sci., Kyoto Imp. Univ. Ser. A., v. 18, no. 5, pp. 295-309.

TURNER, S. F., and FOSTER, M. D., 1934, *A Study of Salt-Water Encroachment in the Galveston Area, Texas*, Trans. Am. Geophys. Union, v. 15, pp. 432-435.

VACHER, H. L., 1974, *Groundwater Hydrology of Bermuda*, Public Works Dept., Govt. of Bermuda, 87 pp.

VANDENBERG, A., 1975, *Simultaneous Pumping of Fresh and Salt Water from a Coastal Aquifer*, J. Hydrology, v. 14, no. 1/2, Jan.

VISHER, F. N., and MINK, J. F., 1964, *Ground-Water Resources in Southern Oahu, Hawaii*, U. S. Geol. Survey Water-Supply Paper 1778, 133 pp.

VOLKER, R. E., and RUSHTON, K. R., 1982, *An Assessment of the Importance of Some Parameters for Sea-Water Intrusion in Aquifers and a Comparison of Dispersive and Sharp-Interface Modelling Approaches*, J. Hydrology, v. 56, no. 3/4, Apr.

WALTERS, K. L., 1971, *Reconnaissance of Sea-Water Intrusion Along Coastal Washington*, Washington Dept. Water Resources, Water Supply Bull. 32, 208 pp.

WANG, F. C., 1965, *Approximate Theory for Skimming Well Formulation in the Indus Plain of West Pakistan*, J. Geophys. Res., v. 70, pp. 5055-5063.

WATSON, L. V., 1964, *Development of Ground Water in Hawaii*, J. Hydraulics Div., Am. Soc. Civil Engrs., v. 90, no. HY6, pp. 185-202.

WENTWORTH, C. K., 1939, *The Specific Gravity of Sea Water and the Ghyben-Herzberg Ratio at Honolulu*, Bull. Univ. of Hawaii, v. 18, no. 8, Jun.

WENTWORTH, C. K., 1942, *Storage Consequences of the Ghyben-Herzberg Theory*, Trans. Am. Geophys. Union, v. 23, pp. 683-693.

WENTWORTH, C. K., 1946, *Laminar Flow in the Honolulu Aquifer*, Trans. Am. Geophys. Union, v. 27, no. 4, pp. 540-548, Aug.

WENTWORTH, C. K., 1947, *Factors in the Behavior of Ground Water in a Ghyben-Herzberg System*, Pacific Sci., v. 1, pp. 172-184.

WENTWORTH, C. K., 1948, *Growth of the Ghyben-Herzberg Transition Zone Under a Rinsing Hypothesis*, Trans. Am. Geophys. Union, v. 29, pp. 97-98.

WENTWORTH, C. K., 1951, *The Process and Progress of Salt-Water Encroachment*, Internat. Assoc. Sci. Hydrology, Publ. 33, General Assembly Brussels, v. 2, pp. 238-248.

WELL DESIGN AND CONSTRUCTION/ DRILLING METHODS

BIBLIOGRAPHY

GIEFER, G. J., 1963, *Water Wells, an Annotated Bibliography*, Water Resources Center, Archives Ser. Rept. no. 13, Berkeley, California, 141 pp.

GENERAL WORKS

AHMAD, N., 1969, *Tubewells, Construction and Maintenance*, Scientific Stores, Lahore, Pakistan, 250 pp.

AHRENS, T. P., 1970, *Basic Considerations of Well Design*, Water Well J., (in four parts), Apr., pp. 45-50, May, pp. 49-52, Jun., pp. 47-51 and Aug., pp. 35-37.

AMERICAN PETROLEUM INSTITUTE, 1973, *Specifications for Casing, Tubing, and Drill Pipe*, A.P.I. Standard 5a, 32nd ed., Dallas, Texas.

AMERICAN PETROLEUM INSTITUTE, 1984, *API Specification for Materials and Testing for Well Cements*, API Spec. 10, API, 211 N. Ervay, Suite 1700, Dallas, TX, 75201, 89 pp.

AMERICAN WATER WORKS ASSOCIATION, 1967, *AWWA Standard for Deep Wells*, AWWA-A100-66, Denver, Colorado, 57 pp., transl. into Spanish by M. R. Llamas and A. Faura, *Normas de la American Water Works Association para Pozos Profundos*, Servicio Geologico, Bull. no. 30, Madrid 1969, 115 pp.

AMERICAN WATER WORKS ASSOCIATION, 1973, *Getting The Most from Your Well Supply*, Proc. AWWA Seminar June 4, 1972, Catalog no. 20122, AWWA, New York, 10016 72 pp.

AMERICAN WATER WORKS ASSOCIATION, 1973, *Ground Water*, AWWA Manual M21, 130 pp.

ANONYMOUS, 1958, *Drilling Mud Data Book*, Baroid Div., National Lead Co., Houston, Texas.

ANONYMOUS, 1966, *Cable Tool Drilling Manual*, Sanderson Cyclone Drill Co., Orville, Ohio, 50 pp.

ANONYMOUS, 1966, *What's New With Mud?*, Baroid Div., National Lead Corp., Spec. Rept., 35 pp.

ANONYMOUS, 1967, *Blaster's Handbook*, E.I. du Pont de Nemours & Co., Inc., Wilmington, Delaware.

ANONYMOUS, 1969, *Principles of Drilling Fluid Control*, ed. by Am. Petroleum Inst. and Univ. of Texas in cooperation with the Am. Assoc. of Oil Well Drilling Contractors, Dallas, 215 pp.

ANONYMOUS, 1969, *Well Drilling Manual*, Koehring, Speedstar Div., Enid, Oklahoma 72 pp.

ANONYMOUS, 1978, *Self-Help Wells*, Irrig. and Drainage Section, FAO, Rome 00100.

ANONYMOUS, 1981, *The Water Well Industry: A Study*, Water Well J., Jan., pp. 79-97. (USA water well statistics)

ANONYMOUS, 1986, *Manual on the Selection and Installation of Thermoplastic Water Well Casing*, National Water Well Assoc., 6375 Riverside Dr., Dublin, Ohio 43017, 64 pp.

AWWA *A100-84, Standard for Water Wells*, AWWA Data Processing Dept., 6666 W. Quincy Ave., Denver, Colorado 80235, 84 pp.

BARCELONA, M.J., GIBB, J.P., and MILLER, R.A., 1984, *Guide to the Selection of Materials for Monitoring Well Construction and Ground-Water Sampling*, Illinois State Water Survey Div., Champaign, IL, PB 84-141779, NTIS, Springfield, Virginia 22161, 87 pp.

BARON, D. M., 1982, *A Well System Can Be Designed to Minimize the Incrusting Tendency*, The Johnson Drillers Journal, v. 54, no. 1.

BENNETT, T. W., 1970, *On the Design and Construction of Infiltration Galleries*, Ground Water, v. 8, no. 3, May-Jun.

BENNISON, E. W., 1953, *Fundamentals of Water Well Operation and Maintenance*, J. Am. Water Works Assoc., v. 45, pp. 252-258.

BERNHART, A. P., 1973, *Protection of Water-Supply Wells from Contamination by Wastewater*, Ground Water, v. 11, no. 3, pp. 9-15.

BLACK, J. H., 1977, *Polyurethane Foam — A Useful New Borehole Grout*, J. Hydrology, v. 32, no. 1/2, Jan.

BLAIR, A. H., 1970, *Well Screens and Gravel Packs*, Ground Water, v. 8, no. 1.

Well Design

BLANKWAARDT, BOB, 1985, *Hand Drilled Wells*, NWWA, 6375 Riverside Dr., Dublin, Ohio 43017.

BOSTOCK, C. A., SIMPSON, E. S., AND ROEFS, T. G., 1977, *Minimizing Costs in Well Field Design in Relation to Aquifer Models*, Water Resources Res., v. 13, no. 2, pp. 420-426.

BOWMAN, ISAIAH, 1911, *Well-Drilling Methods*, U. S. Geol. Survey Water Supply Paper 257.

BRAKENSIEK, D. L., and others, 1979, *Field Manual for Research in Agricultural Hydrology*, USDA Agric. Handbook 224. (well drilling, logging)

BRANTLY, J. E., 1948, *Rotary Drilling Handbook*, Palmer Publications, Los Angeles, 565 pp.

BRANTLY, J. E., 1961, *Hydraulic Rotary-Drilling System* in *History of Petroleum Engineering*, D. V. Carter, (ed.), Am. Petroleum Inst., pp. 271-452.

BRANTLY, J. E., 1961, *Percussion-Drilling System*, in Carter, D. V., (ed.), *History of Petroleum Engineering*, Am. Petroleum Inst., pp. 133-269.

BRUSH, R. E., 1980, *Wells Construction, Hand Dug and Hand Drilled*, Peace Corps Appropriate Technology for Development Series, Peace Corps, Information Collection and Exchange, 806 Connecticut Ave., N.W., Washington, D. C. 20525, 282 pp.

BURKLUND, P.W., and RABER, ELLEN, 1983, *Method to Avoid Ground-Water Mixing Between Two Aquifers During Drilling and Well Completion Procedures*, Ground Water Monitoring Rev., v. 3, no. 4, pp. 48-55.

CALIFORNIA DEPARTMENT OF WATER RESOURCES, 1982, *Water Well Standards: State of California*, Bull. 74-81, P. O. Box 388, Sacramento, California 95802.

CAMPBELL, M. D., AND LEHR, J. H., 1972, *Water Well Technology*, McGraw-Hill Book Co., New York 10020, 681 pp.

CARLSTON, J. E., 1961, *Notes on the Early History of Water-Well Drilling in the United States*, Econ. Geol. v. 38, no. 2, pp. 119-136.

CATES, W. H., 1955, *Water Well Casing Manual*, U. S. Steel Pub. ADCWS-281-55, 44 pp.

CEDERSTROM, D. J., and TIBBITTS, G. C., Jr., 1961, *Jet Drilling In Fairbanks Area, Alaska*, U. S. Geol. Survey Water-Supply Paper 1539-B, 28 pp.

CLARK, L., and TURNER, P.A., 1983, *Experiments to Assess the Hydraulic Efficiency of Well Screens*, Ground Water, v. 21, no. 3, pp. 270-281.

CODE, W. E., 1949, *Rotary Method of Drilling Large Diameter Wells Using Reverse Circulation*, Water Well J., July-Aug.

CSALLANY, S., and WALTON, W. C., 1963, *Yields of Shallow Dolomite Wells in Northern Illinois*, Illinois Water Survey Rept. of Inv. 46.

CUSHMAN, R. V., and others, 1953, *Geologic Factors Affecting Yield of Rock Wells in Southern New England*, J. New England Water Works Assoc., v. 67, pp. 77-93.

DAVIS, S. N., and TURK, L. J., 1964, *Optimum Depth of Wells in Crystalline Rock*, Ground Water, v. 2, no. 2.

DECKER, M. G., 1986, *Cable Tool Fishing*, National Water Well Publ. Co., Dublin, Ohio 43017, 72 pp.

DHV CONSULTING ENGINEERS, 1979, *Shallow Wells*, P. O. Box 85, 3800 AB Amersfoort, The Netherlands. (Tanzania well drilling experience)

EBERLE, MICHAEL and PERSONS, J. L., 1978, *Appropriate Well Drilling Technologies: A Manual for Developing Countries*, National Water Well Assoc., 500 W. Wilson Bridge Rd., Worthington, Ohio 43085, 96 pp.

ERICKSON, C. R., and WRIGHT, R. C., 1957 *Maintenance of Rock Wells*, J. Am. Water Works Assoc., Jun.

ERICSON, W.A., BRINKMAN, J.E., and DARR, P.S., 1985, *Types and Usages of Drilling Fluids Utilized to Install Monitoring Wells Associated with Metals and Radionuclide Ground Water Studies*, Ground Water Monitoring Rev., v. 5, no. 1, pp. 30-33.

FEULNER, A. J., 1964, *Galleries and Their Use for Development of Shallow Ground-Water Supplies, with Special Reference to Alaska*, U. S. Geol. Survey Water Supply Paper 1809-E, 16 pp.

GARG, S. P., and LAL, J., 1971, *Rational Design of Well Screens*, J. Irrigation and Drainage Div., Am. Soc. Civil Engrs. v. 97, no. IR1, pp. 131-147

Well Design

GATLIN, CARL, 1960, *Petroleum Engineering Drilling and Well Completions*, Prentice-Hall, Inc., Englewood Cliffs, New Jersey.

GIBB, J. P., 1973, *Wells and Pumping Systems for Domestic Water Supplies*, Illinois State Water Survey Cir. 117, Urbana, 17 pp.

GIBSON, U. P., and SINGER, R. D., 1971, *Water Well Manual*, Premier Press, P. O. Box 4428, Berkeley, California 94704, 156 pp.

GIDLEY, H. K., and MILLER, J. H., 1960, *Performance Records of Radial Collector Wells in Ohio River Valley*, J. Am. Water Works Assoc., v. 52, pp. 1206-1210.

GORDON, R. W., 1958, *Water Well Drilling with Cable Tools*, Bucyrus-Erie Co., South Milwaukee, Wisconsin.

GOSSETT, O. C., 1958, *Reverse Circulation Rotary Drilling*, Water Well J., Nov.

GRIDLEY, H. K., 1952, *Installation and Performance of Radial Wells in Ohio River Gravel*, J. Am. Water Works Assoc., Dec.

GRIDLEY, H. K., and MILLAR, J. H., 1960, *Performance Records of Radial Collector Wells in Ohio River Valley*, J. Am. Water Works Assoc., v. 52, no. 9, pp. 1206-1210, Sept.

GRIDLEY, H. K., and PAPADOPULOS, I. S., 1962, *Flow of Ground Water to Collector Wells*, Proc. Am. Soc. Civil Engrs., v. 88, no. HY5.

HANTUSH, M. S., 1961, *Economical Spacing of Interfering Wells*, in *Groundwater in Arid Zones*, Internat. Assoc. Sci. Hydrology, Pub. no. 57, pp. 350-364.

HEISS, H. W., Jr., 1980, *Manual of Recommended Safe Operating Procedures and Guidelines for Water Well Contractors and Pump Installers*, National Water Well Assoc., 6375 Riverside Dr., Dublin, Ohio 43017, 86 pp.

HELWEG, O. J., 1982, *Evaluating and Improving Existing Ground-Water Systems*, Ground Water, v. 20, no. 4, pp. 402-409. (well efficiency)

HUNTER BLAIR, A., 1970, *Well Screens and Gravel Packs*, Ground Water, v. 8, no. 1, pp. 10-21.

JANN, R. H., 1966, *Method for Deep Well Alignment Tests*, J. Am. Water Works Assoc., v. 58, pp. 440, Apr.

JOHNSON, A. I. and others, 1966, *Laboratory Study of Aquifer Properties and Well Design for an Artificial-Recharge Site*, U. S. Geol. Survey Water-Supply Paper 1615-H, 42 pp.

JOHNSON, R. C. Jr., KURT, C. E., and DUNHAM, G. F., Jr., 1980, *Experimental Determination of Thermoplastic Casing Collapse Pressures*, Ground Water, v. 18, no. 4, Jul.-Aug., pp. 346-350.

JOHNSON, R. C., Jr., KURT, C. E., and DUNHAM, G. F., Jr., 1980, *Well Grouting and Casing Temperature Increases*, Ground Water, v. 18, no. 1, pp. 7-13.

JOHNSTON, C. N., 1951, *Irrigation Wells and Well Drilling*, Agric. Exp. Sta. Circ. 404, Univ. California, Berkeley, 32 pp.

JONES, E. E., Jr., 1979, *Well Disinfection — How Fast? How Sure?*, Water Well J., v. 33, no. 11, pp. 56-59.

KOJICIC, BOZIDAR, 1979, *Guidelines for Drilling Water Wells*, Drinking Water Program Div. UNICEF, New York, New York 11017.

KOOPMAN, F. C., and others, 1962, *Use of Inflatable Packers in Multiple-Zone Testing of Water Wells*, U. S. Geol. Survey Research, Prof. Paper 450-B, pp. B108-109.

KOVACS, GYORGY, and UJFALUDI, LASZLO, 1983, *Movement of Fine Grains in the Vicinity of Well Screens*, Hydrological Science J., v. 28, no. 2.

KRUSE, GORDON, 1960, *Selection of Gravel Packs for Wells in Unconsolidated Aquifers*, Tech. Bull. 66, Colorado State Univ., Fort Collins, Colorado.

KURT, C. E., 1979, *Collapse Pressure of Thermoplastic Water Well Casings*, Ground Water, v. 17, no. 6, pp. 550-556.

KURT, C. E., and JOHNSON, R. C., Jr., 1982, *Permeability of Grout Seals Surrounding Thermoplastic Well Casing*, Ground Water, v. 20, no. 4, pp. 415-419.

MACHIS, ALFRED, 1946, *Experimental Observations on Grouting Sands and Gravels*, Proc. Am. Soc. Civil Engrs., pp. 1207, 1218, 1226, and 1227, Nov.

MACKANESS, F. G. (ed.), 1968, *Manual of Water Well Construction Practices*, Oregon Drilling Assoc., Salem, Oregon, 84 pp.

MATLOCK, W. G., 1970, *Small Diameter Wells Drilled by Jet-Percussion Method*, Ground Water, v. 8, no. 1, pp. 6-9, Jan.-Feb.

MAURER, W. C., 1968, *Novel Drilling Techniques*, Pergamon Press Inc., Elmsford, New York and Oxford, England, 124 pp.

MEYER, G., and WYRICK, G. G., 1966, *Regional Trends in Water Well Drilling in the United States*, U. S. Geol. Survey Circ. 533, 8 pp.

MOEHRL, K. E., 1964, *Well Grounding and Well Protection*, J. Am. Water Works Assoc., v. 56, pp. 423-431.

MOGG, J. L., 1963, *The Technical Aspects of Gravel Well Construction*, J. New Eng. Water Works Assoc., v. 77, pp. 155-164.

MOGG, J. L., 1972, *Practical Construction and Incrustation Guide Lines for Water Wells*, Ground Water, v. 10, no. 2, pp. 6-11.

MOGG, J. L., 1974, *Designing an Efficient Well Can Be Easy*, Water and Wastes Engineering, v. 11, no. 1, Jan.

MOLZ, F. J., and KURT, C. E., 1979, *Grout-Induced Temperature Rise Surrounding Wells*, Ground Water, v. 17, no. 3, pp. 264-269.

MOORE, P. L., 1974, *Drilling Practices Manual*, Petroleum Publishing, Tulsa, Oklahoma, 448 pp.

MOSS, ROSCOE, Jr., 1958, *Water Well Construction in Formation Characteristics of the Southwest*, J. Am. Water Works Assoc., v. 50, p. 777, Jun.

MOSS, ROSCOE, Jr., 1964, *Design of Casings and Screens for Water Production and Injection Wells*, A.P.I. Pacific Coast Dist. Biennial Symp. on Treatment and Control of Injection Wells, Anaheim, California, 25 pp.

MURRAY, A. A., and ECKEL, J. E., 1961, *Foam Agents and Foam Drilling*, Oil and Gas J., pp. 125-128, Feb.

NAKAMOTO, D., McLAREN, F.R., and PHILLIPS, P.J., 1986, *Multiple Completion Monitor Wells*, Ground Water Monitoring Rev., v. 6, no. 2, pp. 50-55.

NATIONAL WATERWELL AND DRILLING ASSOC. OF AUSTRALIA, 1985, *The Australian Driller's Guide*, NWDAA, P.O. Box 187, St. Ives, NSW 2075, Australia, 1500 pp.

NATIONAL WATERWELL AND DRILLING ASSOCIATION OF AUSTRALIA, 1985, *The Complete Driller's Manual*, NWDAA, P.O. Box 187, St Ives, NSW 2075, Australia, 296 pp.

NATIONAL WATER WELL ASSOCIATION, 1971, *Water Well Driller's Beginning Training Manual*, NWWA, Dublin, Ohio, 84 pp.

NATIONAL WATER WELL ASSOCIATION, 1976, *Manual of Water Well Construction Practices*, NWWA, Dublin, Ohio 43017, 156 pp., also issued by U. S. Environmental Protection Agency, Publ. no. EPA-570/9-75-001, reprinted by Premier Press (1981) P. O. Box 4428, Berkeley, California 94704.

NATIONAL WATER WELL ASSOCIATION/PLASTIC PIPE INSTITUTE, 1980, *Manual on the Selection and Installation of Thermoplastic Water Well Casing*, NWWA, Dublin, Ohio 43017.

NATIONAL WATER WELL ASSOCIATION OF AUSTRALIA, 1980, *Drillers Training and Reference Manual*, NWWA of Australia, P. O. Box 91, St. Ives, 2075 NSW Australia, 284 pp.

NEW ENGLAND AQUARIUM, 1984, *Survey of The Toxicity and Chemical Composition of Used Drilling Muds*, Edgerton Research Lab., Boston, MA, EPA-600/3-84-071, NTIS, PB84-207661, 125 pp.

NOBLE, D. G., 1963, *Well Points for Dewatering*, Ground Water, v. 1, no. 3, pp. 21-26.

NUZMAN, C. E., 1978, *Groundwater and Well Efficiency*, Doerr Metal Products, Larned, Kansas 67550.

OIL AND GAS JOURNAL, 1959, *Driller's Handbook*, Reprints from Oil and Gas J., Tulsa, Oklahoma.

PATCHICK, P. F., 1966, *Quicksand and Water Wells*, Ground Water, v. 4, no. 2, pp. 32-46.

PILLSBURY, A. F., and CHRISTIANSEN, J. E., 1947, *Installing Ground-Water Piezometers by Jetting for Drainage Investigations*, Agric. Eng., v. 28, pp. 409-410.

PLUMB, C. E., and WELSH, J. L., 1955, *Abstract of Laws and Recommendations Concerning Water Well Construction and Sealing in the United States*, Water Quality Invest. Rept. 9, California Div. Water Resources, Sacramento, 391 pp.

POEHLMAN, JIM, 1985, *Drilling With Foam*, Ground Water Age, v. 19, no. 7.

REINKE, J. W., and KILL, D. L., 1970, *Modern Design Techniques for Efficient High Capacity Irrigation Wells*, Winter Meeting, Am. Soc. Agric. Engrs., Paper 70-732, Dec., 23 PP.

REPUBLIC STEEL CORPORATION, 1965, *Standard and Line Pipe*, Cleveland, Ohio, 52 pp.

RITCHIE, E. A., 1976, *Cathodic Protection Wells and Ground-Water Pollution*, Ground Water, v. 14, pp. 146-149.

ROSCOE MOSS COMPANY, *The Engineer's Manual for Well Design*, 4360 Worth St., Los Angeles, California 90031, (also available from NWWA).

SCHWALEN, H. C., 1925, *The Stovepipe or California Method of Well Drilling as Practiced in Arizona*, Bull. 112, Univ. Arizona Agric. Exp. Sta., Tucson, pp. 103-154.

SCHWARTZ, D. H., 1969, *Successful Sand Control Design for High Rate Oil and Water Wells*, J. Petroleum Tech., v. 21, pp. 1193-1198.

SERVICIO GEOLOGICO DE OBRAS PUBLICAS, INSTITUTO NACIONAL DE COLONIZACION, CENTRO DE ESTUDIOS INVESTIGACION Y APLICACIONES DEL AGUA, 1968, *Primer Seminario de Técnicas Modernas para la Construcción de Pozos Ponencias*, C.E.I.A.A., Barcelona, 529 pp.

SHUTER, EUGENE and PEMBERTON, R. R., 1979, *Inflatable Straddle Packers and Associated Equipment for Hydraulic Fracturing and Hydrologic Testing*, PB-288 403, NTIS, Springfield, Virginia 22161.

SMITH, H. F., 1958, *Gravel Packing Water Wells*, Water Well J., v. 8, nos. 1 and 2.

SMITH, L. A., 1941, *Deep Wells in Sandstone Rock*, Water Works Eng., v. 94, pp. 710-712.

SMITH, R. C., 1963, *Relation of Screen Design of Mechanically Efficient Wells*, J. Am. Water Works Assoc., v. 55, no. 5, pp. 609-614.

SOLIMAN, M. M., 1965, *Boundary Flow Considerations in the Design of Wells*, J. Irrigation and Draining Div., Am. Civil Engrs., v. 91, no. IR1, pp. 159-177.

SPOONER, P.A., HUNT, G.E., and others, 1984, *Compatibility of Grouts with Hazardous Wastes*, JRB Associates, Inc., McLean, VA, PB84-139732, NTIS, Springfield, Virginia 22161, 153 pp.

SPIRIDONOFF, S. V., 1964, *Design and Use of Radial Collector Wells*, J. Am. Water Works Assoc., Jun.

STANLEY, ANITA, 1980, *Newest Water Well Drill — The Supersonic Flame Jet*, Water Well J., v. 34, no. 9, pp. 67-68.

STATE OF CALIFORNIA, 1984, *Water Well Standards*, CA, Dept. of Water Resources, P.O. Box 388, Sacramento, California 95802.

STONE, R., 1954, *Infiltration Galleries*, Proc. Am. Soc. Civil Engrs. v. 80, sep 472, 12 pp.

STONER, R. F., and others, 1979, *Economic Design of Wells*, Quarterly J. Engineering Geology, no. 12, pp. 63-78.

STOW, G. R. S., 1963, *Modern Water-Well Drilling Techniques in Use in the United Kingdom*, Ground Water, v. 1, no. 3, pp. 3-12.

THEIS, C. V., 1957, *The Spacing of Pumped Wells*, U. S. Geol. Survey, Ground Water Notes, Hydraulics, no. 31, open-file rept.

TODD, D. K., 1955, *Discussion of Infiltration Galleries*, Proc. Am. Soc. Civil Engrs., v. 81, sep. 647, pp. 7-9.

U. S. BUREAU OF MINES, 1968, *Horizontal Boring Technology: A State of the Art Study*, Info. Cir. 8392, 86 pp.

U. S. DEPARTMENT OF THE ARMY, 1965, *Well Drilling Operations*, Tech. Manual TM5-297, AFM-85-23, U.S. Gov't Printing Office Washington, D.C., 264 pp., AFM-85-23.

U. S. ENVIRONMENTAL PROTECTION AGENCY, 1980, *Manual of Water Well Construction Practices*, Pub. No. EPA-570/9-75-001, Washington, D.C. 20460

U. S. ENVIRONMENTAL PROTECTION AGENCY, 1983, *Manual of Individual Water Supply Systems*, Supt. Doc., Dept 36-BN, Stock no. 055-000-00229-1, Washington, DC 20402, 156 pp.

U. S. PUBLIC HEALTH SERVICE, 1962, *Manual of Individual Water Supply Systems*, Publ. no. 24, 121 pp.

U. S. SOIL CONSERVATION SERVICE, 1969, *Engineering Field Manual for Conservation Practices*, 995 pp.

VAADIA, YOASH, and SCOTT, V. H., 1958, *Hydraulic Properties of Perforated Well Casings*, Proc. Am. Soc. Civil Engrs., Irrig. and Drainage Div., Paper 1505, Jan.

VAN ECK, O. J., 1978, *Plugging Procedures for Domestic Wells*, Iowa Geol. Survey Tech. Info. Series 7, Iowa City 52242.

WALKER, W. H., 1974, *Tube Wells, Open Wells and Optimum Ground-Water Resource Development*, Ground Water, v. 12, no. 1, pp. 10-15.

WALTON, W. C., and CSALLANY, S., 1962, *Yields of Deep Sandstone Wells in Northern Illinois*, Illinois. State Water Survey Rept. of Invest. 43.

Well Design

WALZ, D.H., and TOWNSEND, H.M., 1985, *How to Select the Right Screen*, Groundwater Age, v. 19, no. 7.

WATT, S. B., and WOOD, W. E., 1976, *Hand Dug Wells and Their Construction*, Intermediate Technology Publications, Ltd., London WC2E 8HN, England, 234 pp.

WELCHERT, W. T., and FREEMAN, B. N., 1973, *Horizontal Wells*, J. Range Management, v. 26, pp. 253-256.

WILLIAMS, E. B., 1981, *Fundamental Concepts of Well Design*, Ground Water, v. 19, no. 5, pp. 527-542.

YARE, B. S., 1975, *The Use of a Specialized Drilling and Ground-Water Sampling Technique for Delineation of Hexavalent Chromium Contamination in an Unconfined Aquifer, Southern New Jersey Coastal Plain*, Ground Water, v. 13, no. 2, pp. 151-154.

ZDENEK, F. F. and ALLRED, R. E., 1979, *Correct Methods are Essential to Well Development*, The Johnson Drillers Journal, Jan.-Feb.

WELL MAINTENANCE/CORROSION/IRON BACTERIA/INCRUSTATION/ STIMULATION

BADER, J. S., 1966, *Device for Removing Debris from Wells*, U. S. Geol. Survey Water-Supply Paper 1822, pp. 43-46

BARNES, IVAN and CLARKE, F. E., 1969, *Chemical Properties of Ground Water and Their Corrosion and Encrustation Effects on Wells*, U. S. Geol. Survey Prof. Paper 498-D, 58 pp.

BENNISON, E. W., 1953, *Fundamentals of Water Well Operation and Maintenance*, J. Am. Water Works Assoc., Mar.

BLAKELEY, L. E., 1945, *The Rehabilitation, Cleaning, and Sterilization of Water Wells*, J. Am. Water Works Assoc., v. 37, pp. 101-114.

BRACILOVIC, DRAGOMIR, and others, 1975, *Chlorination for Iron Removal in Well Water*, Water and Sewage Works, v. 122, no. 1, Jan.

BRIGGS, G. F., 1949, *Corrosion and Incrustation of Well Screens*, J. Am. Water Works Assoc., Jan.

BROWN, E. D., 1942, *Restoring Well Capacity with Chlorine*, J. Am. Water Works Assoc., v. 34, no. 5, pp. 698-702.

CHRISTIAN, R. D., 1975, *Distribution, Cultivation, and Chemical Destruction of Gallionella from Alabama Ground Water*, Unpubl. research thesis, Univ. of Alabama, University Station, Alabama.

CLARK, J. B., 1949, *A Hydraulic Process for Increasing the Productivity of Wells*, Trans. Am. Inst. Mining Engineers, v. 186.

CLARKE, F. E., 1964, *Selection of Metal Components for Long-Term Development of Egypt's Corrosive Ground Waters*, U. S. Geol. Survey Open-file Rept., 39 pp.

CLARKE, F. E., 1979, *The Corrosive Water Wells of Egypt's Western Desert*, U. S. Geol. Survey Water-Supply Paper 1757-0.

CLARKE, F. E., 1980, *Corrosion and Encrustration in Water Wells*, FAO Irrigation and Drainage Paper 34, FAO, Via delle Terme di Caracalla, 00100 Rome, Italy, 95 pp.

CLARKE, F. E., and BARNES, IVAN, 1965, *Study of Water-Well Corrosion, Chad Basin, Nigeria*, U. S. Geol. Survey Open-file Rept.

CLARKE, F. E. and BARNES, IVAN, 1969, *Evaluation and Control of Corrosion and Encrustation in Tube Wells of the Indus Plains, West Pakistan*, U. S. Geol. Survey Water-Supply Paper 1608-L, 61 pp.

CULLIMORE, D. R., 1980, *Iron Bacteria — Controlling Them with Disinfectants*, The Johnson Driller's Journal, v. 52, no. 4, pp. 6-8.

CULLIMORE, D. R., and McCANN, A. E., 1977, *The Identification, Cultivation, and Control of Iron Bacteria in Ground Water*; in Skinner and Shewan, (eds.), *Aquatic Microbiology*, Academic Press, New York.

DeWITT, M. M., 1947, *How to Acidize Water Wells*, American City, v. 62, no. 10, pp. 92-93.

EBAUGH, R. M., 1950, *Water Well Redevelopment by Explosives*, J. Am. Water Works Assoc., Feb.

ERICKSON, C. R., 1961, *Cleaning Methods for Deep Wells and Pumps*, J. Am. Water Works Assoc., v. 53, pp. 155-162.

FARBIN, A. O., 1948, *Column Life of Deep Well Pumps*, J. Am. Water Works Assoc., no. 40, p. 415.

GASS, T. E., and others, 1981, *Manual of Water Well Maintenance and Rehabilitation Technology*, Nat. Environmental Research Center, Ada, Oklahoma.

GROOM, C. H. AND BROWNING, J. T., 1947, *Water Well Acidizing*, Water Well J., v. 1, no. 1, pp. 9-11, Urbana, Illinois.

HACKETT, GLEN, and LEHR, J.H., 1986, *Iron Bacteria Occurrence, Problems and Control Methods in Water wells*, National Water Well Assoc., Dublin, Ohio 43017, 79 pp.

HARDER, E. C., 1915, *Iron-Bacteria*, Science, N.S., v. 42, pp. 310-311.

HARDER, E. C., 1919, *Iron-Depositing Bacteria and Their Geologic Relations*, U. S. Geol. Survey Prof. Paper 113, 85 pp.

HELWEG, J., and others, 1983, *Improving Well and Pump Efficiency*, Am. Water Works Assoc., Denver, Colorado 80235, 158 pp.

HEM, J. D., and CROPPER, W. H., 1959, *Survey of Ferrous-Ferric Equilibria and Redox Potentials*, U. S. Geol. Survey Water-Supply Paper 1459-A, 30 p.

JENSEN, O. F., Jr. and RAY, W., 1965, *Photographic Examination of Wells*, J. Am. Water Works Assoc., v. 57, pp. 441-447.

KELLY, G. J. and KEMP, R. G., 1974, *The Corrosion of Ground Water Pumping Equipment*, Australian Water Resources Council Project Rept. 72R, 5 appendices.

KELLY, S. F., 1939, *Photographing Rock Walls and Casings of Boreholes*, Trans. Am. Geophys. Union, v. 20, pp. 269-271.

KLEBER, J. P., 1950, *Well Cleaning with Calgon*, J. Am. Water Works Assoc., v. 42, pp. 481-484.

KOENIG, LOUIS, 1960, *Economic Aspects of Water Well Stimulation*, J. Am. Water Well Assoc., v. 52, p. 631, May.

KOENIG, LOUIS, 1960, *Effects of Stimulation on Well Operating Costs and its Performance on Old and New Wells*, J. Am. Water Well Assoc., v. 52, p. 1499, Dec.

KOENIG, LOUIS, 1960, *Survey and Analysis of Well Stimulation Performance*, J. Am. Water Works Assoc., v. 52, no.3, p. 333, Mar.

KOENIG, LOUIS, 1961, *Relation Between Aquifer Permeability and Improvement Achieved by Well Stimulation*, J. Am. Water Works Assoc., v. 53, p. 652, May.

LARSON, T. E., 1947, *Corrosion in Vertical Turbine Pumps*, J. Water and Sewage Works, v. 94, no. 4.

LARSON, T. E., 1975, *Corrosion by Domestic Waters*, Bull. 59, Illinois State Water Survey, Urbana, 48 pp.

LEWIS, R. F., 1965 *Control of Sulfate-Reducing Bacteria*, J. Am. Water Works Assoc., v. 57, no. 8, Aug.

McCOMBS, J., and FIEDLER, A. G., 1928, *Methods of Exploring and Repairing Leaky Artesian Wells*, U. S. Geol. Survey Water Supply Paper 596, pp. 1-32.

MILAEGER, R. E., 1942, *Development of Deep Wells by Dynamiting*, J. Am. Water Works Assoc., v. 34, pp. 684-690.

MOGG, J. L., 1972, *Practical Corrosion and Incrustation Guide Lines for Water Wells*, Ground Water, v. 10, no. 2, pp. 6-11.

MYLANDER, H. A., 1952, *Well Improvement by Use of Vibratory Explosives*, J. Am. Water Works Assoc., v. 44, pp. 39-48.

ONGERTH, H. J., 1942, *Sanitary Construction and Protection of Wells*, J. Am. Water Works Assoc., v. 34, pp. 671-677.

SHAFER, D. C., 1974, *The Right Chemicals are Able to Restore or Increase Well Yield*, The Johnson Drillers Journal, Jan.-Feb., p. 1, Mar.-April. p. 4.

SMITH, D. K., 1969, *Fiberglass Plastic Casing Overcomes Corrosion Problems in Water Wells in West Pakistan*, Trans. Soc. Min. Eng., A.I.M.E., v. 224, no. 1, pp. 24-28, Mar.

SMITH, STUART, 1982, *Sulfur Bacterial Problems Revisited*, Water Well J., Feb., pp. 44-45.

SMITH, STUART, and TUOVINEN, OLLI, 1985, *Environmental Analysis of Iron-Precipitating Bacteria in Ground Water and Wells*, Ground Water Monitoring Rev., v. 5, no. 4, pp. 45-53.

STETZENBACH, L.D., KELLEY, L.M., and SINCLAIR, N.A., 1986, *Isolation, Identification, and Growth of Well-Water Bacteria*, Ground Water, v. 24, no. 1, pp. 6-11.

STOTT, G. A., 1973, *The Tenacious Iron Bacteria*, The Johnson Drillers Journal, Jul.-Aug., p. 4.

STRAMEL, G. J., 1965, *Maintenance of Well Efficiency*, J. Am. Water Works Assoc., v. 57, pp. 996-1010.

TRAINER, F. W., and EDDY, J. E., 1964, *A Periscope for the Study of Borehole Walls, and its Use in Ground Water Studies in Niagara County, New York*, U. S. Geol. Survey Prof. Paper 501-D, pp. 203-206.

UHLIG, H. H., 1961, *The Corrosion Handbook*, John Wiley, New York and London.

VAN BEEK, C. G. E. M., 1984, *Restoring Well Yield in the Netherlands*, J. Am. Water Works Assoc., Oct., pp. 66-72.

VAN BEEK, C. G. E. M., and VAN DER KOOIJ, D., 1982, *Sulfate-Reducing Bacteria in Ground Water from Clogging and Non-Clogging Shallow Wells in the Netherlands River Region*, Ground Water, v. 20, no. 3, pp. 298-302.

WALKER, W. H., 1967, *When not to Acidize*, Ground Water, v. 5, no. 2, pp. 36-40, Apr.

WHITE, H. L., 1942, *Rejuvenating Wells with Chlorine*, Civil Eng., v. 12, pp. 263-265.

WILLIAMSON, W. H., and WOOLLEY, D. R., 1980, *Hydraulic Fracturing to Improve the Yield of Bores in Fractured Rock*, Australian Water Res. Council, Tech. Paper 55, Austr. Govt. Publishing Service, Box 84, Canberra ACT 2600, Australia.

ZINGG, W. M., 1953, *The How and When of Water Well Acidizing*, Plant Engineering J., Aug.

PUMPING EQUIPMENT

AMERICAN WATER WORKS ASSOCIATION, 1961, *(American Standard Specifications for Vertical Turbine Pumps (ASA B58.1; AWWA E101)*, J. Am. Water Works Assoc., v. 53, p. 333, Mar.

ANONYMOUS, 1954, *The Vertical Pump*, Johnston Pump Co., Pasadena, California.

ANONYMOUS, 1962, *Vertical Turbine Pump Facts*, Vertical Turbine Pump Assoc., Pasadena, California.

ANONYMOUS, 1970, *Application Manual for Large Submersible Pumps*, Red Jacket Mfg. Co., Davenport, Iowa.

ANONYMOUS, 1970, *Pump Fundamentals*, Goulds Pumps, Inc., Seneca Falls, New York.

ANONYMOUS, 1977, *Hand Pumps*, Internat. Reference Center for Community Water Supply, Tech. Paper Ser. No. 10, IRC P. O. Box 140, 2260 AC, Leidschendam, The Netherlands, 230 pp.

ERICKSON, C. R., 1960, *Submersible Water Well Pumps*, J. Am. Water Works Assoc., Sept.

HELWEG, O. J., SCOTT, V. H., and SCALMANINI, J. C., 1983, *Improving Well and Pump Efficiency*, Am. Water Works Assoc., Data Processing Dept., 6666 West Quincy Ave., Denver, Colorado 80235, 168 pp. (software programs).

HICKS, T. G., 1957, *Pump Selection and Application*, McGraw-Hill Book Co., New York, 422 pp.

HICKS, T. G., 1958, *Pump Operation and Maintenance*, McGraw-Hill Book Co., Inc., New York, 310 pp.

KARASSIK, I. J., and others, 1986, *Pump Handbook*, McGraw-Hill Book Co., Hightstown, New Jersey, 08520 1,280 pp.

KILL, D. L., 1974, *Cost Comparisons and Practical Applications of Air-Lift Pumping*, The Johnson Drillers Journal, Sep.-Oct.

PACEY, ARNOLD, 1978, *Hand Pump Maintenance*, Intermediate Technology Publications, Ltd. 9 Kings St., London WC2E 8HN, 138 pp.

SOMMERFELDT, T. G., and CAMPBELL, D. E., 1975, *A Pneumatic System to Pump Water from Piezometers*, Ground Water, v. 13, no. 3, pp. 293.

TRESCOTT, P. C., and PINDER, G. F., 1970, *Air Pump for Small-Diameter Piezometers*, Ground Water, v. 8, no. 3, May-Jun.

WASSON, R. H., 1968, *Industrial Pump Manual*, v. 1 and 2, Fairbanks, Morse & Co., Kansas City, Kansas.

WORLD HEALTH ORGANIZATION, 1979, *Handpumps Testing and Evaluation to Support Selection and Development of Handpumps for Rural Water Supply Programmes*, WHO Internat. Reference Center for Community Water Supply, P. O. Box 140, 2260 AC Leidschendam, The Netherlands, 55 p.

ARTIFICIAL RECHARGE, RIVER INFILTRATION, STORAGE/WASTE WATER RECLAMATION

BIBLIOGRAPHY

ANONYMOUS, 1982, *Percolation Techniques for Water Quality Control and Waste Treatment, 1977-June 1982, (255 Citations from the Selected Water Resources Abstracts Data Base)*, PB82-868134, NTIS, Springfield, Virginia 22161, 358 pp.

ANONYMOUS, 1985, *Artificial Recharge of Aquifers 1977-1984, (Citations from the Selected Water Resources Abstracts Data Base)*, PB85-852259/WEP, NTIS, Springfield, Virginia 22161, 172 pp.

BOURGUET, L., 1971, *Inventaire Internationale des Aménagements d'Alimentation Artificielle*, Depouillement et Synthèse des Réponses, Bull. Internat. Assoc. Sci. Hydrology, v. 16, no. 3, pp. 51-102.

CARLILE, B. L., and STEWART, J. M., 1977, *Land Application of Waste Water — A Bibliography*, (460 abstracts) PB-269 511/2WP, NTIS, Springfield, Virginia 22161.

INTERNATIONAL ASSOCIATION OF SCIENTIFIC HYDROLOGY, 1970, *Artificial Groundwater Recharge — International Survey of Existing Water Recharge Facilities*, Internat. Assoc. Sci. Hydrology, Pub. 87, 762 pp.

KEITH, W. J., and others, 1982, *Bibliography on Ground-Water Recharge in Arid and Semiarid Areas*, Univ. Arizona Water Resources Research Center, Tucson, Arizona, 149 pp.

KNAPP, G. L., 1973, *Artificial Recharge of Groundwater — A Bibliography*, WRSIC 73-202, Water Resources Sci. Info. Center, Washington, D. C., 309 pp.; also PB 221479, NTIS, Springfield, Virginia 22161, 309 pp.

SIGNOR, D. C., and others, 1970, *Annotated Bibliography on Artificial Recharge of Ground Water, 1955-67*, U. S. Geol. Survey Water-Supply Paper 1990, 141 pp.

TODD, D. K., 1959, *Annotated Bibliography on Artificial Recharge of Ground Water Through 1954*, U. S. Geol. Survey Water-Supply Paper 1477, 115 pp.

GENERAL WORKS

ABERBACH, S. H., and SELLINGER, A., 1967, *Review of Artificial Ground-Water Recharge in the Coastal Plain of Israel*, Internat. Assoc. Sci. Hydrology, Symp. Haifa, v. 12, no. 1, pp. 65-77.

AMERICAN WATER WORKS ASSOCIATION, 1967, *Artificial Ground Water Recharge*, Task Group Rept., J. Am. Water Works Assoc., v. 59, p. 103, Jan.

ANDERSON, G. S., 1977, *Artificial Recharge Experiments on the Ship Creek Alluvial Fan, Anchorage, Alaska*, U. S. Geol. Surv. Water-Resours. Inv. 77-30, 50 pp.

ANONYMOUS, 1957, *Artificial Recharge Experiments at McDonald Well Field, Amarillo, Texas*, State Bd. of Water Engrs. Bull. 570.

ANONYMOUS, 1985, *The Future of Water Reuse*, Proc. 3rd Symp. on Water Reuse, San Diego, Calif., Aug. 1984, AWWA Research Foundation, 6666 W. Quincy Ave., Denver, Colorado 80235, 1830 pp.

ARNOLD, C. E., and others, 1949, *Report upon the Reclamation of Water from Sewage and Industrial Wastes in Los Angles County, California*, Los Angeles County Flood Control District, Los Angeles, 159 pp.

ARONOVICI, V. S. and others, 1972, *Basin Recharge of the Ogallala Aquifer*, J. Irrig. Drain. Div., Am. Soc. Civil Engrs., v. 98, no. IR1, pp.65-76.

ARONSON, D. A., 1980, *The Meadowbrook Artificial-Recharge Project in Nassau County, Long Island, New York*, Long Island Water Resources Bull. 14, 23 p.

ARONSON, D. A., 1980, *Use of Highly Treated Wastewater to Recharge the Ground-Water Reservoir in Nassau County, Long Island, New York*, in Am. Soc. Civil Engineers, Proc. of the ASCE Environmental Engineering Div., National Conf. on Environmental Engineering, p. 214-220.

ARONSON, D. A., and PRILL, R. C., 1977, *Analysis of Recharge Potential of Storm-Water Basins on Long Island, New York*, U. S. Geol. Surv. J. Res., v. 5, no. 3, pp. 307-318.

ASANO, TAKASHI, (ed.), 1980, *Proc. Ground-Water Recharge Symp. Pomona, California, Sept. 6-7, 1979*, State Water Resources Control Board, Office of Water Recycling, P. O. Box 100, Sacramento, California 95801.

ASANO, TAKASHI, 1984, *Artificial Recharge of Ground Water*, Butterworth Publishers, 80 Montvale Ave., Stoneham, Massachusetts 02180, 800 pp.

ASANO, TAKASHI and WASSERMANN, K. L., 1980, *Ground-Water Recharge Operations in California*, J. Am. Water Works Assoc., Jul., pp. 380-385.

BABCOCK, H. M., and CUSHING, E. M., 1942, *Recharge to Ground Water from Floods in a Typical Desert Wash, Pinal County, Arizona*, Trans. Am. Geophys. Union, v. 23, pp. 49-56.

BAFFA, J. J., 1970, *Injection Well Experience at Riverhead, N.Y.*, J. Am. Water Works Assoc., v. 62, no. 1, Jan.

BAFFA, J. J., 1975, *Artificial Groundwater Recharge and Wastewater Reclamation*, J. Am. Water Works Assoc., v. 67, no. 9, pp. 471-476.

BAFFA, J. J., and others, 1958, *Developments in Artificial Ground-Water Recharge*, J. Am. Water Works Assoc., v. 50, no. 7, p. 812-871.

BAIER, D. C., and WESNER, G. M., 1971, *Reclaimed Waste Water for Ground-Water Recharge*, Water Resources Bull., v. 7, no. 5, pp. 991-1001.

BANKS, H. O., and others, 1954, *Artificial Recharge in California*, California Div. of Water Resources, Sacramento, 41 pp.

BARKSDALE, H. C., and DE BUCHANANNE, G. D., 1946, *Artificial Recharge of Productive Ground-Water Aquifers in New Jersey*, Econ. Geol., v. 41, no. 7, pp. 726-737.

BAUMANN, P., 1952, *Ground-Water Movement Controlled Through Spreading*, Trans. Am. Soc. Civil Engrs., v. 117, pp. 1024-1074.

BAUMANN, P., 1955, *Ground Water Phenomena Related to Basin Recharge*, Proc. Am. Soc. Civil Engrs., v. 81, sep. 806, 25 pp.

BAUMANN, P., 1963, *Theoretical and Practical Aspects of Well Recharge*, Trans. Am. Soc. Civil Engrs., v. 128, pt. I, pp. 739-764.

BAUMANN, P., 1965, *Technical Development in Ground Water Recharge,* in Chow, V.T., (ed.), Advances in Hydroscience , v. 2, pp. 209-279, Academic Press, New York.

BAXTER, K.M., and CLARK, L., 1984, *Effluent Recharge: The Effects of Effluent Recharge on Groundwater Quality,* Water Research Centre, Stevenage (England), TR-199, NTIS, PB84-201805, 61 pp, (effects of recharging sewage effluent to major aquifers).

BEAR, J., and BRAESTER, C., 1966, *Flow from Infiltration Basins to Drains and Wells,* J. Hydraulics Div., Am. Soc. Civil Engrs., v. 92, no. HY5, pp. 115-134.

BEAR, J., and JACOBS, M., 1965, *On the Movement of Water Bodies Injected into Aquifers,* J. Hydrology, v. 3, pp. 37-57.

BEHNKE, J. J., 1969, *Clogging in Surface Spreading Operations for Artificial Ground-Water Recharge,* Water Resources Res. v. 5, no. 4., p. 870.

BEREND, J. E., and others, 1967, *Use of Storm Runoff for Artificial Recharge,* Trans. Am. Soc. Agric. Engrs. v. 10, no. 5.

BETTAQUE, R. H. G., 1958, *Studien zur Künstlichen Grundwasseranreicherung,* Pub. Inst. Siedlungswasserwirtschaft, v. 2, Techn. Hochschule, Hanover, 105 pp.

BIANCHI, W. C., and MUCKEL, D. C., 1970, *Ground-Water Recharge Hydrology,* ARS 41-161, Agric. Research Service, U. S. Dept. Agric., 62 pp.

BIANCHI, W. C. and others, 1978, *A Case History to Evaluate the Performance of Water-Spreading Projects,* J. Am. Water Works Assoc., v. 70, pp. 176-180.

BIEMOND, C., 1957, *Dune Water Flow and Replenishment in the Catchment Area of the Amsterdam Water Supply,* J. Inst. Water Engrs., v. 11, pp. 195-213.

BITTINGER, M. W., and TRELEASE, F. J., 1960, *The Development and Dissipation of a Ground-Water Mound Beneath a Spreading Basin,* Winter Meeting of Am. Soc. Agric. Engrs., Memphis, Tennessee, Dec. 4-7; also Trans. Am. Soc. Agric. Engrs., v. 8, pp. 103-104, 106, 1965.

BLISS, E. S., and JOHNSON, C. E., 1952, *Some Factors Involved in Ground-Water Replenishment,* Trans. Am. Geophys. Union, v. 33, pp. 547-558.

BOGGESS, D. H., and RIMA, D. R., 1962, *Experiments in Water Spreading at Newark, Delaware*, U. S. Geol. Survey Water-Supply Paper 1594-B, 15 pp.

BOND, J. G., and others, 1972, *Delineation of Areas for Terrestrial Disposal of Waste Water*, Water Resources Res., v. 8, no. 6, pp. 1560-1573.

BOUWER, HERMAN, 1974, *Design and Operation of Land Treatment Systems for Minimum Contamination of Ground Water*, Ground Water, v. 12, no. 3, pp. 140-147.

BOUWER, HERMAN, 1974, *Renovating Municipal Wastewater by High-Rate Infiltration of Groundwater Recharge*, J. Am. Water Works Assoc., v. 66, pp. 159-162.

BOUWER, HERMAN, 1976, *Zoning Aquifers for Tertiary Treatment of Wastewater*, Ground Water, v. 14, no. 6, pp. 386-392.

BOUWER, HERMAN, and others, 1972, *Renovating Secondary Sewage by Ground Water Recharge with Infiltration Basins*, U. S. Water Conservation Lab., Office of Research and Monitoring, Proj. No. 16060 DRV, U. S. Environmental Protection Agency.

BOUWER, HERMAN, and others, 1978, *Land Treatment of Wastewater in Today's Society*, Civil Engrng, v. 48, no.1, pp. 78-81.

BRASHEARS, M. L., Jr., 1946, *Artificial Recharge of Ground Water on Long Island, New York*, Econ. Geol., v. 41, pp. 503-516.

BRASHEARS, M. L. Jr., 1953, *Recharging Ground-Water Reservoirs with Wells and Basins*, Min. Eng., v. 5, pp. 1029-1032.

BRAUNSTEIN, JULES, (ed.), 1974, *Underground Waste Management and Artificial Recharge*, 2 vols., Am. Assoc. Petrol. Geol., Tulsa, Oklahoma, 931 pp.

BRICE, H. D., and others, 1959, *A Progress Report on the Disposal of Storm Water at an Experimental Seepage Basin near Mineola, New York*, U. S. Geol. Survey Open-file Rept., 34 p.

BROCK, R. R., 1974, *Hydrodynamics of Artificial Ground-Water Recharge*, No. PB-235831, NTIS Springfield, Virginia 22161.

BROWN, D. L., and SILVEY, W. D., 1977, *Artificial Recharge to a Freshwater-Sensitive Brackish-Water Sand Aquifer, Norfolk, Virginia*, U. S. Geol. Survey Prof. Paper 939, 53 pp.

BROWN, R. F., and SIGNOR, D. C., 1972, *Groundwater Recharge*, Water Resources Bull., v. 8, pp. 132-149.

BROWN, R. F., and SIGNOR, D. C., 1973, *Artificial Recharge — State of the Art*, Underground Waste Management and Artificial Recharge, v. 2, Am. Assoc. Petrol. Geol., Tulsa, Oklahoma, pp. 668-686.

BROWN, R. F., and SIGNOR, D. C., 1974, *Artificial Recharge — State of the Art*, Ground Water, v. 12, pp. 152-160.

BUCHAN, S., 1955, *Artificial Replenishment of Aquifers*, J. Inst. Water Engrs., v. 9, pp. 111-163.

BUSH, P. W., 1983, *Connector Well Experiment to Recharge the Floridan Aquifer East Orange Co., Florida*, U. S. Geol. Survey Water Supply Paper W 2210, 26 pp.

CALIFORNIA DEPT. OF WATER RESOURCES, 1961, *Feasibility of Reclamation of Water from Wastes in the Los Angeles Metropolitan Area*, Bull. 80, Sacramento, 183, pp.

CALIFORNIA DEPT. OF WATER RESOURCES, 1973, *Waste Water Reclamation*, Bull. 189, The Resources Agency, Sacramento, 43 pp.

CALIFORNIA STATE WATER RESOURCES CONTROL BOARD., DEPT. OF WATER RESOURCES AND DEPT. OF HEALTH, 1978, *A "State-of-the-Art" Review of Health Aspects of Wastewater Reclamation for Groundwater Recharge*, Water Information Center, Plainview, New York, 240 pp.

CEDERSTROM, D. J., 1947, *Artificial Recharge of a Brackish Water Well*, The Commonwealth, Virginia Chamber of Commerce, Richmond, Dec.

CHRISTIANSEN, J. E., and MAGISTAD, O. C., 1945, *Report for 1944 — Laboratory Phases of Cooperative Water-Spreading Study*, U. S. Regional Salinity Lab., Riverside, California, 74 pp.

CLYDE, G. D., 1951, *Utilization of Natural Underground Water Storage Reservoirs*, J. Soil and Water Conserv., v. 6, pp. 15-19.

CLYMA, WAYNE, 1964, *Artificial Groundwater Recharge by a Multiple-Purpose Well*, Texas Agric. Exp. Sta., Misc. Pub. 712.

COHEN, PHILIP, and DURFOR, C. N., 1966, *Design and Construction of a Unique Injection Well on Long Island, New York*, U. S. Geol. Survey Prof. Paper 550-D, p. D253-D257.

COHEN, PHILIP, and DURFOR, C. N., 1967, *Artificial-Recharge Experiments Utilizing Renovated Sewage-Plant Effluent — A Feasibility Study at Bay Park, New York*, Internat. Assoc. Sci., Hydrology Pub. 72, p. 194-199.

COORDINATING COMMITTEE ON ENVIRONMENTAL QUALITY, 1973, *Recycling Municipal Sludges and Effluents on Land*, Proc. Joint Conf., Washington, D.C.: Nat. Assoc. of State Universities and Land-Grant Colleges.

COTTON, F. S., and others, 1977, *Artificial Recharge of Ground Water in the San Joaquin-Central Coast Area*, California Dept. of Water Resources, San Joaquin District, State of California, Documents Section, P. O. Box 20191, Sacramento, California 95802.

DAVIS, G. H., and others, 1964, *Use of Ground-Water Reservoirs for Storage of Surface Water in the San Joaquin Valley, California*, U. S. Geol. Survey Water-Supply Paper 1618, 125 pp.

DEUTSCH, MORRIS, 1967, *Artificial Recharge Induced by Interaquifer Leakage*, Internat. Assoc. Sci. Hydrology, Pub. 72, pp. 159-172.

DVORACEK, M. J., and PETERSON, S. H., 1971, *Artificial Recharge in Water Resources Management*, J. Irrig. Drain. Div., Am. Soc. Civil Engrs., v. 97, no. IR2, pp. 219-232.

DVORACEK, M. J., and SCOTT, V. H., 1963, *Ground-Water Flow Characteristics Influenced by Recharge Pit Geometry*, Trans. Am. Soc. Agric. Engrs., v. 6, pp. 262-265.

EHRLICH, G. G., and others, 1972, *Microbiological Aspects of Ground-Water Recharge — Injection of Purified Chlorinated Sewage Effluent*, U. S. Geol. Survey Prof. Paper 800-B, Geol. Survey Research 1972 — Chapter B, pp. 241-247.

ELLIS, B. G., 1973, *The Soil as a Chemical Filter,* in *Recycling Treated Municipal Wastewater and Sludge Through Forest and Cropland*, W. E. Sopper and L. T. Kardos (eds.), Pennsylvania State University Press.

ENGELEN, G. B., and ROEBERT, A. J., 1974, *Chemical Water Types and Their Distribution in Space and Time in the Amsterdam Dune-Water Catchment Area with Artificial Recharge*, J. Hydrology, v. 21, no. 4, Apr.

ERICKSON, E. T., 1949, *Using Runoff for Ground Water Recharge*, J. Am. Water Works Assoc., v. 41, pp. 647-649.

ESMAIL, O. J., and KIMBLER, O. K., 1967, *Investigation of the Technical Feasibility of Storing Fresh Water in Saline Aquifers*, Water Resources Res., v. 3, no. 3, pp. 683-695.

FERRIS, J. G., 1950, *Water Spreading and Recharge Wells*, Proc. Indiana Water Conserv. Conf., Ind. Dept. Conserv., Div. Water Resources, Indianapolis, pp. 52-59.

FETTER, C. W., and HOLZMACHER, R. H., 1974, *Groundwater Recharge with Treated Wastewater*, J. Water Poll. Control Fed., v. 46, no. 2, Feb., pp. 260-270.

FLACK, J. E., (ed.), 1973, *Proceedings of the Symposium on Land Treatment of Secondary Effluent*, Info. Ser. no. 9, Environmental Resources Center, Colorado State University, Fort Collins, 257 pp.

FRANKEL, J. R., 1967, *Economics of Artificial Recharge for Municipal Water Supply*, Symp. Haifa, Internat. Assoc. Sci. Hydrology Pub. no. 72, also Resources for the Future, Inc., Reprint no. 62, Mar. 1967.

FREEMAN, V. M., 1936, *Water-Spreading as Practiced by the Santa Clara Water-Conservation District, Ventura County, California*, Trans. Am. Geophys. Union, v. 17, pp. 465-471.

FOXWORTHY, B. L., 1970, *Hydrological Conditions and Artificial Recharge Through a Well in the Salem Heights Area of Salem, Oregon*, U. S. Geol. Survey Water-Supply Paper 1594-F, 56 pp.

GILLESPIE, J. B., HARGADINE, G. D., and STOUGH, M. J., 1977, *Artificial-Recharge Experiments near Lakin, Western Kansas*, Kansas Water Resour. Board Bull. 20, 91 pp.

GODSY, E. M., and EHRLICH, G. G., 1978, *Reconnaissance for Microbial Activity in the Magothy Aquifer, Bay Park, New York, Four Years after Artifical Recharge*, J. Research of the U. S. Geol. Survey, v. 6, no. 6, Nov.-Dec.

GOLDSHMID, J., 1974, *Water-Quality Aspects of Ground-Water Recharge in Israel*, J. Am. Water Works Assoc., v. 66, no. 3, Mar.

GOSS, D. W., and others, 1973, *Fate of Suspended Sediment During Basin Recharge*, Water Resources Res., v. 9, no. 3, pp. 668-675, Jun.

GOTASS, H. B., and others, 1953, *Final Report on Field Investigation and Research on Waste Water Reclamation and Utilization in Relation to Underground Water Pollution*, California State Water Poll. Control Board, Pub. 6, 124 pp.

GOUDEY, R. F., 1931, *Reclamation of Treated Sewage*, J. Am. Water Works Assoc., v. 23, pp. 230-240.

GREEN, D. W., and others, 1966, *Storage of Fresh Water in Underground Reservoirs Containing Saline Water — Phase 1*, Kansas State Univ. Water Resources Res. Inst. Contrib. no. 3, Phase 2, Contrib, no. 36, 1970.

GREENBERG, A. E., and GOTAAS, H. B., 1952, *Reclamation of Sewage Water*, Am. J. Public Health, v. 42, pp. 401-410.

GROVE, D. B. and WOOD, W. W., 1979, *Prediction and Field Verification of Subsurface-Water Quality Changes During Artificial Recharge, Lubbock, Texas*, Ground Water, v. 17, no. 3, May-Jun. pp. 250-257.

GUYTON, W. F., 1946, *Artificial Recharge of Glacial Sand and Gravel with Filtered River Water at Louisville, Kentucky*, Econ. Geol., v. 41, pp. 644-658.

HAJEK, B. F., 1969, *Chemical Interactions of Wastewater in a Soil Environment*, Water Poll. Control Federation, v. 41, no. 10, pp. 1775-1786.

HANOR, J. S., 1980, *Aquifers as Processing Plants for the Processing of Modification of Injected Water*, Louisiana Water Resources Res. Inst. Bull. 11, Louisiana State Univ., Baton Rouge 70803.

HANTUSH, M. S., 1967, *Growth and Decay of Ground-Water Mounds in Response to Uniform Percolation*, Water Resources Res., v. 3, no. 1.

HARGIS, D. R., and PETERSON, F. L., 1974, *Effects of Well Injection on a Basaltic Ghyben-Herzberg Aquifer*, Ground Water, v. 12, no. 1, pp. 4-9.

HARMESON, R. H., and others, 1968, *Coarse Media Filtration for Artificial Recharge*, J. Am. Water Works Assoc., v. 60, no. 12, pp. 1396-1404.

HARPAZ, YOAV, 1965, *Field Experiments in Recharge and Mixing Through Wells*, TAHAL, P.N. 483, Tel Aviv.

HARPAZ, YOAV, 1971, *Artificial Ground-Water Recharge by Means of Wells in Israel*, J. Hydraulics Div., ASCE, v. 97, no. HY 12, Dec.

HARPAZ, YOAV, and BEAR, JACOB, 1964, *Investigations on Mixing of Waters in Underground Storage Operations*, Internat. Assoc. Sci. Hydrology, Pub. 64, pp. 132-153.

HASKELL, E. E., Jr., and BIANCHI, W. C., 1965, *Development and Dissipation of Ground Water Mounds Beneath Square Recharge Basins*, J. Am. Water Works Assoc., v. 57, no. 3, pp. 349-353.

HAUSER, V. L., and LOTSPEICH, F. B., 1967, *Artificial Groundwater Recharge Through Wells*, J. Soil and Water Conservation, v. 22, pp. 11-15.

HAUSER, V. L., and LOTSPEICH, F. B., 1968, *Treatment of Playa Lake Water for Recharge Through Wells*, Trans. Am. Soc. Agr. Engrs., v. 11, no. 1, pp. 108-111.

HUISMAN, L., 1957, *The Determination of the Geo-Hydrological Constants for Dune-Water Catchment Area of Amsterdam*, Internat. Assoc. Sci. Hydrology Pub. 44, v. 2, pp. 168-182.

HUISMAN, L., 1983, *Artificial Groundwater Recharge*, Pitman Books Ltd., 128 Long Acre, London WC2E 9AN, England, 336 pp.

HUISMAN, L., and VAN HAAREN, F. W. J., 1966, *Treatment of Water Before Infiltration and Modification of its Quality During its Passage Underground*, Internat. Water Supply Congr., Barcelona, Spec. Paper no. 3.

HUNT, G. W., 1940, *Description and Results of Operations of the Santa Clara Valley Water Conservation District's Project*, Trans. Am. Geophys. Union, v. 21, pp. 13-23.

IDELOVITCH, EMANUEL, and MICHAIL, MEDY, 1984, *Soil-Aquifer Treatment - A new Approach to an Old Method of Wastewater Reuse*, J. Water Pollution Control Fed., v. 56, no. 8, pp. 936-943.

IDELOVITCH, EMANUEL, TERKELTOUB, RICHARD and MICHAIL, MEDY, 1980, *The Role of Groundwater Recharge in Wastewater Reuse: Israel's Dan Region Project*, J. Am. Water Works Assoc., Jul., pp. 391-399.

INTERNATIONAL ASSOCIATION OF SCIENTIFIC HYDROLOGY, 1967, *Artificial Recharge and Management of Aquifers*, Publ. 72, 523 pp.

JANSA, O. V., 1952, *Artificial Replenishment of Underground Water*, Internat. Water Supply Assoc., Second Cong., 105 pp., Paris.

JANSA, O. V., 1954, *Artificial Ground-Water Supplies of Sweden*, Internat. Assoc. Sci. Hydrology Rept. no. 2, Pub. no. 37, pp. 269-275.

JEFFORDS, R. M., 1945, *Recharge to Water-Bearing Formations Along the Ohio Valley*, J. Am. Water Works Assoc., v. 37.

JENKINS, C. T., and HOFSTRA, W. E., 1970, *Availability of Water for Artificial Recharge, Plains Ground Water Management District, Colorado*, Colorado Water Conserv. Bd. Ground-Water Ser. Circ. 13, 16 pp.

JOHNSON, A. H., 1948, *Ground-Water Recharge on Long Island*, J. Am. Water Works Assoc., v. 40, pp. 1159-1166.

JOHNSON, A. I., and others, 1966, *Laboratory Study of Aquifer Properties and Well Design for an Artificial Recharge Site*, U. S. Geol. Survey Water Supply Paper 1615-H.

KATZ, D. L., and COATS, K. H., 1968, *Underground Storage of Fluids*, Ulrich's Books, Inc., Ann Arbor, Michigan, 575 pp.

KATZ, D. L., and TEK, M. R., 1970, *Storage of Natural Gas in Saline Aquifers*, J. Water Resources Res., v. 6, no. 5, pp. 1515-1521, Oct.

KAZMANN, R. G., 1947, *Discussion of Apparent Changes in Water Storage During Floods at Peoria, Illinois*, Trans. Am. Geophys. Union, v. 28.

KAZMANN, R. G., 1978, *Underground Hot Water Storage Could Cut National Fuel Needs 10%*, Civil Engrng., v. 48, no. 5, pp. 57-60.

KELLY, T. E., 1967, *Artificial Recharge at Valley City, North Dakota, 1932 to 1965*, Ground Water, v. 5, no. 2, pp. 20-25.

KIMBLER, O. K., 1970, *Fluid Model Studies of the Storage of Freshwater in Saline Aquifers*, Water Resources Res., v. 6, pp. 1522-1527.

KIMBLER, O. K., KAZMANN, R. G., and WHITEHEAD, W. R., 1975, *Cyclic Storage of Fresh Water in Saline Aquifers*, Louisiana Water Resour. Res. Institute, Bull. 10.

KIMREY, J. O., 1985, *Proposed Artificial Recharge Studies in Northern Qatar*, U. S. Geol. Survey Open-File Rept. 85-343, 31 pp.

KOCH, ELLIS, and others, 1973, *Design and Operation of the Artificial Recharge Plant at Bay Park, New York*, U. S. Geol. Survey Prof. Paper 751-B, 14 pp.

KRONE, R. B., and WALLS, R. W., 1984, *Clarification of Surface Waters for Recharge Through Wells*, California Univ., Davis. Water Resources Center, PB85-225910/WEP, NTIS, Springfield, Virginia 22161, 32 pp.

KU, H. F. H., VECCHIOLI, J., and RAGONE, S. E., 1975, *Changes in Concentration of Certain Constituents of Treated Waste Water During Movement Through the Magothy Aquifer, Bay Park, New York*, U. S. Geol. Survey J. Res., v. 3, no. 1, pp. 89-92.

KUMAR, A., and KIMBLER, O. K., 1970, *Effect of Dispersion, Gravitational Segration, and Formation Stratification on the Recovery of Freshwater Stored in Saline Aquifers*, Water Resources Res., v. 6, pp. 1698-1700.

LANE, D. A., 1934, *Surface Spreading-Operations by the Basin-Method and Tests on Underground Spreading by Means of Wells*, Trans. Am. Geophys. Union, v. 15, pp. 523-527.

LAU, L. S., and others, 1972, *Water Recycling of Sewage Effluent by Irrigation, A Field Study on Oahu*, Hawaii Water Resources Res. Center Tech. Report no. 62.

LAVERTY, F. B., 1946, *Correlating Flood Control and Water Supply, Los Angeles Coastal Plain*, Trans. Am. Soc. Civil Engrs., v. 111, pp. 1127-1158.

LAVERTY, F. B., 1952, *Ground-Water Recharge*, J. Am. Water Works Assoc., v. 44, pp. 677-681.

LAVERTY, F. B., 1954, *Water-Spreading Operations in the San Gabriel Valley*, J. Am. Water Works Assoc., v. 46, pp. 112-122.

LAVERTY, F. B., and others, 1961, *Reclaiming Hyperion Effluent*, J. Sanitary Engrng. Div., Am. Soc. Civil Engrs., v. 87, no. SA6, pp. 1-40.

LEGGETTE, R. M., and BRASHEARS, M. L., Jr., 1938, *Groundwater for Air-Conditioning on Long Island, New York*, Trans. Am. Geophys. Union, v. 19, pp. 412-418.

LEHR, J. H., 1965, *Relation of Shape of Artificial Recharge Pits to Infiltration Rate*, J. Am. Water Works Assoc., v. 56, p. 699, Jun.

LI, A. D., and others, 1969, *Evaluation of the Suitability of Water for Injection into Strata*, Tr. Tatar Neft. (U.S.S.R.), 9, 299 (1966); Chem. Abs. 71, 69786 (1969).

LINDENBERGH, P. C., 1951, *Drawing Water from a Dune Area*, J. Am. Water Works Assoc., v. 43, pp. 713-724.

LINSTEDT, K. D., and BENNETT, E. R., (eds.), 1975, *Research Needs for the Potable Reuse of Municipal Wastewater*. EPA-600/9-75007.

LIPPMANN, M. J., and TSANG, C. F., 1980, *Ground-Water Use for Cooling: Associated Aquifer Temperature Changes*, Ground Water, v. 18, no. 5, Sept.-Oct., pp. 452-458.

MacWHORTER, D. B., and BROOKMAN, J. A., 1972, *Pit Recharge Influenced by Subsurface Spreading*, Ground Water, v. 10, no. 5, pp. 6-11.

MARGAT, J., 1972, *L'Alimentation Artificielle des Nappes Souterraines en France*, Bureau de Recherche Geologique et Minières, 72 SGNS 63 AME, Orleans, France.

MARLETTE, R. R., 1967, *Artificial Recharge Through Injection Wells in a Sandstone Aquifer*, Proc. Internat. Assoc. Sci. Hydrology, Gen. Assembly Bern.

MARMION, K. R., 1962, *Hydraulics of Artificial Recharge in Non-Homogeneous Formations*, Univ. California, Berkeley Water Resources Center Contrib. no. 48, 88 pp. (glass beads sand tank).

MATHER, J. R., 1953, *The Disposal of Industrial Effluent by Woods Irrigation*, Trans. Am. Geophys. Union, v. 34, pp. 227-239.

McCARTY, P. L., ROBERTS, P. V., and DUNLAP, W. J., 1979, *Contaminant Transport in the Ground Water Environment During Recharge of Reclaimed Water*, U.S. EPA Environmental Research Brief, Mar., Robert S. Kerr Env. Res. Lab., Ada., Oklahoma 74820.

McCARTY, P. L., and others, 1980, *Wastewater Contaminant Removal for Groundwater Recharge at Water Factory 21*, MERL 736, Municipal Environmental Res. Lab., U.S. EPA, SBDP/MERL, 4208 Airport Rd., Cincinnati, Ohio 45226, 163 pp

McCORMICK, R. L., 1975, *Filter-Pack Installation and Redevelopment Techniques for Shallow Recharge Shafts*, Ground Water, v. 13, pp. 400-405.

McGAUHEY, P. H., and KRONE, R. B., 1967, *Soil Mantle as a Wastewater Treatment System*, Sanitary Eng. Research Lab. Rep. no. 67-11, Univ. of California, Berkeley, California.

McMICHAEL, F. C. and McKEE, J. E., 1966, *Wastewater Reclamation at Whittier Narrows*, Publ. 33, California State Water Quality Control Board, Sacramento, 100 pp.

McWHORTER, D. B., and BROOKMAN, J. A., 1972, *Pit Recharge Influenced By Subsurface Spreading*, Ground Water, v. 10, no. 5, pp. 6-11.

MEINZER, O. E., 1946, *General Principles of Artificial Ground-Water Recharge*, Econ. Geol. v. 41, pp. 191-201.

MERCADO, A., 1967, *The Spreading Pattern of Injected Water in a Permeability Stratified Aquifer*, Internat. Assoc. Sci. Hydrology (Haifa Symp.), Pub. 72, pp. 23-36.

MERRITT, M. L., and others, 1984, *Subsurface Storage of Freshwater in South Florida: A Prospectus*, U. S. Geol. Survey, Water Resources Inv. Rep. 83-4214, Box 25425, Federal Center, Denver, Colorado 80225

MEYER, C. F., 1976, *Status Report on Heat Storage Wells*, Water Resources Bull., v. 12, no. 2, pp. 237-252.

MITCHELSON, A. T., and MUCKEL, D. C., 1937, *Spreading Water for Storage Underground*, U. S. Dept. Agric. Tech. Bull. 578, Washington, D. C., 80 pp.

MOLZ, F. J., and BELL, L. C., 1977, *Head Gradient Control in Aquifers Used for Fluid Storage*, Water Resources Res., v. 13, no. 4, pp. 795-798.

MOLZ, F. J., WARMAN, J. C., and JONES, T. E., 1978, *Aquifer Storage of Heated Water, Part 1 — A Field Experiment*, Ground Water, v. 16, no. 4.

MOULDER, E. A., 1970, *Freshwater Bubbles: A Possibility for Using Saline Aquifers to Store Water*, J. Water Resources Res., v. 6, no. 5, pp. 1528-1531, Oct.

MOULDER, E. A., and FRAZOR, D. R., 1957, *Artificial Recharge Experiments at McDonald Well Field*, Amarillo, Texax, Bull. Texas Board Water Eng. no. 5701.

MUCKEL, D. C., 1953, *Research in Water Spreading*, Trans. Am. Soc. Civil Engrs., v. 118, pp. 209-219.

MUCKEL, D. C., 1959, *Replenishment of Ground-Water Supplies by Artificial Means*, Tech. Bull. 1195, U. S. Dept. Agric., Washington, D. C.

NAGEL, G., and others, 1982, *Sanitation of Groundwater by Infiltration of Ozone Treated Water*, GWF-Wasser/Abwasser H. 8.

NATIONAL COMMISSION ON WATER QUALITY, 1975, *Disposal of Wastewater Residuals*, Environmental Quality Systems, Inc., Rockville, Maryland.

NELLOR, M. A. H., 1980, *Health Effects of Water Reuse by Ground-Water Recharge*, Univ. Texas Center for Research in Water Resources, Tech. Rept. 175, Austin 78712.

NIGHTINGALE, H. I., and BIANCHI, W. C., 1973, *Ground-Water Recharge for Urban Use: Leaky Acres Project*, Ground Water, v. 11, no. 6, pp. 36-43.

NIGHTINGALE, H. I. and BIANCHI, W. C., 1977, *Environmental Aspects of Water Spreading for Ground-Water Recharge*, U. S. Dept. of Agriculture, Agric. Research Service, Tech. Bull. no. 1568, Washington, D. C. 20250, 21 pp.

NIGHTINGALE, H. I., and BIANCHI, W. C., 1977, *Ground-Water Turbidity Resulting from Artificial Recharge*, Ground Water, v. 15, no. 2, pp. 146-152.

OBERDORFER, J. A., and PETERSON, F. L., 1985, *Waste-Water Injection: Geochemical and Biogeochemical Clogging Processes*, Ground Water, v. 23, no. 6, pp. 753-761.

OLTSHOORN, T. N., 1983, *The Clogging of Recharge Wells*, KIWA, P.O. Box 70, NL 2280 AB, Rijswijk, The Netherlands, 135 pp.

PAGE, A. L., 1974, *Fate and Effects of Trace Elements in Sewage Sludge when Applied to Agricultural Lands — A Literature Review Study*, U. S. Environmental Protection Agency, 670/2-74-005.

PARIZEK, R. R., 1974, *Site Selection Criteria for Wastewater Disposal — Soils and Hydrogeologic Considerations*, Conference on Recycling Treated Municipal Wastewater Through Forest and Cropland, U. S. Environmental Protection Agency, EPA-660/2-74-003, Washington, D. C., pp. 95-130.

PARIZEK, R. R., and MEYERS, E. A., 1970, *Recharge of Ground Water from Renovated Sewage Effluent by Spray Irrigation*, Pennsylvania State Univ., Inst. for Research on Land & Water Resources, Reprint Ser. no. 2, 26 pp.

PARKER, G. G., and others, 1967, *Artificial Recharge and Its Role in Scientific Water Management, with Emphasis on Long Island, New York*, Proc. Nat. Symp. on Ground-Water Hydrology, Am. Water Resources Assoc., pp. 193-213.

PEAVY, H. S., STARK, P. E., and SCHWENDEMAN, T. G., 1981, *The Effects of Rapid Infiltration of Municipal Wastewater on Groundwater Quality*, Montana Water Resources Res. Center, Bozeman. PB82-157900, NTIS Springfield, Virginia 22161, 81 pp.

PETERSON, F., and HARGIS, D., 1971, *Effect of Storm Runoff Disposal and Other Artificial Recharge to Hawaiian Ghyben-Herzberg Aquifers*, Hawaii Water Res. Center Tech. Rept. no. 54.

PETERSON, J. R., and others, 1973, *Chemical and Biological Quality of Municipal Sludge*, in *Recycling Treated Municipal Wastewater and Sludge Through Forest and Cropland*, W. E. Sopper and L. T. Kardos (eds.), Pennsylvania State Univ. Press, pp. 26-36.

PRICE, C. E., 1961, *Artificial Recharge Through a Well Tapping Basalt Aquifers Walla Walla Area, Washington*, U. S. Geol. Survey Water-Supply Paper 1594-A, 33 pp.

PRICE, D., and others, 1965, *Artificial Recharge in Oregon and Washington, 1962*, U. S. Geol. Survey Water-Supply Paper 1594-C, 65 pp.

PRILL, R. C., and ARONSON, D. A., 1973, *Flow Characteristics of a Subsurface-Controlled Recharge Basin on Long Island, New York*, J. Research U. S. Geol. Survey, v. 1, no. 6, Nov.-Dec., pp. 735-744.

PRILL, R. C., and ARONSON, D. A., 1978, *Ponding-Test Procedure for Assessing the Infiltration Capacity of Storm-Water Basins, Nassau County, New York*, U. S. Geol. Survey Water-Supply Paper 2049.

PRILL, R. C., OAKSFORD, E. T., and POTORTI, J. E., 1979, *A Facility Designed to Monitor the Unsaturated Zone During Infiltration of Tertiary-Treated Sewage, Long Island, New York*, PB-102 700, NTIS, Springfield, Virginia 22161.

RAFTER, G. W., 1897 and 1899, *Sewage Irrigation*, U. S. Geol. Survey Water Supply Papers 3 and 22, 100 and 100 pp.

RAGONE, S. E., 1977, *Geochemical Effects of Recharging the Magothy Aquifer, Bay Park, New York, with Tertiary-Treated Sewage*, U. S. Geol. Survey Prof. Paper 751-D, D1-D22.

RAGONE, S. E., KU, H. F. H., and VECCHIOLI, J., 1975, *Mobilization of Iron in Water in the Magothy Aquifer During Long-Term Recharge with Tertiary-Treated Sewage, Bay Park, New York*, U. S. Geol. Surv. J. Res., v. 3, no. 1, pp. 93-98.

RAGONE, S. E., and VECCHIOLI, J., 1975, *Chemical Interaction During Deep Well Recharge, Bay Park, New York*, Ground Water, v. 13, no. 1, pp. 17-23.

RAHMAN, M. A., and others, 1969, *Effect of Sediment Concentration on Well Recharge in a Fine Sand Aquifer*, Water Resources Res., v. 5, no. 3, p. 641.

REBHUN, M., and SCHWARZ, J. 1968, *Clogging and Contamination Processes in Recharge Wells*, Water Resources Res. v. 4, no. 6, pp. 1207-1219.

REED, J. E., and others, 1966, *Induced Recharge of an Artesian Glacial Drift Aquifer at Kalamazoo, Michigan*, U. S. Geol. Survey Water-Supply Paper 1594-D, 62 pp.

REEDER, H. O., 1975, *Injection-Pipe System for Artificial Recharge*, U. S. Geol. Surv. J. Res., v. 3, no. 4, pp. 501-503.

REEDER, H. O., and others, 1976, *Artificial Recharge Through a Well in Fissured Carbonate Rock, West St. Paul, Minnesota*, U. S. Geol. Survey Water-Supply Paper 2004, 80 pp.

REICHARD, E. G., and BREDEHOEFT, J. D., 1984, *An Engineering Economic Analysis of a Program for Artificial Groundwater Recharge*, Water Resources Bull., v. 20, no. 6, pp. 929-939.

REICHENBAUGH, R. C., 1977, *Effects on Ground-Water Quality from Irrigating Pasture with Sewage Effluent near Lakeland, Florida*, U. S. Geol. Survey, Water-Resources Inv. 76-108, 63 pp.

RICHERT, J. G., 1900, *On Artificial Underground Water*, C. E. Fritze's Royal Bookstore, Stockholm, 33 pp.

RICHTER, R. C., and CHUN, R. Y. D., 1959, *Artificial Recharge of Ground Water Reservoirs in California*, J. Irrig. Drain. Div., Am. Soc. Civil Engrs. v. 85, no. IR4, pp. 1-27.

RIPLEY, D. P. and SALEEM, Z. A., 1973, *Clogging in Simulated Glacial Aquifers Due to Artificial Recharge*, Water Resources Res., v. 9, no. 4, pp. 1047-1057.

ROBERTS, P. V., 1980, *Water Reuse for Groundwater Recharge: An Overview*, J. Am. Water Works, Jul., pp. 375-379.

ROBERTS, P. V., McCARTY, P. L., and ROMAN, WILLIAM, 1978, *Direct Injection of Reclaimed Water into an Aquifer*, J. Environmental Engineering Div., Am. Soc. Civil Engrs., v. 104, no. EE5, Oct., p. 933.

ROSENSHEIN, J. S., and HICKEY, J. J., 1977, *Storage of Treated Sewage Effluent and Storm Water in a Saline Aquifer, Pinellas Peninsula, Florida*, Ground Water, v. 15, no. 4, Jul.-Aug., pp. 284-293.

RUSSELL, R. H., 1960, *Artificial Recharge of a Well at Walla Walla*, J. Am. Water Works Assoc., v. 52, p. 1427, Nov.

SANFORD, J. H., 1938, *Diffusing Pits for Recharging Water into Underground Formations*, J. Am. Water Works Assoc., v. 30, pp. 1755-1766.

SANITARY ENGINEERING RESEARCH LABORATORY, 1955, *An Investigation of Sewage Spreading on Five California Soils*, Tech. Bull. 12, Univ. California, Berkeley, 53 pp.

SANITARY ENGINEERING RESEARCH LABORATORY, 1955, *Studies in Water Reclamation*, Tech. Bull. 13, Univ. California, Berkeley, 65 pp.

SASMAN, R. T., 1972, *Thermal Pollution of Ground Water by Artificial Recharge*, Water and Sewage Works, v. 119, no. 12, pp. 52-55.

SAYRE, A. N., and STRINGFIELD, V. T, 1948, *Artificial Recharge of Ground-Water Reservoirs*, J. Am. Water Works Assoc., v. 40, pp. 1152-1158.

SCHAEFER, D. H., and WARNER, J. W., 1975, *Artificial Recharge in the Upper Santa Ana River Area, San Bernardino, California*, U. S. Geol. Survey, Water-Resources Inv. 15-75, 32 pp.

SCHICHT, R. J., 1971, *Feasibility of Recharging Treated Sewage Effluent into a Deep Sandstone Aquifer*, Ground Water, v. 9, no. 6, pp. 29-35.

SCHICHT, R. J., and RISK, N. E., 1974, *Recharging Treated Sewage Effluent Through Cored Aquifer Samples*, Illinois State Water Survey Reprint Ser. no. 252.

SCHIFF, L., 1953, *The Effect of Surface Head on Infiltration Rates Based on the Performance of Ring Infiltrometers and Ponds*, Trans. Am. Geophys. Union, v. 34, pp. 257-266.

SCHIFF, L., 1954, *Water Spreading for Storage Underground*, Agric. Eng., v. 35, pp. 794-800.

SCHIFF, L., 1955, *The Status of Water Spreading for Ground Water Replenishment*, Trans. Am. Geophys. Union, v. 36, pp. 1009-1020.

SCHIFF, L., 1956, *The Darcy Law in its Selection of Water-Spreading Systems for Ground-Water Recharge*, Symposia Darcy, Internat. Assoc. Sci. Hydrology Pub. no. 41, pp. 99-110.

SCHIFF, L., 1957, *The Use of Filters to Maintain High Infiltration Rates in Aquifers for Ground-Water Recharge*, Internat. Assoc. Sci. Hydrology Pub. no. 44, pp. 217-221.

SCHMIDT, C. J. and CLEMENTS, E. V., III, 1977, *Reuse of Municipal Wastewater for Ground-Water Recharge*, U. S. Environmental Protection Agency Rept. EPA-600/2-77-183.

SCHMIDT, C. J., CLEMENTS, E. V., III, and SHELTON, S. P., 1978, *A Survey of Practices and Regulations for Reuse of Water by Ground-Water Recharge*, J. Am. Water Assoc., v. 70, no. 3, pp. 140-147.

SCHNEIDER, A. D., and others, 1971, *Recharge of Turbid Water to the Ogallala Aquifer Through A Dual-Purpose Well*, Texas Agric. Experiment Station Bull. 1112, College Station 77840.

SCHULTZ, T. R., and others, 1976, *Tracing Sewage Effluent Recharge — Tuscon, Arizona*, Ground Water, v. 14, no. 6, pp. 463-470.

SCHWARZ, JEHOSHUA, 1967, *Clogging and Contamination of Wells Recharging Lake Kinnereth Water*, TAHAL, P.N. 550, March.

SCOTT, V. H., and ARON, G., 1967, *Aquifer Recharge Efficiency of Wells and Trenches*, Ground Water, v. 5, no. 3, pp. 6-14.

SEABURN, G. E., and ARONSON, D. A., 1974, *Influence of Recharge Basins on the Hydrology of Nassau and Suffolk Counties, Long Island, New York*, U. S. Geol. Survey Water-Supply Paper 2031, 66 pp.

SIGNOR, D. C., 1973, *Laboratory Facility for Studies Related to Artificial Recharge*, Underground Waste Management and Artificial Recharge, Am. Assoc. Petrol. Geol., Tulsa, Oklahoma, pp. 799-824.

SINCLAIR, W. C., 1977, *Experimental Study of Artificial Recharge Alternatives in Northwest Hillsborough County, Florida*, U. S. Geol. Survey, Water-Resources Inv. 77-13, 52 pp.

SINGH, S. P. and MURTY, V. V. N., 1980, *Storage of Freshwater in Saline Aquifers*, J. Irrigation and Drainage Div., Am. Soc. Civil Engrs., v. 106, no. IR2, pp. 93-104.

SISSON, W. H., 1955, *Recharge Operations at Kalamazoo*, J. Am. Water Works Assoc., v. 47, pp. 914-922.

SLACK., L. J., 1975, *Hydrologic Environmental Effects of Sprayed Sewage Effluent, Tallahassee, Florida*, U. S. Geol. Survey, Water-Resources Inv. 55-75.

SMITH, C. G., Jr., and HANOR, J. S., 1975, *Underground Storage of Treated Water: A Field Test*, Ground Water, v. 13, no. 5, pp. 410-417.

SMITH, E. D., and others, 1979, *A Study of the Reuse of Reused Water*, Ground Water, v. 17, no. 4, Jul.-Aug., pp. 366-374.

SNIEGOCKI, R. T., 1960, *Effects of Viscosity and Temperature — Ground Water Recharge and Conservation*, (discussion), J. Am. Water Works Assoc., v. 52, no. 12.

SNIEGOCKI, R. T., 1963, *Geochemical Aspects of Artifical Recharge in the Grand Prairie Region, Arkansas*, U. S. Geol. Survey Water-Supply Paper 1615-E, 41 pp.

SNIEGOCKI, R. T., 1963, *Problems in Artificial Recharge Through Wells in the Grand Prairie Region Arkansas*, U. S. Geol. Survey Water-Supply Paper 1615-E.

SNIEGOCKI, R. T., and others, 1965, *Testing Procedures and Results of Studies of Artificial Recharge in the Grand Prairie Region, Arkansas*, U. S. Geol. Survey Water-Supply Paper 1615-G, 56 pp.

SONDEREGGER, A. L., 1918, *Hydraulic Phenomena and the Effect of Spreading of Flood Water in the San Bernardino Basin, Southern California*, Trans. Am. Soc. Civil Engrs., v. 82, pp. 802-851.

SOPPER, W. E., 1967, *Renovation of Municipal Sewage Effluent for Groundwater Recharge Through Forest Irrigation*, Internat. Conf. on Water for Peace, Paper no. 571, 11 pp.; also School of Forest Resources, The Pennsylvania State University.

SOPPER, W. E., and KARDOS, L. T., 1973, *Recycling Treated Municipal Wastewater and Sludge Through Forest and Cropland*, Pennsylvania State Univ. Press, University Park, Pennsylvania.

SOPPER, W. E., and KARDOS, L. T., 1974, *Conference on Recycling Treated Municiple Wastewater Through Forest and Cropland*, Rept. EPA-660/2-74-003, U. S. Environmental Protection Agency, Washington, D. C., 463 pp.

STEELINK, and others, 1981, *Organic Pollutants in Ground-Recharged Water*, PB81-205122, NTIS, Springfield, Virginia 22161, 31 pp.

STEINBRUEGGE, G. W., and others, 1954, *Groundwater Recharge by Means of Wells*, Agric. Exp. Sta., Univ. Arkansas, Fayetteville, 119 pp.

STONE, R., and GARBER, W. F., 1952, *Sewage Reclamation by Spreading Basin Infiltration*, Trans. Am. Soc. Civil Engrs., v. 117, pp. 1189-1217.

STONE, R. and ROWLAND, J., 1980, *Long-Term Effects of Land Application of Domestic Wastewater; Mesa, Arizona: Irrigation Site*, PB80-203144, NTIS, Springfield, Virgina 22161.

SUNDSTROM, R. V., and HOOD, H. W., 1952, *Results of Artificial Recharge of the Ground-Water Reservoir at El Paso, Texas*, Texas Board Water Engrs. Bull. 5206, Austin, 19 pp.

SUTER, M., 1956, *High-Rate Recharge of Ground Water by Infiltration*, J. Am. Water Works Assoc., v. 48, pp. 355-360.

SUTER, M., 1956, *The Peoria Recharge Pit: its Development and Results*, Proc. Am. Soc. Civil Engrs., v. 82, no. IR3, 17 pp.

SUTER, M., and HARMESON, R. H., 1960, *Artificial Ground-Water Recharge at Peoria, Illinois*, Bull. 48, Illinois State Water Survey, Urbana, 48 pp.

TASK GROUP ON ARTIFICIAL GROUND WATER RECHARGE, 1963, *Artificial Ground Water Recharge*, J. Am. Water Works Assoc., v. 55, pp. 705-709.

TASK GROUP ON ARTIFICIAL GROUND WATER RECHARGE, 1963, *Design and Operation of Recharge Basins*, J. Am. Water Works Assoc., v. 55, pp. 697-704.

TASK GROUP ON ARTIFICIAL GROUND WATER RECHARGE, 1965, *Experience with Injection Wells for Artificial Ground Water Recharge*, J. Am. Water Works Assoc., v. 57, pp. 629-639.

TAYLOR, O. J., and LUCKEY, R. R., 1972, *A New Technique for Estimating Recharge Using a Digital Model*, Ground Water, v. 10, no. 6, pp. 22-26, Nov.-Dec.

THOMAS, H. E., 1978, *Cyclic Storage, Where Are You Now?*, Ground Water, v. 16, no. 1, pp. 12-17.

TIBBETTS, F. H., 1936, *Water-Conservation Project in Santa Clara County*, Trans. Am. Geophys. Union, v. 17, pp. 458-465.

TODD, D. K., 1961, *The Distribution of Ground Water Beneath Artificial Recharge Areas*, Internat. Assoc. Sci. Hydrology Publ. 57, pp. 254-262.

TODD, D. K., 1964, *Economics of Ground Water Recharge by Nuclear and Conventional Means*, Rept. UCRL-7850, Lawrence Radiation Lab., pp. 1-135, Livermore, California, Feb.; also Proc. Am. Soc. Civil Engrs., v. 91, 1965, HY 4, pp. 249-270.

TODD, D. K., 1965, *Nuclear Craters for Ground Water Recharge*, J. Am. Water Works Assoc., v. 57, p. 429, Apr.

TODD, D. K., 1965, *Economics of Ground Water Recharge*, J. Hydraulics Div., Am. Soc. Civil Engrs., v. 91, no. HY4, pp. 249-270.

TOUPS, J. M., 1969, *Use of Reclaimed Water for Sea Water Intrusion Barrier*, Bull. Calif. Water Poll. Control Assoc., v. 5, no. 3, p. 5.

TROUT, T. T., and others, 1976, *Environmental Effects of Land Application of Anaerobically Digested Municipal Sewage Sludge*, Trans. Am. Soc. Agric. Engrs., v. 19, no. 2, pp. 266-270.

UNITED NATIONS DEPT. OF ECONOMICS AND SOCIAL AFFAIRS, 1975, *Ground-Water Storage and Artificial Recharge*, Natural Resources/Water Ser. 2, United Nations, New York, 270 pp.

UNKLESBAY, A. G., and COOPER, H. H., Jr., 1946, *Artificial Recharge of Artesian Limestone at Orlando, Florida*, Econ. Geol., v. 41, pp. 293-307.

U. S. ENVIRONMENTAL PROTECTION AGENCY, 1975, *Land Application of Wastewater*, Rept. EPA 903-9-75-017, U. S. EPA, Philadelphia, Pensylvania, 94 pp.

VALLIANT, J. C., 1962, *Artificial Recharge of Surface Water to the Ogallala Formation in the High Plains of Texas*, High Plains Research Foundation, Bull. 1, 17 pp.

VECCHIOLI, JOHN, 1972, *Experimental Injection of Tertiary-Treated Sewage in a Deep Well at Bay Park, Long Island, New York — A Summary of Early Results*, J. New England Water Works Assoc., v. 86, no. 2, pp. 87-103.

VECCHIOLI, JOHN, and GIAIMO, A. A., 1972, *Corrosion of Well-Casing and Screen Metals in Water from the Magothy Aquifer and in Injected Reclaimed Water, Bay Park, Long Island, New York*, U. S. Geol. Survey Prof. Paper 800-B. Geol. Survey Research 1972-Chapter B, pp. 247-253.

VECCHIOLI, JOHN, KU, H. F. H., and SULAM, D. J., 1980, *Hydraulic Effects of Recharging the Magothy Aquifer, Bay Park, New York, with Tertiary-Treated Sewage*, U. S. Geol. Survey Prof. Paper 751-F, 21 pp.

VECCHIOLI, JOHN, and others, 1975, *Wastewater Reclamation and Recharge, Bay Park, New York*, J. Environ. Eng. Div., Am. Soc. Civ. Eng., v. 101 (EE2), pp. 201-214.

Artificial Recharge

VENHUISEN, K. D., 1967, *The Storage Capacity in the Dunewater Catchment Area of Amsterdam and its Effect on the Water Quality*, Internat. Assoc. Sci. Hydrology, Pub. 72, pp. 109-123.

WARNER, D. L., and LEHR, J. H., 1977, *An Introduction to the Technology of Subsurface Wastewater Injection*, Rept. EPA-600/2-77-240, U. S. Environmental Protection Agency, Ada, Oklahoma, 344 pp.

WATER RESEARCH ASSOCIATION, 1971, *Proceedings September 1970 Artificial Groundwater Recharge Conference*, 2 vols., Medmenham, England, 481 pp.

WATKINS, F. A., Jr., 1978, *Effectiveness of Pilot Connector Well in Artificial Recharge of the Floridan Aquifer, Western Orange County, Florida*, U. S. Geol. Surv. Water-Resources Inv., 77-112, 33 pp.

WEAVER, R. J., 1971, *Recharge Basins for Disposal of Highway Storm Drainage; Theory, Design Procedure, and Recommended Engineering Practice* Final Report on Research Project 301. New York Dept. of Transp., Eng. Res. and Dev. Bureau, Research Rept. 69-2, Albany, New York 12226.

WEGENSTEIN, M., 1954, *La Recharge de Nappes Souterraines au Moyen de Puits Centraux et Galeries d'Alimentation Horizontales*, Internat. Assoc. Sci. Hydrology Pub. no. 37, pp. 232-237.

WELSCH, W. F., 1957, *Water Supply Problems in Nassau County, Long Island*, presented at Albany, New York at Ann. Convocation of New York Dist. Personnel of Water Resources Div., U. S. Geol. Survey.

WHETSTONE, G. A., 1954, *Mechanism of Ground-Water Recharge*, Agric. Eng., v. 35, pp. 646-647, 650.

WILLIAMS, R. E., and others, 1969, *Feasibility of Reuse of Treated Wastewater for Irrigation, Fertilization and Ground-Water Recharge in Idaho*, Idaho Bur. Mines and Geol. Pamph. 143, 110 pp.

WILSON, L. G., 1971, *Observations on Water Content Changes in Stratified Sediments During Pit Recharge*, Ground Water, v. 9, no. 3, p. 29-41.

WIPPLINGER, O., 1958, *The Storage of Water in Sand, and Investigation of the Properties of Natural and Artificial Sand Reservoirs and of Methods of Developing such Reservoirs*, South West Africa Adm., Windhoek. 107 pp.

WITHERSPOON, P. A., and others, 1967, *Interpretation of Aquifer Gas Storage Conditions from Water Pumping Tests*, American Gas Association, Inc., New York, New York, 273 pp.

YOUNG, R., EKERN, P., and LAU, L. S., 1972, *A Study of Wastewater Reclamation by Irrigation*, J. Water Poll. Control Fed., v. 44, no. 9.

RIVER INFILTRATION

BAURNE, GORAN, 1984, *Trap-Dams: Artificial Subsurface Storage of Water*, Water International, v. 9, no. 1, pp. 2-9.

BOULTON, N. S., 1942, *The Steady Flow of Ground Water to a Pumped Well in the Vicinity of a River*, The Philosophical Magazine, v. 7, pp. 34-50.

CHATURVEDI, M. C., and SRIVASTAVA, V. K., 1979, *Induced Ground-Water Recharge in the Ganges Basin*, Water Resources Res., v. 15, no. 5, Oct., p. 1156.

COOPER, H. H., Jr., and RORABAUGH, M. I., 1963, *Groundwater Movements and Bank Storage due to Flood Stages in Surface Streams*, U. S. Geol. Survey Water Supply Paper 1536-J, pp. 343-366.

DANIELSON, J. A., and RAZIQ QUAZI, A., 1972, *Stream Depletion by Wells in the South Platte Basin — Colorado*, Water Resources Bull., v. 8, no. 2, pp. 359-366.

DE WIEST, R. J. M., 1963 and 1964, *Replenishment of Aquifers Intersected by Streams*, J. Am. Soc. Civil Engrs., Hydraulics Div., Nov., 1963, pp. 165-191; Sept. 1964, pp. 161-168.

DIRECTO, L. S., and LINDAHL, M. E., 1969, *River Water Quality for Artificial Recharge*, J. Am. Water Works Assoc., v. 61, p. 175, Apr.

GLOVER, R. E., and BALMER, G. G., 1954, *River Depletion Resulting from Pumping a Well near a River*, Trans. Am. Geophys. Union, v. 35, pp. 468-470.

HANTUSH, M. S., 1959, *Analysis of Data from Pumping Wells near a River*, J. Geophys. Res., v. 64, pp. 1921-1932.

HANTUSH, M. S., 1965, *Wells near Streams with Semipervious Beds*, J. Geophysical Research, v. 70, pp. 2829-2838.

HANTUSH, M. S., 1967, *Depletion of Flow in Right-Angle Stream Bends by Steady Wells*, Water Resources Res., v. 3, no. 1, pp. 235-240.

HERBERT, ROBIN, 1969, *Solving Multiwell, River Ground-Water Flow Problems*, J. Hydrology, v. 4, art. 420, pp. 30-38.

JENKINS, C. T., 1968, *Techniques for Computing Rate and Volume of Stream Depletion by Wells*, Ground Water, v. 6, pp. 37-46.

JENKINS, C. T., 1970, *Computation and Volume of Stream Depletion by Wells*, Techniques of Water Resources Inv. of the U. S. Geol. Survey, Book 4, Chapter D-1.

KAZMANN, R. G., 1948, *River Infiltration as a Source of Ground-Water Supply*, Trans. Am. Soc. Civil Engrs., v. 113, pp. 404-424.

KAZMANN, R. G., 1948, *The Induced Infiltration of River Water to Wells*, Trans. Am. Geophys. Union, v. 29, pp. 85-92.

KAZMANN, R. G., 1960, *Discussion of Paper by M. S. Hantush, Analysis of Data from Pumping Wells near a River*, J. Geophys. Res., v. 65, pp. 1625-1626.

KLAER, F. H., Jr., 1953, *Providing Large Industrial Water Supplies by Induced Infiltration*, Min. Eng., v. 5, pp. 620-624.

MIKELS, F. C., 1952, *Report on Hydrogeological Survey for City of Zion, Illinois*, Ranney Method Water Supplies, Inc., Columbus, Ohio.

MIKELS, F. C., and KLAER, F. H., Jr., 1956, *Application of Ground Water Hydraulics to the Development of Water Supplies by Induced Infiltration*, Symposia Darcy, Internat. Assoc. Sci. Hydrology, Pub. 41, pp. 232-242.

MILOJEVIC, M., 1963, *Radial Collector Wells Adjacent to the River Bank*, J. Hydraulics Div., Amer. Soc. Civil Engrs., v. 89, no. HY6, pp. 133-151.

MOORE, J. E., and JENKINS, C. T., 1966, *An Evaluation of the Effect of Groundwater Pumpage on the Infiltration Rate of a Semipervious Streambed*, Water Resources Res., v. 2, no. 4, pp. 691-696.

NORRIS, S. E. and EAGON, H. B., Jr., 1971, *Recharge Characteristics of a Water-Course Aquifer System at Springfield, Ohio*, Ground Water, v. 9, no. 1, pp. 30-41.

OLMSTED, F. H., and HELY, A. G., 1962, *Relation Between Ground Water and Surface Water in Brandywine Creek Basin, Pennsylvania*, U. S. Geol. Survey Prof. Paper 417-A, 21 pp.

ROPER, R. M., 1939, *Ground-Water Replenishment by Surface Water Diffusion*, J. Am. Water Works Assoc., v. 31, pp. 165-179.

RORABAUGH, M. I., 1951, *Stream-Bed Percolation in Development of Water Supplies*, Trans. General Assembly Brussels, Internat. Assoc. Sci. Hydrology, v. 2, pp. 165-174; also U. S. Geol. Survey Ground Water Note 25, 1951.

RORABAUGH, M. I., 1956, *Ground-Water Resources of the Northeastern Part of the Louisville Area, Kentucky*, U. S. Geol. Survey Water-Supply Paper 1360-B.

SONTHEIMER, HEINRICH, 1980, *Experience with Riverbank Filtration along the Rhine River*, J. Am. Water Works Assoc., Jul., pp. 386-390.

THEIS, C. V., 1941, *The Effect of a Well on the Flow of a Nearby Stream*, Trans. Am. Geophys. Union, 22nd Ann. Meeting, pt. 3.

TODD, D. K., 1955, *Ground-Water Flow in Relation to a Flooding Stream*, Proc. Am. Soc. Civil Engrs., v. 81, sep. 628, 20 pp.

TODD, D. K., and BEAR, JACOB, 1959, *River Seepage Investigation*, Water Resources Center Contrib. 20, Univ. of California, Berkeley.

TRAINER, F. W., and SALVAS, E. H., 1962, *Ground Water Resources of the Massena-Waddington Area, St. Lawrence County, New York, with Emphasis on the Effect of Lake St. Lawrence on Ground Water*, New York Water Res. Comm. Bull. GW-47.

WALTON, W. C., 1963, *Estimating the Infiltration Rate of a Streambed by Aquifer-Test Analysis*, Internat. Assoc. Sci. Hydrology, Gen. Assembly Berkeley.

WALTON, W. C., 1969, *Recharge from Induced Streambed Infiltration under Varying Ground-Water Level Conditions*, U. S. Dept. Agric., Agric. Res. Service, ARS-41-147.

WALTON, W. C., and others, 1967, *Recharge from Induced Streambed Infiltration under Varying Groundwater-Level and Stream-Stage Conditions*, Minn. Water Resources Res. Center Bull. 6.

WILSON, L. G., and DE COOK, K. J., 1968, *Field Observations on Changes in the Subsurface Water Regime During Influent Seepage in the Santa Cruz River*, Water Resources Res., v. 4, no. 6, pp. 1219-1235.

WRIGHT, C. E., 1982, *Surface Water and Groundwater Interaction*, Studies and Reports in Hydrology 29, UNIPUB, New York, New York 10036, 123 pp.

BASIN STUDIES/FLOW SYSTEMS/ VARIABLE DENSITY FLOW/WATER BALANCES

ACKROYD, E. A., and others, 1967, *Groundwater Contribution to Streamflow and its Relation to Basin Characteristics in Minnesota*, Minn. Geol. Survey Rept. of Invest. 6.

ANDERSON, R. Y., and KIRKLAND, D. W., 1980, *Dissolution of Salt Deposits by Brine Density Flow*, Geology. v. 8, no. 2, Feb.

BAKER, D. M., 1950, *Safe Yield of Ground-Water Reservoirs*, General Assembly of Brussels, Internat. Assoc. Sci. Hydrology, v. 2, pp. 160-164.

BANKS, H. O., 1953, *Utilization of Underground Storage Reservoirs*, Trans. Am. Soc. Civil Engrs., v. 118, pp. 220-234.

BERRY, F. A. F., 1973, *High Fluid Potentials in California Coast Ranges and Their Tectonic Significance*, Bull. Am. Assoc. Petrol. Geol., v. 57, no. 7, pp. 1219-1249.

BETHKE, C. M., 1983, *Fluid Flow and Heat Transport in Compacting Sedimentary Basins*, Geol. Soc. Am. Abstracts with Programs, v. 15, p. 526.

BOGOMOLOV, G. V., and PLOTNIKOV, N. A., 1956, *Classification des Ressources d'Eaux Souterraines et Evaluation de leurs Reserves*, Symposia Darcy, Internat. Assoc. Sci. Hydrology, v. 2, pp. 263-271.

BOKE, R. L., and STONER, D. S., 1953, *The Application of Hydrologic Techniques to Ground-Water Problems in California's Central Valley Project*, Proc. Ankara Symp. on Arid Zone Hydrology, UNESCO, Paris, pp. 134-139.

BOND, D. C., 1972, *Hydrodynamics in Deep Aquifers of the Illinois Basin*, Illionis Geol. Survey Circ. 470, 72 pp.

BOND, D. C., 1973, *Deduction of Flow Patterns in Variable-Density Aquifers from Pressure and Water-Level Observations*, in *Underground Waste Management and Artificial Recharge*, Jules Braunstein, ed., Am. Assoc. Petrol. Geol., Tulsa, Oklahoma, pp. 357-378.

BOYLE, J. M. and SALEEM, Z. A., 1979, *Determination of Recharge Rates Using Temperature-Depth Profiles in Wells*, Water Resources Res., v. 15, no. 6, Dec., p. 1616.

BROWN, C. B., 1944, *Report on an Investigation of Water Losses in Streams Flowing East out of the Black Hills, South Dakota*, U. S. Dept. Agr., Soil Conserv. Serv., Sedimentation Sec., Spec. Rept. 8.

BRUNE, GUNNAR, 1970, *How Much Underground Water Storage Capacity does Texas Have?*, Water Res. Bull., v. 6, no. 4, pp. 588-600.

CLENDENEN, F. B., 1954, *A Comprehensive Plan for the Conjunctive Utilization of a Surface Reservoir with Underground Storage for Basin-Wide Water Supply Development Solano Project, California*, PhD. Eng. Thesis, Univ. California, Berkeley, 160 pp.

CLENDENEN, F. B., 1955, *Economic Utilization of Ground Water and Surface Storage Reservoirs*, Paper, San Diego, California meeting Am. Soc. Civil Engrs., Feb.

CLIFFORD, M. J., 1973, *Hydrodynamics of the Mount Simon Sandstone, Ohio and Adjoining Areas*, in *Underground Waste Management and Artificial Recharge*, Jules Braunstein, ed., Am. Assoc. of Petrol. Geol., Tulsa, Oklahoma, pp. 349-356.

CONKLING, H., 1934, *The Depletion of Underground Water Supplies*, Trans. Am. Geophys. Union, v. 15, pp. 531-539.

CONKLING, H., 1946, *Utilization of Ground-Water Storage in Stream System Development*, Trans. Am. Soc. Civil Engrs. v. 111, pp. 275-354.

DICKINSON, GEORGE, 1953, *Geological Aspects of Abnormal Reservoir Pressures in the Gulf Coast, Louisiana*, Bull. Am. Assoc. Petrol. Geol., v. 37, no. 2, pp. 410-432.

DOLCINI, A. J., and others, 1957, *The California Water Plan*, Bull. 3, California Dept. Water Resources, Sacramento, 246 pp.

DOWNEY, J. S., 1984, *Hydrodynamics of the Williston Basin in the Northern Great Plains*, in *Geohydrology of the Dakota Aquifer* (D. G. Jorgensen, ed.), National Water Well Assoc., Dublin, Ohio, pp. 92-98.

EDELMAN, J. H., 1972, *Groundwater Hydraulics of Extensive Aquifers*, Internat. Institute for Land Reclamation and Improvement/ILRI, Bull. no. 13, P. O. Box 45, Wageningen, The Netherlands.

FETH, J. H., 1964, *Hidden Recharge*, Ground Water, v. 2, no. 4.

GARVEN, GRANT, 1982, *The Role of Groundwater Flow in Genesis of Stratabound Ore Deposits: A Quantitative Analysis*: Unpubl. Ph.D. Dissert., Univ. British Columbia, 304 pp.

GARVEN, GRANT, 1985, *The Role of Regional Fluid Flow in the Genesis of the Pine Point Deposit, Western Canada Sedimentary Basin*, Economic Geol., v. 80, pp. 307-324.

GARVEN, GRANT and FREEZE, R. A., 1984, *Theoretical Analysis of the Role of Groundwater Flow in the Genesis of Stratabound Ore Deposits: 1. Mathematical and Numerical Model*: Am. Jour. Sci., v. 284, pp. 1085-1124.

GARVEN, GRAND and FREEZE, R. A., 1984, *Theoretical Analysis of the Role of Groundwater flow in the Genesis of Stratabound Ore Deposits: 2. Quantitative Results*, Am. Jour. Sci., v. 284, pp. 1125-1174.

GELHAR, L. W., and others, 1972, *Density Induced Mixing in Confined Aquifers*, U. S. Environmental Protection Agency Water Poll. Control Research Series Publ. 16060 ELJ 03/72.

GLEASON, G. B., 1947, *South Coastal Basin Investigation - Overdraft on Ground Water Basins*, Bull. 53, California Div. Water Resources, Sacramento, 256 pp.

GOLDSCHMIDT, J. J., 1959, *On the Water Balances of Several Mountain Underground Water Catchments in Israel and Their Flow Patterns*, Hydrological Service of Israel, Jerusalem, Israel, 10 pp.

GREGORY, K. J., (ed.), 1984 *Background to Palaeohydrology - A Perspective*, Wiley-Interscience, One Wiley Drive, Somerset, New Jersey 08873, 486 pp.

HAGMAIER, J. L., 1971, *The Relation of Uranium Occurrences to Ground Water Flow Systems*, Wyoming Geol. Assoc. Earth Science Bull., v. 4, no. 2, pp. 19-24, Jun.

HALEY, J. M., and others, 1955, *Santa Clara Vally Investigation*, Bull. 7, California State Water Resources Board, Sacramento, 154 pp.

HANOR, J. S., 1979, *The Sedimentary Genesis of Hydrothermal Fluids*, in H. L. Barnes, (ed.), *Geochemistry of Hydrothermal Ore Deposits*, 2nd ed.: New York, Wiley, pp. 137-172.

HANSHAW, B. B., BACK, WILLIAM, and DEIKE, R. G., 1971, *A Geochemical Hypothesis for Dolomitization by Ground Water*, Econ. Geol., v. 66, pp. 710-724.

HITCHON, BRIAN, 1969, *Fluid Flow in the Western Canada Sedimentary Basin 1. Effect of Topography*, Water Resources Res., v. 5, no. 1, p. 186; *2. Effect of Geology*, Water Resources Res., v. 5, no. 2, p. 460-469

HITCHON, BRIAN, 1971, *Origin of Oil: Geological and Geochemical Constraints: Washington*, Am. Chem. Soc. Advances in Chemistry Ser., no. 103, pp. 30-66.

HITCHON, BRIAN, 1971, *Hydrodynamics and Hydrocarbon Occurrences Surat Basin, Queensland, Australia*, Water Resources Res., v. 7, no. 3, pp. 658-676.

HOAG, G. E., BRUELL, C. J., and MARLEY, M. C., 1984, *Study of the Mechanisms Controlling Gasoline Hydrocarbon Partitioning and Transport in Groundwater Systems*, Conn. Univ., Storrs. Inst. Water Resources, PB85-242907/WEP, NTIS, Springfield, Virginia 22161, 61 pp.

HOLLYDAY, E. F., and SEABER, P. R., 1968, *Estimating Cost of Ground-Water Withdrawal for River Basin Planning*, Ground Water, v. 6, no. 4, pp. 15-24.

HUBBERT, M. K., 1953, *Entrapment of Petroleum under Hydrodynamic Conditions*, Bull. Am. Assoc. Petroleum Geol. v. 37, no. 8.

INGERSON, I. M., 1941, *The Hydrology of Southern San Joaquin Valley, California, and its Relation to Imported Water-Supplies*, Trans. Am. Geophys. Union, v. 22, pp. 20-45.

JACKSON, R. E., GILLILAND, J. A., and ADAMOWSKI, K., 1973, *Time Series Analysis of the Hydrologic Regimen of a Groundwater Discharge Area*, Water Resources Res., v. 9, no. 5, pp. 1411-1419.

JAQUET, N. G., 1976, *Ground-Water and Surface-Water Relationships in the Glacial Province of Northern Wisconsin — Snake Lake*, Ground Water, v. 14, no. 4, pp. 194-199.

JORGENSEN, D. G., GOGEL, TONY, and SIGNOR, D. C., 1982, *Determination of Flow in Aquifers Containing Variable - Density Water*, Ground Water Monitoring Review, v. 2, no. 2, pp. 40-45.

Basin Studies

KANEHIRO, B. Y., and PETERSON, F. L., 1977, *Groundwater Recharge and Coastal Discharge for the Northwest Coast of the Island of Hawaii-A Computerized Water Budget Approach*, Water Resources Research Center, University of Hawaii, Honolulu, Hawaii, 83 pp.

KAZMANN, R. G., 1951, *The Role of Aquifers in Water Supply*, Trans. Am. Geophys. Union, v. 32, pp. 227-230.

KAZMANN, R. G., 1956, *Safe Yield in Ground Water Development, Reality or Illusion?*, Proc. Am. Soc. Civil Engrs., v. 82, no. IR3, 12 pp.

KENT, D. C., and SENDLEIN, L. V. A., 1972, *A Basin Study of Ground-Water Discharge from Bedrock into Glacial Drift; Part 1 — Definition of Ground-Water System*, Ground Water, v. 10, no. 4, pp. 24-34; *Part 2 — Estimate of Recharge*, Ground Water, v. 10, no. 5, pp. 24-32.

KONOPLYANTSEV, A. A., and KOVALEVSKIY, V. S., 1968, *Ground-Water Regime and Balance Problems in the Soviet Union*, Nature and Resources, v. 4, no. 3, Sept.

KUDELIN, B. I., 1958, *The Principles of Regional Estimation of Underground Water Natural Resources and the Water Balance Problem*, Internat. Assoc. Sci. Hydrology, Assembly of Toronto, v. II, pp. 150-168.

KUDELIN, B. I., and others, 1971, *The Problem of Groundwater Discharge into the Seas*, (transl. by F. W. Trainer), US IHD Bull. no. 19, Oct., National Academy of Sciences, Washington, D. C., pp. 717-721.

LaFLEUR, R. G., (ed.), 1984, *Groundwater as a Geomorphic Agent*, Proc. 13th Annual Geomorphology Symp. at RPI, Sept. 1982, Allen and Unwin, Inc., 50 Cross St., Winchester, Massachusetts 01890, 390 pp.

LOEHNBERG, ALFRED, 1957, *Water Supply and Drainage in Semi-Arid Countries*, Trans. Am. Geophys. Union, v. 38, no. 4.

LOFGREN, B. E., 1973, *Monitoring Ground Movement in Geothermal Areas*, Am. Soc. Civil Engrs. Hydraulics Div. Specialty Conf. Proc., Bozeman, Montana, pp. 437-447.

LOWRY, R. L., Jr. and JOHNSON, A. F., 1942, *Consumptive Use of Water for Agriculture*, Trans. Am. Soc. Civil Engrs., v. 107, pp. 1243-1266.

LULL, H. M., and MUNNS, E. N., 1950, *Effect of Land Use Practices on Ground Water*, J. Soil and Water Conserv., v. 5, pp. 169-179.

MANN, J. F., 1961, *Factors Affecting the Safe Yield of Ground-Water Basins*, J. Irrigation and Drainage Div., Am. Soc. Civil Engrs., pp. 63-69, Sept.

McDONALD, H. R., 1955, *The Irrigation Aspects of Ground-Water Development*, Proc. Am. Soc. Civil Engrs., v. 81, sep. 707, 17 pp.

McNEAL, R. P., 1965, *Hydrodynamics of the Permian Basin*, in Young, A., and J. E. Galley, eds., *Fluids in Subsurface Environments*, Am. Assoc. Petroleum Geologists, Memoir 4, pp. 308-326.

MEINZER, O. E., 1932, *Outline of Methods for Estimating Ground-Water Supplies*, U. S. Geol. Survey Water-Supply Paper 638-C, pp. 99-144.

MEINZER, O. E., 1945, *Problems of the Perennial Yield of Artesian Aquifers*, Econ. Geol. v. 40, pp. 159-163.

MEINZER, O. E., and STEARNS, N. D., 1928, *A Study of Groundwater in the Pomperaug Basin, Connecticut, with Special Reference to Intake and Discharge*, U. S. Geol. Survey Water Supply Paper 597-B.

MIFFLIN, M. D., 1968, *Delineation of Ground-Water Flow System in Nevada*, Desert Res. Inst., Univ. of Nevada, Tech. Rept. Series H-W, Hydrology and Water Resources, Publ. no. 4.

MOENCH, A. F., SAUER, V. B., and JENNINGS, M. B., 1974, *Modification of Routed Streamflow by Channel Loss and Base Flow*, Water Resources Res., v. 10, no. 5, pp. 963-968.

NORRIS, S. E., 1974, *Regional Flow System and Ground Water Quality in Western Ohio*, U. S. Geol. Survey, J. Research 5.

NORRIS, S. E., and FIDLER, R. E., 1973, *Availability of Water from Limestone and Dolomite Aquifers in Southwest Ohio and the Relation of Water Quality to the Regional Flow System*, U. S. Geol. Survey Water-Resources Inv. 17-73, 42 pp.

RASMUSSEN, W. C., and ANDREASEN, G. E., 1959, *Hydrologic Budget of the Beaverdam Creek Basin, Maryland*, U. S. Geol. Survey Water Supply Paper 1427, 106 pp.

RUSHTON, K. R. and WARD, CATHERINE, 1979, *The Estimation of Ground-Water Recharge*, J. Hydrology, v. 41, no. 3/4, May.

SCHICHT, R. J., and WALTON, W. C., 1961, *Hydrologic Budgets for Three Small Watersheds in Illinois*, Illinois State Water Survey Rept. of Invest. 40, Urbana, Illinois, 40 pp.

SHARP, J. M., Jr., 1978, *Energy and Momentum Transport Model of the Quachita Basin and Its Possible Impact on Formation of Economic Mineral Deposits*, Econ. Geol., v. 73, pp. 1057-1068.

SIMPSON, T. R., 1952, *Utilization of Ground Water in California*, Trans. Am. Soc. Civil Engrs., v. 117, pp. 923-934.

SKLASH, M. G. and FARVOLDEN, R. N., 1979, *The Role of Ground Water in Storm Runoff*, J. Hydrology, v. 43, no. 1/4, Oct.

STEELE, K. F., *Groundwater Quality and Mineral Deposits Relationships in the Ozark Mountains*, PB85-211779/WEP, NTIS, Springfield, Virginia 22161.

STRINGFIELD, V. T., 1951, *Geologic and Hydrologic Factors Affecting Perennial Yields of Aquifers*, J. Am. Water Works Assoc., v. 43, pp. 803-816.

SUTCLIFFE, J. V., and RANGELEY, W. R., 1960, *An Estimation of the Long Term Yield of a Large Aquifer at Teheran*, Assembly of Helsinki, Internat. Assoc. Sci. Hydrology, Pub. 52, pp. 264-271.

SWENSON, F. A., 1968, *New Theory of Recharge to the Artesian Basin of the Dakotas*, Bull. Geol. Soc. Am., v. 79, pp. 163-182.

TAYLOR, S., 1964, *The Problem of Groundwater Recharge with Special Reference to the London Basin*, J. Inst. Water Engrs., 18, pp. 247-254.

THOMAS, R. O., 1955, *General Aspects of Planned Ground-Water Utilization*, Proc. Am. Soc. Civil Engrs., v. 81, sep. 706, 11 pp.

THORNTHWAITE, C. W., and MATHER, J. R., 1955, *The Water Balance*, Drexel Inst. Techn., Centerton, New Jersey, 86 pp.

THORNTHWAITE, C. W., and MATHER, J. R., 1955, *The Water Budget and its Use in Irrigation* in *Water, the Yearbook of Agriculture*, U. S. Dept. of Agric., pp. 346-358.

THORNTHWAITE, C. W., and MATHER, J. R., 1957, *Instructions and Tables for Computing Potential Evapotranspiration and the Water Balance*, Drexel Inst. Techn. Pub. in Climatology, v. 10, no. 3, Centerton, New Jersey, 311 pp.

THORNWAITE, C. W., and others, 1958, *Three Water Balance Maps of Eastern North America*, Resources for the Future, Inc., Washington, D. C.

TÓTH, J. A., 1970, *A Conceptual Model of the Groundwater Regime and the Hydrogeologic Environment*, J. Hydrology, v. 10, no. 2, pp. 164-176.

TÓTH, J. A., 1962, *A Theory of Ground-Water Motion in Small Drainage Basins in Central Alberta, Canada*, J. Geophys. Res., v. 67, pp. 4375-4387, Oct.

TÓTH, J. A., 1963, *Theoretical Analysis of Groundwater Flow in Small Drainage Basins*, J. Geophys. Res., v. 68, pp. 4795-4812.

TÓTH, J. A., 1978, *Gravity-Induced Cross-Formational Flow of Formation Fluids, Red Earth Region, Alberta, Canada: Analysis, Patterns and Evolution*, Water Resources Res., v. 14, no. 5, pp. 805-843.

TÓTH, J. A., and MILLAR, R. F., 1983, *Possible Effects of Erosional Changes of the Topographic Relief on Pore Pressures at Depth*: Water Resources Res., v. 19, pp. 1585-1597.

TROXELL, H. C., 1953, *The Influence of Ground-Water Storage on the Runoff in the San Bernardino and Eastern San Gabriel Mountains of Southern California*, Trans. Am. Geophys. Union, v. 34, pp. 552-562,

TURNER, S. F., and HALPENNY, L. C., 1941, *Ground-Water Inventory in the Upper Gila Valley, New Mexico and Arizona, Scope of Investigation and Methods Used*, Trans. Am. Geophys. Union, v. 22, pp. 738-744.

UBELL, K., 1966, *Investigations into Groundwater Balance by Applying Radioisotope Tracers,* Isotopes in Hydrology Symp. Vienna, Internat. Atomic Energy Agency, SM 83/36, pp. 521-530.

VAN EVERDINGEN, R. O., 1968, *Studies of Formation Waters in Western Canada: Geochemistry and Hydrodynamics*: Canadian J. Earth Sci, v. 5, pp. 523-543.

WALTON, W. C., 1964, *Future Water-Level Declines in Deep Sandstone Wells in Chicago Region*, Ground Water, v. 2, pp. 13-20.

WALTON, W. C., 1964, *Potential Yield of Aquifers and Ground Water Pumpage*, J. Am. Water Works Assoc., v. 56, pp. 172-186, Feb.

WALTON, W. C., 1965, *Groundwater Recharge and Runoff in Illinois*, Illinois State Water Survey Rept. of Invest. 48.

WENTWORTH, C. K., 1951, *The Problem of Safe Yield in Insular Ghyben-Herzberg Systems*, Trans. Am. Geophys. Union, v. 32, pp. 739-742.

WILLIAMS, C. C., and LOHMAN, S. W., 1947, *Methods Used in Estimating the Ground-Water Supply in Wichita, Kansas, Well-Field Area*, Trans. Am. Geophys. Union, v. 28, pp. 120-131.

WILLIAMS, C. C., and LOHMAN, S. W., 1949, *Geology and Ground-Water Resources of a Part of Southern-Central Kansas, with Special Reference to the Wichita Municipal Water Supply*, Kansas Geol. Survey Bull. 79.

WILLIAMS, R. E., 1970, *Groundwater Flow Systems and Accumulation of Evaporite Minerals*, Bull. Am. Assoc. Petrol. Geol., v. 54, no. 7, pp. 1290-1295.

WILLIAMS, R. E., and ALLMAN, D. W. , 1969, *Factors Affecting Infiltration and Recharge in a Loess Covered Basin*, J. Hydrology, v. 8, no. 3, Jul.

ZEIZEL, A. J., and others, 1962, *Ground-Water Resources of DuPage County, Illinois*, Illinois State Water Survey Cooperative Ground-Water Rept. 2.

ZEKSTER, I. S., and others, 1973, *The Problem of Direct Ground-Water Discharge to the Seas*, J. Hydrology, v. 21, no. 1, Sept.

PHREATOPHYTES/
EVAPOTRANSPIRATION

BIBLIOGRAPHY

ROBINSON, T. W., and JOHNSON, A. I., 1961, *Selected Bibliography on Evaporation and Transpiration*, U. S. Geol. Survey Water-Supply Paper 1539-R, 25 pp.

GENERAL WORKS

BLANEY, H. F., 1954, *Consumptive Use of Ground Water by Phreatophytes and Hydrophytes*, Internat. Assoc. Sci. Hydrology, General Assembly Rome, Pub. 37, v. 2, pp. 53-62.

BLANEY, H. F. and CRIDDLE, W. D., 1962, *Determining Consumptive Use and Irrigation Water Requirements*, U. S. Dept. of Agriculture Tech. Bull. 1257, 59 pp.

BOUWER, HERMAN, 1975, *Predicting Reduction In Water Losses from Open Channels by Phreatophyte Control*, Water Resources Res., v. 11, pp. 96-101.

CRUFF, R. W., and THOMPSON, T. H., 1967, *A Comparison of Methods of Estimating Potential Evapotranspiration from Climatological Data in Arid and Subhumid Environments*, U. S. Geol. Survey Water-Supply Paper 1839-M.

CULLER, R. C., 1970, *Water Conservation by Removal of Phreatophytes*, EOS, Trans. Am. Geophys. Union, v. 51, no. 10, Oct.

CULLER, R. C., JONES, J. E., and TURNER, R. M., 1972, *Quantitative Relationship Between Reflectance and Transpiration of Phreatophytes Gila River Test Site*: 4th Annual Earth Res. Aircraft Prog. Status Review, National Aeronautics and Space Adm., pp. 83-1 to 83-9.

DANIEL, J. F., 1976, *Estimating Groundwater Evapotranspiration from Streamflow Records*, Water Resources Res., v. 12, no. 3, pp. 360-364.

Phreatophytes

GARDNER, W. R., and FIREMAN, M., 1958, *Laboratory Studies of Evaporation from Soil Columns in the Presence of a Water Table*, Soil Sci., v. 85, pp. 244-249.

GATEWOOD, J. S., and others, 1950, *Use of Water by Bottom-Land Vegetation in Lower Safford Valley, Arizona*, U. S. Geol. Survey Water-Supply Paper 1103, 210 pp.

HANSON, R. L., 1972, *Subsurface Hydraulics in the Area of the Gila River Phreatophyte Project*, U. S. Geol. Survey Prof. Paper 655-F, 27 pp.

HANSON, R. L., KIPPLE, F. P., and CULLER, R. C., 1972, *Changing the Consumptive Use on the Gila River Flood Plain, Southeastern Arizona*, Proc. Am. Soc. Civil Engrs. Irrigation and Drainage Div. Specialties Conf., Sept. 1972, pp. 309-330.

HELLWIG, D. H. R., 1973, *Evaporation of Water from Sand*, J. Hydrology, v. 18, pp. 93-118.

HORTON, R. E., 1933, *The Role of Infiltration in the Hydrologic Cycle*, Trans. Am. Geophys. Union, v. 14, pp. 446-460.

JENSEN, M. E., and HAISE, H. R., 1963, *Estimating Evapotranspiration from Solar Radiation*, Proc. Am. Soc. Civil Engrs. J. Irrigation and Drainage Div., v. 89 (IR 4), pp. 15-41.

LEPPANNEN, O. E., 1980, *Barren Area Evapotranspiration Estimates Generated from Energy Budget Measurements in the Gila River Valley of Arizona*, USGS, WRD, Gulf Coast Hydroscience Ctr., NSTL Station Mississippi 39529, 80-1003 NTIS, Springfield, Virginia 22161, 36 pp.

LEWIS, D. C., and BURGY R. H., 1964, *The Relationship Between Oak Tree Roots and Groundwater in Fractured Rock as Determined by Tritium Tracing*, J. Geophys. Res., v. 69, pp. 2579-2588.

LOWRY, R. L. and JOHNSON, A. F., 1942, *Consumptive Use of Water for Agriculture*, Trans. Am. Soc. of Civil Engrs Transactions, v. 107, pp. 1243-1266.

PENMAN, H. L., 1948, *Natural Evapotranspiration from Open Water, Bare Soil, and Grass*, Proc. Royal Soc. of London, v. 193, pp. 120-145.

PENMAN, H. L., 1956, *Estimating Evapotranspiration*, Trans. Am. Geophys. Union, v. 37, pp. 43-46.

RIPPLE, C. D., and others, 1972, *Estimating Steady-State Evaporation Rates from Bare Soils under Conditions of High Water Table*, U. S. Geol. Survey Water-Supply Paper 2019-A, 39 pp.

ROBINSON, T. W., 1958, *Phreatophytes*, U. S. Geol. Survey Water-Supply Paper 1423, 84 pp.

ROBINSON, T. W., 1970, *Evapotranspiration by Woody Phreatophytes in the Humboldt River Valley near Winnemucca, Nevada*, U. S. Geol. Survey Prof. Paper 491-D.

SIBBONS, J. L. H., 1962, *A Contribution to the Study of Potential Evapotranspiration*, Geografiska Annaler, v. 44, pp. 279-292.

SMITH, G. E. P., 1924, *The Effect of Transpiration of Trees on the Groundwater Supply*, (abstract), J. Wash. Acad. Sci., v. 14.

TANNER, C. B., and PELTON, W. L., 1960, *Potential Evapotranspiration Estimates by the Approximate Energy Balance Method of Penman*, J. Geophys. Res. v. 65, pp. 3391-3413.

VAN HYLCKAMA, T. E. A., 1974, *Water Use by Saltcedar as Measured by the Water Budget Method*, U. S. Geol. Survey Prof. Paper 491-E, 30 pp.

VEIHMEYER, F. J. and BROOKS, F. A., 1954, *Measurement of Cumulative Evaporation of Bare Soil*, Trans. Amer. Geophys. Union, v. 35, pp. 601-607.

WEIST, W. G., Jr., 1971, *Geology and Ground-Water System, Gila River Phreatophyte Project*, U. S. Geol. Survey Prof. Paper 655-D, 22 pp.

WEIST, W. G., Jr., 1971, *Measurement of Spatial and Temporal Changes in Vegetation from Color IR Film*, Proc., Internat. Workshop on Earth Resources Survey Systems, Ann Arbor, Michigan, May 3-15, pp. 513-524.

WHITE, W. N., 1932, *A Method of Estimating Ground-Water Supplies Based on Discharge by Plants and Evaporation from Soil, Results of Investigations in Escalante Valley, Utah*, U. S. Geol. Survey Water-Supply Paper 659-A.

BASE FLOW/BANK STORAGE

BARNES, B. S., 1939, *The Structure of Discharge Recession Curves,* Trans. Am. Geophys. Union, v. 20, pp. 721-725.

BIRTLES, A. B., 1978, *Identification and Separation of Major Base Flow Components from a Stream Hydrograph,* Water Resources Res., v. 14, no. 5, Oct.

CHERNAYA, T. M., 1964, *Comparative Evaluation of Graphical Methods of Separation of Ground-Water Components of Streamflow Hydrographs,* Soviet Hydrology, v. 5, pp. 454-465.

CROSS, W. P., and HEDGES, R. E., 1959, *Flow Duration of Ohio Streams,* Ohio Div. of Water Bull. 31.

FARVOLDEN, R. N., 1964, *Geologic Controls on Groundwater Storage and Base Flow,* J. Hydrology, v. 50, no. 3, pp. 219-250.

FREEZE, R. A., 1972, *Role of Subsurface Flow in Generating Surface Runoff,* Water Resources Res., v. 8, pp. 609-623, 1272-1283.

GILL, M. A., 1985, *Bank Storage Characteristics of a Finite Aquifer Due to Sudden Rise and Fall of River Level,* J. Hydrology, v. 76, no. 1/2.

GRUNDY, F., 1951, *The Ground-Water Depletion Curve, its Construction and Uses,* General Assembly of Brussels, Internat. Assoc. Sci. Hydrology, v. 2, pp. 213-217.

HALL, F. R., 1968, *Base-Flow Recessions — A Review,* Water Resources Res., v. 4, no. 5, pp. 973-983.

HOUK, I. E., 1921, *Rainfall and Runoff in the Miami Valley, State of Ohio,* Miami Conserv. Dist. Tech. Repts., pt. VIII.

ISIHARA, T. and TAGAKI, F., 1965, *A Study on the Variation of Low Flow,* Disaster Prevention Res. Inst., Kyoto, Japan, Bull. 95, pp. 75-98.

JOHNSTON, R. H., 1971, *Base Flow as an Indicator of Aquifer Characteristics in the Coastal Plain of Delaware,* U. S. Geol. Survey Prof. Paper 750-D, pp. 212-215.

KEPPEL, R. V., and RENARD, K. G., 1962, *Transmission Losses in Ephemeral Stream Beds,* J. Hydraulics Div., Am. Soc. Civil Engrs., v. 88, no. HY3, pp. 59-68.

KNISEL, Jr. and WALTER, G., 1963, *Baseflow Recession Analysis for Comparison of Drainage Basins and Geology*, J. Geophys. Res., v. 68, no. 12, pp. 3649-53.

KUNKLE, G. R., 1962, *The Base Flow-Duration Curve, a Technique for the Study of Groundwater Discharge from a Drainage Basin*, J. Geophys. Res., v. 67, no. 4, pp. 1543-1554.

KUNKLE, G. R., 1965, *Computations of Ground-Water Discharge to Streams During Floods or to Individual Reaches During Baseflow, by Use of Specific Conductance*, U. S. Geol. Survey Prof. Paper 525-D, pp. 207-210.

LANGBEIN, W. B., 1940, *Some Channel Storage and Unit Hydrograph Studies*, Trans. Am. Geophys. Union, v. 21, pp. 620-627.

MENDENHALL, W. C., 1905, *The Hydrology of San Bernardino Valley, California*, U. S. Geol. Survey Water-Supply Paper 142.

MERRIAM, C. F., 1948, *Ground-Water Records in River-Flow Forecasting*, Trans. Am. Geophys. Union, v. 29, pp. 384-386.

MERRIAM, C. F., 1951, *Evaluation of Two Elements Affecting the Characteristics of the Recession Curve*, Trans. Am. Geophys. Union, v. 32, pp. 597-600.

MEYBOOM, P. 1961, *Estimating Ground-Water Recharge from Stream Hydrographs*, J. Geophys. Res., v. 66, no. 4, pp. 1203-1214.

MINSHALL, N. E., 1967, *Precipitation and Base Flow Variability*, Internat. Assoc. Sci. Hydrology Publ. 76, pp. 137-145.

NANEY, J. W., and others, 1978, *Predicting Base Flow Using Hydrogeologic Parameters*, Water Resour. Bull., v. 14, no. 3, pp. 640-649.

NEWCOMB, R. C., and BROWN, S. G., 1961, *Evaluation of Bank Storage Along the Columbia River Between Richland and China Bar, Washington*, U. S. Geol. Survey Water-Supply Paper 1539-1, 13 pp.

NORUM, D. I., and LUTHIN, J. N., 1968, *The Effects of Entrapped Air and Barometric Fluctuations in the Drainage of Porous Mediums*, Water Resources Res., v. 4, pp. 417-424.

PINDER, G. F., and JONES, J. F., 1969, *Determination of the Ground-Water Component of Peak Discharge from the Chemistry of Total Runoff*, Water Resources Res., v. 5, no. 2, p. 438.

PIPER, A. M., and others, 1939, *Geology and Groundwater Hydrology of the Mokelumne Area, California,* U. S. Geol. Survey Water-Supply Paper 780.

PLUHOWSKI, E. J., and SPINELLO, A. G., 1978, *Impact of Sewage Systems on Stream Base Flow and Ground-Water Recharge on Long Island, New York,* U. S. Geol. Surv. J. Res., v. 6, no. 2, pp. 263-271.

POGGE, E. C., 1968, *Effects of Bank Seepage on Flood Hydrographs,* Kansas State Univ. Water Resources Res. Inst. Proj. Completion Rept., Contrib. no. 33.

RIGGS, H. C., 1963, *The Base-Flow Recession Curve as an Indicator of Ground Water,* Internat. Assoc. Sci. Hydrology Publ. 63, pp. 352-363.

SINGH, K. P., 1968, *Some Factors Affecting Base Flow,* Water Resources Res. v. 4, no. 5, pp. 985-999.

SINGH, K. P., 1969, *Theoretical Baseflow Curves,* J. Hydraulics Div., Am. Soc. Civil Engrs., v. 95, no. HY6, Nov.

SINGH, K. P., and STALL, J. B., 1971, *Derivation of Base Flow Recession Curves and Parameters,* Water Resources Res., v. 7, pp. 292-303.

SNYDER, W. M., 1968, *Subsurface Implications from Surface Hydrograph Analysis,* Proc. Second Seepage Symp., Phoenix, Arizona.

TODD, D. K., 1955, *Ground-Water Flow in Relation to a Flooding Stream,* Proc. Am. Soc. Civil Engrs., v. 81, sep. 628, 20 pp.

TSCHINKEL, H. M., 1963, *Short-Term Fluctuation in Streamflow as Related to Evaporation and Transpiration,* J. Geophysical Research, v. 68, pp. 6459-6469.

VISOCKY, A. P., 1970, *Estimating the Ground-Water Contribution to Storm Runoff by the Electrical Conductance Method,* Ground Water, v. 8, no. 2, pp. 5-11.

WERNER, P. W., and SUNDQUIST, K. J., 1951, *On the Ground-Water Recession Curve for Large Watersheds,* Gen. Assembly Brussels, Internat. Assoc. Sci. Hydrology, v. 2, pp. 202-212.

ZITTA, V. L., and WIGGERT, J. M., 1971, *Flood Routing in Channels with Bank Seepage,* Water Resources Res., v. 7, no. 5, pp. 1341-1345.

GROUNDWATER MODELS

BIBLIOGRAPHY

JOHNSON, A. L., 1963, *Selected References on Analog Models for Hydrologic Studies*, App. F, Proc. Symp. on Transient Ground Water Hydraulics, Colorado State Univ., July 25-27.

GENERAL WORKS

ANGUITA BARTOLOME, F., and others, 1972, *Teoria Basica de Modelos Analogicos y Digitales de Acuiferos*, Servicio Geológico de Obras Publicas, Madrid, Bull. no. 37, Informaciones y Estudios, Oct. 178 pp.

APPEL, C. A., and BREDEHOEFT, J. D., 1976, *Status of Ground-Water Modeling in the U. S. Geological Survey*, U. S. Geol. Survey Cir. 737, 9 pp.

BOONSTRA, J., and DE RIDDER, N. A., 1983, *Numerical Modelling of Groundwater Basins*, Water Resources Publ. Distrib. Div., P. O. Box 2841, Littleton, Colorado 80161, 236 pp.

CARSLAW, H. S., and JAEGER, J. C., 1959, *Conduction of Heat in Solids*, Oxford, England, 510 pp.

CHANG, S., and YEH, W. W-G., 1970, *A Proposed Algorithm for Solution of the Large-Scale Inverse Problem in Groundwater*, Water Resources Res., v. 12, no. 3, pp. 365-374.

DOMENICO, P. A., 1972, *Concepts and Models in Groundwater Hydrology*, McGraw Hill Book Co., New York, 405 pp.

INTERNATIONAL ASSOCIATION OF SCIENTIFIC HYDROLOGY, 1968, *The Use of Analog and Digital Computers in Hydrology*, Publs. 80 and 81, 755 pp.

IRMAY, S., 1964, *Theoretical Models of Flow Through Porous Media*, Rilem Symp. on the Transfer of Water in Porous Media, Paris, April 7-10.

Models

JAVANDEL, I., DOUGHTY, C., and TSANG, C. F., (eds.), 1984, *Groundwater Transport - Handbook of Mathematical Models*, AGU Water Res. Monograph Ser. 10, Am. Geophys. Union, 1000 Florida Ave. NW, Washington, DC 20009.

JENKINS, C. T., 1968, *Electric-Analog and Digital-Computer Model Analysis of Stream Depletion by Wells*, Ground Water, v. 6, no. 6, pp. 27-35.

KLEINECKE, DAVID, 1971, *Hydrological Adaptations of Petroleum Reservoir Simulation Techniques*, PB-205 313, NTIS, Springfield, Virginia 22161.

KUNKEL, FRED, 1973, *Data Requirements for Modelling a Ground-Water System in an Arid Region*, PB 2-19588, NTIS, Springfield, Virgina 22161, 24 pp.

MERCER, J. W., and FAUST, C. R., 1980, *Ground-Water Modeling: An Overview*, Ground Water, v. 18, no. 2, Mar.-Apr., pp. 108-115.

MEYER, C. F., 1971, *Using Experimental Models to Guide Data Gathering*, J. Hydraulics Div., Am. Soc. Civil Engrs., v. 97, no. HY10, pp. 1681-1697.

MEYER, C. F., 1972, *Surrogate Modeling*, Water Resources Res., v. 8, no. 1, pp. 212-216.

MIDO, K. W., 1979, *A Breakthrough in Cutting the Cost of Ground Water Basin Modeling*, California Dept. of Water Resources, P. O. Box 6598, Los Angeles, California 90055.

PETERS, H. J., 1972, *Criteria for Ground-Water Level Data Networks for Hydrologic and Modeling Purposes*, Water Resources Res., v. 8, no. 1, Feb. pp. 194-200.

PRICKETT, T. A., 1975, *Modeling Techniques for Ground-Water Evaluation*, Chapter 1 in Ven Te Chow, *Advances of Hydroscience*, v. 10, pp. 1-143, Academic Press, Inc., New York, 10003.

PRICKETT, T. A., 1979, *Ground Water Computer Models - State of the Art*, Ground Water, v. 17, no. 2, pp. 167-174.

PRICKETT, T. A., and LONNQUIST, C. G., 1968, *Comparison Between Analog and Digital Simulation Techniques for Aquifer Evaluation*, Symp. of Tucson, Arizona, Internat. Assoc. Sci. Hydrology; also IASH/UNESCO Symp. on Use of Analog and Digital Computers in Hydrology, v. 2, pp. 625-634.

REILLY, T. E., and HARBAUGH, A. W., 1980, *Simulating the Movement of Ground Water on Long Island, New York — A Comparison of Modeling Techniques*, U. S. Geol. Survey Water Resources Inv. 80-14, 40 pp.

ROBINOVE, C. J., 1962, *Ground-Water Studies and Analog Models*, U. S. Geol. Survey Circ. 468, 12 pp.

RUSHTON, K. R., and ASH, J. E., 1974, *Ground Water Modelling Using Interactive Analogue and Digital Computers*, Ground Water, v. 12, no. 5, pp. 296-300.

RUSHTON, K. R., and RATHOD, K. S., 1979, *Modelling Rapid Flow in Aquifers*, Ground Water, v. 17, no. 4, Jul.-Aug., pp. 351-358.

RUSHTON, K. R., and REDSHAW, S. C., 1980, *Seepage and Ground-Water Flow: Numerical Analysis by Analog and Digital Methods*, John Wiley & Sons, Inc., One Wiley Drive, Somerset, New Jersey 08873, 339 pp.

RUSHTON, K. R., and WEDDERBURN, L. A., 1973, *Starting Conditions for Aquifer Simulations*, Ground Water, v. 11, no. 1, pp. 37-42, Jan.-Feb.

SAGAR, BUDHI, and others, 1975, *A Direct Method for the Identification of the Parameters of Dynamic Nonhomogeneous Aquifers*, Water Resources Res., v. 11, no. 4, pp. 563-570.

SALEEM, Z. A., 1973, *Groundwater Flow in Multiaquifer Systems: Numerical and Analytical Models*, N. Mexico Water Resources Res. Inst. Rept. No. 17, Las Cruces, New Mexico.

SCHNEIDER, A. D., and LUTHIN, J. N., 1978, *Simulations of Ground-Water Mound Perching in Layered Media*, Trans. Am. Soc. Agric. Engrs., v. 21, no. 5, Sept.-Oct.

SCHWARTZ, F. W., and DOMENICO, P. A., 1973, *Simulation of Hydrochemical Patterns in Regional Groundwater Flow*, Water Resources Res., v. 9, no. 3, pp. 707-720, Jun.

SHESTAKOB, V. M., 1968, *On the Technique for Solving Hydrological Problems Using Solid and Network Models*, Internat. Assoc. Sci. Hydrology Publ. 77, pp. 353-360.

TAYLOR, O. J., 1971, *A Shortcut for Computing Stream Depletion by Wells Using Analog or Digital Methods*, Ground Water, v. 9, no. 2, pp. 9-11.

Models

THOMAS, R. G., 1973, *Groundwater Models*, Irrigation and Drainage Paper 21, FAO, Rome, 191 pp.

THRAILKILL, JOHN, 1974, *Pipe Flow Models of a Kentucky Limestone Aquifer*, Ground Water, v. 12, no. 4, pp. 202-205.

UNESCO, 1969, *The Use of Analog and Digital Computers in Hydrology*, Proceedings Tucson Symposium.

U. S. OFFICE OF SURFACE MINING, 1981, *Ground Water Model Handbook*, U. S. Dept. of the Interior, Western Tech. Service Center, Denver, CO, 254 pp.

VAN EVERDINGER, R. O., and BHATTACHARYA, B. K., 1963, *Data for Ground-Water Model Studies*, Geol. Survey of Canada, Paper 63, 31 pp.

VERRUIJT, A., 1984, *Numerical Methods in Geomechanics*, Technological University of Delft, Geotechnical Laboratory, P.O. Box 5048, 2600 GA, Delft, The Netherlands.

WALTON, W. C., 1984, *Practical Aspects of Ground Water Modeling*, NWWA, 6375 Riverside Dr., Dublin, Ohio 43017, 566 pp.

WANG, H. F., and ANDERSON, M P., 1984, *Introduction to Groundwater Modeling: Finite Difference and Finite Element Methods*, W. H. Freeman and Co., 600 Market St., San Francisco, California 94104, 237 pp.

SAND TANK MODELS

BABBITT, H. E., and CALDWELL, D. H., 1948, *The Free Surface Around, and Interference Between Gravity Wells*, Univ. Illinois Bull., v. 45, no. 30.

BAUMANN, PAUL, 1951, *Ground-Water Movement Controlled Through Spreading*, Proc. Am. Soc. Civil Engrs. v. 77, sep. no. 86.

CAHILL, J. M., 1973, *Hydraulic Sand-Model Studies of Miscible-Fluid Flow*, U. S. Geol. Survey J. Research, v. 1, pp. 243-250.

CHILDS, E. C., and other, 1953, *The Measurement of the Hydraulic Permeability of Saturated Soil in Situ*, pt. 2, Proc. Royal Soc. London, ser. A., v. 216, no. 1124.

D'ANDRIMONT, R., 1905, *Note Préliminaire sur une Nouvelle Méthode pour Etudier Experimentalement l'Allure des Nappes Aquiferes dans les Terrains Permeables en Petit*, Annales Soc. Geol. Belgique, v. 32, Liége, pp. M115-M120.

D'ANDRIMONT, R., 1906, *Sur la Circulation de l'Eau des Nappes Aquiféres Contenues dans des Terrains Permeables en Petit*, Annales, Soc. Geol. Belgique, v. 33, Liége, pp. M21-M33.

DAY, P. R., and LUTHIN, J. N., 1954, *Sand-Model Experiments on the Distribution of Water-Pressure Under an Unlined Canal*, Proc. Soil Sci. Soc. Am., v. 18, no. 2.

HARDER, J. A., and others, 1953, *Laboratory Research on Sea Water Intrusion into Fresh Ground-Water Sources and Methods of its PreventionFinal Report*, Sanitary Eng. Research Lab., Univ. California, Berkeley, 68 pp.

KIRKHAM, D., 1940, *Pressure and Streamline Distribution in Waterlogged Land Overlying an Impervious Layer*, Proc. Soil Sci. Soc. Am., v. 5, pp. 65-68.

LEHR, J. H., 1963, *Ground-Water Flow Models Simulating Subsurface Conditions*, J. Geological Education, v. 11, pp. 124-132.

MUSKAT, M., 1935, *Seepage of Water Through Dams with Vertical Faces*, Physics, v. 6.

SEMCHINOVA, M. M., 1953, *Comparison of Experimental Data with Theory for the Case of Unsteady Flow Located on a Horizontal Water Table* (in Russian), Inghenerny Sbornik, Inst., Mech., Acad. Sci., U.S.S.R., v. 15, pp. 195-200.

STALLWORTH, T. W., 1950, *Quickly Constructed Model Facilities Seepage Studies*, Civil Eng., v. 20, no. 7, pp. 45-46.

VAIDHIANATHAN, V. I., and others, 1934, *A Hydrodynamical Investigation of the Subsoil Flow from Canal Beds by Means of Models*, Proc. Indian Acad. Sci., sec. A, v. 1.

WRIGHT, D. E., 1968, *Nonlinear Flow Through Granular Media*, Proc. Am. Soc. Civil Engrs., J. Hydraulics Div., v. 94, HY4, pp. 851-871.

WYCKOFF, R. D., and others, 1932, *Flow of Liquids Through Porous Media Under the Action of Gravity*, Physics, v. 3.

ANALOG MODELS

ARAVIN, V. I., 1941, *Experimental Investigation of Unsteady Flow of Ground Water* (in Russian), Trans. Sci. Res. Inst. Hydrotechnics, U.S.S.R., v. 30, pp. 79-88.

BEAR, JACOB, 1960, *Scales of Viscous Analogy Models for Ground Water Studies*, J. Hydraulics Div. Am. Soc. Civil Engrs., pp. 11-23, Feb.

CECEN, K., and OMAY, E., 1973, *Three Dimensional Viscous Flow Analogy*, Proc. Fifteenth Congress of the International Assoc. for Hydraulic Research, State Hydraulic Works of Turkey, Istanbul, v. 3, pp. C10-1 to C10-8 (viscous fluid).

COLLINS, M. A., GELHAR, L. W., and WILSON, J. L., III, 1972, *Hele-Shaw Model of Long Island Aquifer System*, Hydraulics Div., Proc. Am. Soc. Civil Engrs., v. 98, HY9, pp. 1701-1714 (viscous fluid).

COLUMBUS, N., 1966, *The Design and Construction of Hele-Shaw Models*, Ground Water, v. 4, no. 2, pp. 16-22.

DE JONG, G. DE JOSSELIN, 1961, *Moire Patterns of the Membrane Analogy for Groundwater Movement Applied to Multiple Fluid Flow*, J. Geophys. Res., v. 66, pp. 3625-3628.

DE JONG, G. DE JOSSELIN, 1962, *Een Eenvoudige Methode voor het Nabootsen van een Oneindig Potentiaalveld*, Water, v. 46, pp. 185-186.

DIETZ, D. N., 1941, *Een Modelproef ter Bestudeering van Niet-Stationnaire Bewegingen van het Grondwater*, Water, v. 25, The Hague, pp. 185-188, (viscous flow).

DIETZ, D. N., 1944, *Ervaringen met Modelonderzoek in de Hydrologie*, Water, v. 28, The Hague, pp. 17-20, (viscous flow).

FREEZE, R. A., 1970, *Moire Pattern Techniques in Ground-Water Hydrology*, Water Resources Res., v. 6, pp. 634-641, (graphical technique).

GUNTHER, E., 1940, *Losung von Grundwasseraufgaben mit Hilfe der Stromung in Dunnen Schichten*, Wasserkraft und Wasserwirtschaft, v. 3, no. 3, pp. 49-55.

GUNTHER, E., 1940, *Untersuchung von Grundwasserstromungen durch Analoge Stromungen zaher Flussigkeiten*, Forschung auf dem Gebiete des Ingenieur-wesens, v. 11, pp. 76-88.

GUPTA, S. P., VARNON, J. E., and GREENKORN, R. A., 1973, *Viscous Finger Wavelength Degeneration in Hele-Shaw Models*, Water Resources Res., v. 9, no. 4, pp. 1039-1046.

HANSEN, V. E., 1952, *Complicated Well Poblems Solved by the Membrane Analogy*, Trans. Am. Geophys. Union, v. 33, pp. 912-916.

HELE-SHAW, H. S., 1897, *Experiments on the Nature of the Surface Resistance in Pipes and on Ships*, Trans. Inst. Naval Architects, v. 39, pp. 145-156.

HELE-SHAW, H. S., 1898, *Investigation of the Nature of Surface Resistance of Water and of Stream-Line Motion under Certain Experimental Conditions*, Trans. Inst., Naval Architects, v. 40, pp. 21-46.

HELE-SHAW, H. S., 1899, *Stream-Line Motion of a Viscous Film*, Rept. 68th Meeting British Assoc. for the Advancement Sci., pp. 136-142.

KIMBLER, O. K., 1970, *Fluid Model Studies of the Storage of Freshwater in Saline Aquifers*, J. Water Resources Res., v. 6, no. 5, pp. 1522-1527, Oct.

KRAIJENHOFF VAN DE LEUR, D. A., 1962, *Some Effects of the Unsaturated Zone on Non-Steady Free-Surface Groundwater Flow as Studied in a Scaled Granular Model*, J. Geophys. Res. v. 67, pp. 4347-4362, Oct.

KRUL, W. F. J. M., and LIEFRINCK, F. A., 1946, *Recent Ground-Water Investigations in the Netherlands*, Elsevier Publishing Co., New York, 78 pp. (viscous flow).

MARINO, M. A., 1967, *Hele-Shaw Model Study of the Growth and Decay of Ground-Water Ridges*, J. Geophys. Research, v. 72, pp. 1195-1205.

MOORE, A. D., 1949, *Fields from Fluid Flow Mappers*, J. Appl. Phys., v. 20.

SANTING, G., 1951, *Infiltratie en Modelonderzoek*, Water, v. 35, no. 21, pp. 234-238, no. 22, pp. 243-246, The Hague.

SANTING, G., 1951, *Modele pour l'Etude des Problemes de l'Ecoulement Simultane des Eaux Souterraines Souces et Saluées*, General Assembly of Brussels, Internat. Assoc. Sci. Hydrology, v. 2, pp. 184-193.

SANTING, G., 1958, *A Horizontal Scale Model, Based on the Viscous Flow Analogy, for Studying Ground-Water in an Aquifer Having Storage*, General Assembly of Toronto, Internat. Assoc. Sci. Hydrology, Publ. no. 43, pp. 105-114.

SANTING, G., 1958, *Recente Ontwikkelingen op het Gebied van de Spleet-Modellen voor het Onderzoek van Grondwaterstromingen*, Water, no. 15, Jul.

SEVENHUYSEN, R. J., 1970, *Blotting Paper Models Simulating Groundwater Flow*, J. Hydrology, v. 10, pp. 276-281.

STERNBERG, Y., and SCOTT, V., 1963, *The Hele-Shaw Model as a Tool in Ground-Water Research*, Natl. Water Well Assoc. Conf., San Francisco, Sept.

TODD, D. K., 1954, *Unsteady Flow in Porous Media by Means of a Hele-Shaw Viscous Fluid Model*, Trans. Am. Geophys. Union, v. 35, pp. 905-916.

TODD, D. K., 1955, *Flow in Porous Media Studied in Hele-Shaw Channel*, Civil Eng., v. 25, no. 2, p. 85.

TODD, D. K., 1956, *Laboratory Research with Ground-Water Models*, Symposia Darcy, Internat. Assoc. Sci. Hydrology, Pub. 41, pp. 199-206.

VARRIN, R. D., and FANG, H. Y., 1967, *Design and Construction of a Horizontal Viscous Flow Model*, Ground Water, v. 5, no. 3, pp. 35-41.

WILLIAMS, D. E., 1966, *Viscous Model Study of Ground-Water Flow in a Wedge-Shaped Aquifer*, Water Resources Res. v. 2, no. 3, pp. 479-495.

YEN, B. C., and HSIE, C. H., 1972, *Viscous Flow Model for Groundwater Movement*, Water Resources Res., v. 8, no. 5, pp. 1299-1306.

ZANGER, C. N., 1953, *Theory and Problems of Water Percolation*, U. S. Bur. Reclamation Eng. Mon. 8. (viscous flow).

ELECTRIC ANALOG MODELS

ANDERSON, T. W., 1972, *Electrical-Analog Analysis of the Hydrologic System, Tucson Basin, Southeastern Arizona*, U. S. Geol. Survey Water-Supply Paper 1939-C, 34 pp.

BATURIC-RUBCIC, J., 1969, *The Study of Nonlinear Flow Through Porous Media by Means of Electrical Models*, J. Hydraulic Res., v. 7, no. 1, pp. 31-65, Delft, The Netherlands.

BEDINGER, M. S., and others, 1970, *Methods and Applications of Electrical Simulation in Ground-Water Studies in the Lower Arkansas and Verdigris River Valleys, Arkansas and Oklahoma*, U. S. Geol. Survey Water-Supply Paper 1971, 71 pp.

BENNETT, G. D., and others, 1968, *Electric-Analog Studies of Brine Coning Beneath Fresh-Water Wells in the Punjab Region, West Pakistan*, U. S. Geol. Survey Water-Supply Paper 1608.

BERMES, B. J., 1960, *An Electric Analog Model for Use in Quantitative Studies*, U. S. Geol. Survey Mimeographed Rept.

BOTSET, H. G., 1946, *The Electrolytic Model and its Application to the Study of Recovery Problems*, Trans. Am. Inst. Min. and Metal. Engrs. v. 165, pp. 15-25.

BOUWER, HERMAN, 1962, *Analyzing Ground-Water Mounds by Resitance Network*, J. Irrigation and Drainage Div., Am. Soc. Civil Engrs., pp. 15-36, Sept.

BOUWER, HERMAN, 1967, *Analyzing Subsurface Flow Systems with Electric Analogs*, Water Resources Res. v. 3, pp. 897-907.

DEBRINE, B. E., 1970, *Electrolytic Model Study for Collector Wells under River Beds*, J. Water Resources Res., v. 6, no. 3, pp. 971-978.

DE JONG, G. DE JOSSELIN, 1962, *Electrische Analogie Modellen voor het Oplossen van Geo-Hydrologische Problemen*, Water, v. 46, pp. 43-45.

FELIUS, G. P., 1954, *Recherches Hydrologiques par des Modeles Electriques*, Gen. Assembly Rome, Internat. Assoc. Sci. Hydrology, v. 2, pp. 162-169.

GETZEN, R. T., 1977, *Analog-Model Analysis of Regional Three-Dimensional Flow in the Ground-Water Reservoir of Long Island, New York*, U. S. Geol. Survey Prof. Paper 982.

HARBAUGH, A. W., and GETZON, R. T., 1978, *Stream Simulation in an Analog Model of the Ground-Water System on Long Island, New York*, U. S. Geol. Survey Water-Resources Inv. 77-58, 20 pp.

HERBERT, ROBIN, 1968, *Analyzing Pumping Tests by Resistance Network Analogue*, Ground Water, v. 6, no. 2, pp. 12-19.

Models

HERBERT, ROBIN, 1968, *Time Variant Ground-Water Flow by Resistance Network Analogues*, J. Hydrology, v. 6, pp. 237-264.

HERBERT, ROBIN, 1970, *Modelling Partially Penetrating Rivers on Aquifer Models*, Ground Water, v. 8, no. 2, pp. 29-37.

HORNER, W. L., and BRUCE, W. A., 1950, *Electrical-Model Studies of Secondary Recovery*, in *Secondary Recovery of Oil in the United States*, 2nd ed., Am. Petrol. Inst., Washington, D. C., pp. 195-203.

HURST, W., 1941, *Electrical Models as an Aid in Visualizing Flow in Condensate Reservoirs*, The Petroleum Engr., v. 12, no. 10, pp. 123-124, 127, 129.

JORGENSEN, D. G., 1975, *Analog-Model Studies of Ground-Water Hydrology in the Houston District, Texas*, Texas Water Developement Board Rep. 190, 84 pp. (resistor capacitor).

KARPLUS, W. J., 1958, *Analog Simulation*, McGraw-Hill Book Co., Inc., New York, 434 pp.

KHAN, I. A., 1980, *Wastewater Disposal Through a Coastal Aquifer*, Water Resources Bull., v. 16, no. 4, pp. 608-614. (electric analog modeling of waste water injection well system in Hawaii).

KORN, G. A., and KORN, T. M., 1952, *Electronic Analog Computers*, McGraw-Hill Book Co., Inc., New York.

LEE, B. D., 1948, *Potentiometric-Model Studies of Fluid Flow in Petroleum Reservoirs*, Trans. Am. Inst. Min. and Metal. Engrs., v. 174, pp. 41-66.

LIEBMANN, G., 1950, *Solution of Partial Differential Equations with a Resistance Network Analogue*, British J. Applied Physics, v. 1, pp. 92-103.

LIEBMANN, G., 1954, *Resistance Network Analogues with Unequal Meshes or Sub-Divided Meshes*, British J. Applied Physics, v. 5, pp. 362-366.

LUTHIN, J. N., 1953, *An Electrical Resistance Network Solving Drainage Problems*, Soil Sci., v. 75, pp. 259-74.

MACK, L. E., 1957, *Evaluation of a Conducting-Paper Analog Field Plotter as an Aid in Solving Ground-Water Problems*, State Geol. Survey of Kansas, Bull. no. 127, part 2, 47 pp. (resistance paper).

MEIN, R. G., and TURNER, A. K., 1968, *A Study of the Drainage of Irrigated Sand Dunes Using an Electrical Resistance Analogue*, J. Hydrology, v. 6, no. 1, Jan.

MUNDORFF, M. J., and others, 1972, *Electric Analog Studies of Flow to Wells in the Punjab Aquifer of West Pakistan*, U. S. Geol. Survey Water-Supply Paper 1608-N, 28 pp.

MUSKAT, M, 1949, *The Theory of Potentiometric Models*, Trans. Am. Inst. Min. and Metal. Engrs., v. 179, pp. 216-221.

OPSAL, F. W., 1955, *Analysis of Two-and Three-Dimensional Ground-Water Flow by Electrical Analogy*, The Trend in Eng. at the Univ. Washington, v. 7, no. 2, Seattle, pp. 15-20, 32.

PATTEN, E. P., Jr., 1965, *Design, Construction and Use of Electric Analog Model*, in Wood, L. A., and Gabrysch, R. K., *Analog Model Study of Ground water in Houston District*, Texas Water Commission Bull. no. 6508, 103 pp. (resistor capacitor).

PRICKETT, T. A., 1967, *Designing Pumped Well Characteristics into Electric Analog Models*, Ground Water, v. 5, no. 4.

RUSHTON, K. R., and BANNISTER, R. G., 1970, *Aquifer Simulation on Slow Time Resistance-Capacitance Networks*, Ground Water, v. 8, no. 4.

SCHICHT, R. J., 1965, *Ground-Water Development in East St. Louis Area, Illinois*, Illinois State Water Survey Rept. of Invest. 51.

SKIBITZKE, H. E., 1961, *Electronic Computers as an Aid to the Analysis of Hydrologic Problems*, Internat. Assoc. Sci. Hydrology Pub. 52, Comm. Subterranean Waters, Gentbrugge, Belgium, p. 347.

SKIBITZKE, H. E., and DaCOSTA, J. A., 1962, *The Ground-Water Flow System in the Snake River Plain, Idaho — An Idealized Analysis*, U. S. Geol. Survey Water-Supply Paper 1536-D, pp. 47-67.

SKIBITZKE, H. E., 1963, *The Use of Analogue Computers for Studies in Ground-Water Hydrology*, J. Inst. Water Engineers, v. 17, pp. 216-230.

SPIEKER, A. M., 1968, *Effect of Increased Pumping of Ground Water in the Fairfield-New Baltimore Area, Ohio — A Prediction by Analog-Model Study*, U. S. Geol. Survey Prof. Paper 605-C, 34 pp.

STALLMAN, R. W., 1963, *Calculation of Resistance and Error in an Electric Analog of Steady Flow Through Nonhomogeneous Aquifers*, U. S. Geol. Survey Water-Supply Paper 1544G.

STALLMAN, R. W., 1963, *Electric Analog of Three-Dimensional Flow to Wells and its Application to Unconfined Aquifers*, U. S. Geol. Survey Water Supply Paper 1536-H, pp. 205-242.

Models

TINLIN, R. M., 1972, *Analysis and Application of a Passive Electronic Analog Model to the Hydrologic Regime of a Watershed*, Univ. Arizona Ph. D. Thesis.

TODD, D. K. and BEAR, JACOB, 1959, *River Seepage Investigation*, Univ. California, Berkeley Water Resources Res. Center Contrib. No. 20, 163 pp. (electrolytic tank).

VREEDENBURGH, C. G. J., and STEVENS, O., 1936, *Electric Investigation of Underground Water Flow Nets*, Proc. Internat. Conf. Soil Mech. and Foundation Eng., v. 1, Harvard Univ., Cambridge, Massachusetts, pp. 219-222.

WALTON, W. C., 1964, *Electric Analog Computers and Hydrogeologic System Analysis in Illinois*, Ground Water, v. 2, no. 4.

WALTON, W. C., 1966, *Pre-Feasibility Report on Chicago Tunnel Drainage Project*, typewritten report prepared for Harza Engineering Co., Chicago, Illinois.

WALTON, W. C., and ACKROYD, E. A., 1966, *Effects of Induced Streambed Infiltration on Water Levels in Wells During Aquifer Tests*, Minnesota Water Res. Center, Bull. 2.

WALTON, W. C., and PRICKETT, T. A., 1963, *Hydrogeologic Electric Analog Computers*, J. Hydraulics Div., Am. Soc. Civil Engrs., pp. 67-91, Nov.

WINSLOW, J. D., and NUZMAN, C. E., 1966, *Electronic Simulation of Ground-Water Hydrology in the Kansas River Valley near Topeka, Kansas*, State Geol. Survey of Kansas Special Distribution Pub. No. 29, 24 pp. (resistor network).

WOLF, A., 1948, *Use of Electrical Models in Study of Secondary Recovery Projects*, The Oil and Gas J., v. 46, no. 50, pp. 94-98.

WORSTELL, R. V., and LUTHIN, J. N., 1959, *A Resistance Network Analog for Studying Seepage Problems*, Soil Science, v. 88, pp. 267-269.

WYCKOFF, R. D., and REED, D. W., 1935, *Electrical Conduction Models for the Solution of Water Seepage Problems*, Physics, v. 6, pp. 395-401.

ZEE, C. H., 1955, *Flow into a Well by Electric and Membrane Analogy*, Proc. Am. Soc. Civil Engrs., v. 81, sep. 817, 21 pp.

MATHEMATICAL MODELS

AFSHAR, A., and MARINO, M. A., 1976, *Mathematical Model for Simulating Soil Moisture Flow Considering Evapotranspiration,* Water Science & Engineering Papers No. 6002, Dept. of Water Science and Engineering, Univ. California, Davis, California. (finite difference).

AGUADO, E., SITAR, N., and REMSON, I., 1977, *Sensitivity Analysis in Aquifer Studies,* Water Resources Res., v. 13, no. 4, pp. 733-737.

AMERMAN, C. R., 1976, *Waterflow in Soils: A Generalized Steady-State, Two-Dimensional Porous Media Flow Model,* Agric. Research Service, U. S. Dept. of Agric. Rept. ARS-NC-30, 62 pp. (finite difference).

ANDERSON, M. P., 1979, *Using Models to Simulate the Movement of Contaminants Through Groundwater Flow Systems,* CRC Critical Reviews in Environmental Control, v. 1, no. 2, p. 97, Nov.

BACHMAT, YEHUDA, and others, 1978, *Utilization of Numerical Groundwater Models for Water Resource Management,* EPA-600/ 8-78-012. Robert S. Kerr Environmental Res. Laboratory, U. S. EPA, Ada, Oklahoma 74820, 178 pp.

BACHMAN, Y., and others, 1980, *Ground-Water Management: The Use of Numerical Models,* American Geophys. Union Water Resources Monograph 5, 2000 Florida Ave., N.W., Washington, D. C. 20009, 127 pp.

BAGCHI, S., and GOODMAN, A. S., 1979, *Emergency Water Supplies from Ground Water in Humid Regions,* (model study), Water Resources Bull., v. 15, p. 536.

BAIR, E. S., and O'DONNELL, T. P., 1984, *Uses of Numerical Modeling in the Design and Licensing of Dewatering and Depressurizing Systems,* Ground Water, v. 21, no. 4, pp. 411-420.

BAIR, E. S., and PARIZEK, R. R., 1981, *Numerical Simulation of Potentiometric Surface Changes Caused by a Proposed Open-Pit Anthracite Mine,* Ground Water, v. 19, no. 2, Mar.-Apr., pp. 190-200.

BENNETT, G. D., KONTIS, A. L., and LARSON, S. P., 1982, *Representation of Multiaquifer Well Effects in Three- Dimensional Ground-Water Flow Simulation,* Ground Water, v. 20, no. 3, pp. 334-341.

BIBBY, R. and SUNADA, D. K., 1971, *Mathematical Model of Leaky Aquifer*, J. Irrigation and Drainage Div., Am. Soc. Civil Engrs., pp. 387-395.

BIRTLES, A. B., and WILKINSON, W. B., 1975, *Mathematical Simulation of Groundwater Abstraction from Confined Aquifers for River Regulation*, Water Resources Res., v. 11, no. 4, pp. 571-580.

BOONSTRA, J. A., and DE RIDDER, N. A., 1981, *Numerical Modelling of Groundwater Basins: A User-Oriented Manual*, Publ. 29 Internat. Inst. Land Reclamation and Improvement, Wageningen 6700 AA, The Netherlands.

BREDEHOEFT, J. D.,and PINDER, G. F., 1970, *Digital Analysis of Aerial Flow in Multiaquifer Ground Water Systems: A Quasi Three-Dimensional Model*, Water Resources Res., v. 6, no. 3, pp. 883-888.

BREDEHOEFT, J. D., and PINDER, G. F., 1972, *The Application of Transport Equations to Groundwater Systems*, in T. D. Cook, (ed.), *Underground Waste Management and Environmental Implications*, Am. Assoc. Petrol. Geol. Mem. 18, pp. 199-201.

BREDEHOEFT, J. D., and PINDER, G. F., 1973, *Mass Transport in Flowing Groundwater*, Water Resources Res., v. 9, no. 1, pp. 194-210.

BREDEHOEFT, J. D., and YOUNG, R. A., 1970, *The Temporal Allocation of Ground Water, A Simulation Approach*, Water Resources Res., v. 6, pp. 3-21.

BROCK, R. R., 1976, *Hydrodynamics of Perched Mounds*, J. Hydraulics Div., Proc. Am. Soc. Civil Engrs., v. 102, HY8, pp. 1083-1100. (finite difference).

BRUTSAERT, W. F., 1971, *A Functional Iteration Technique for Solving the Richards Equation Applied to Two-Dimensional Infiltration Problems*, Water Resources Res., v. 7, no. 6, pp. 1583-1596.

BRUTSAERT, W. F., and others, 1971, *Computer Analysis of Free Surface Well Flow*, J. Irrigation and Drainage Div., Am. Soc. Civil Engrs., v. 97, no. IR 3, pp. 405-420.

CABRERA, G., and MARINO, M. A., 1976, *A Finite Element Model of Contaminant Movement in Groundwater*, Water Resour. Bull., v. 12, no. 2, pp. 317-335.

CALIFORNIA DEPT. OF WATER RESOURCES, 1974, *Mathematical Modeling of Water Quality for Water Resources Management, 2 vols.*, The Resources Agency, Sacramento, 304 pp.

CHEREMISINOFF, P. N., GIGLIELLO, K. A., and O'NEILL, T.K., 1985, *Groundwater Leachate: Modeling, Monitoring, Sampling,* Technomic Publ. Co., 851 New Holland Ave., Box 3535, Lancaster, Pennsylvania 17604, 154 pp.

CHIDLEY, T. R. E., and LLOYD, J. W., 1977, *A Mathematical Model Study of Freshwater Lenses,* Ground Water, v. 15, pp. 215-222.

CHRISTENSEN, B. A., 1980, *Mathematical Methods for Analysis of Fresh Water Lenses in the Coastal Zones of the Floridan Aquifer,* PB-293 067/5WP, NTIS, Springfield, Virginia 22161.

CODELL, R. B., KEY, K. T., and WHELAN, G., 1983, *A Collection of Mathematical Models for Dispersion in Surface Water and Groundwater,* NUREG-0868, NTIS, Springfield, Virginia 22161, 280 pp.

COE, J. J., 1974, *Mathematical Modeling of Water Quality for Water Resources Management - v. 1, Development of the Ground-Water Quality Model,* District Rept. Dept. of Water Resources State of California, (finite difference).

COOLEY, R. L., 1971, *A Finite Difference Method for Unsteady Flow in Variably Saturated Porous Media; Application to a Single Pumping Well,* Water Resources Res., v. 7, no. 6, pp. 1607-1625.

COOLEY, R. L., 1972, *Numerical Simulation of Flow in an Aquifer Overlain by a Water Table Aquitard,* Water Resources Res., v. 8, no. 4, pp. 1046-1050.

COOLEY, R. L., 1974, *Finite Element Solutions for the Equations of Ground-Water Flow,* Tech. Rep. Series H-W. Hydrology and Water Resources Pub. no. 18, Center for Water Resources Research, Desert Research Institute, Univ. of Nevada

COOLEY, R. L., and SINCLAIR, P. J., 1976, *Uniqueness of a Model of Steady-State Groundwater Flow,* J. Hydrology, v. 31, no. 3/4, pp. 245-269.

COOLEY, R. L., 1977, *A Method of Estimating Parameters and Assessing Reliability for Models of Steady-State Ground-Water Flow,* Water Resources Res., v. 13, no. 2, p. 318-324.

COOLEY, R. L., and PETERS, JOHN, 1972, *Finite Element Solution of Steady-State Potention Flow Problems,* in *Hydrologic Engineering Methods for Water Resources Development.* v. 10: Principles of Ground-Water Hydrology Hydrologic Engineering Center, Corps of Engineers U. S. Army, Davis, California.

Models

CUNNINGHAM, A. B., 1977, *Modeling and Analysis of Hydraulic Interchange of Surface and Ground Water*, Tech. Rep. Ser. Hydrol. and Water Resources Publ. no. 34, Desert Research Institute, P. O. Box 60220, Reno, Nevada 89506.

DAVIS, L. A., 1980, *Computer Analysis of Seepage and Ground-Water Response Beneath Tailings Impoundments*, PB80-178619, NTIS, Springfield, Virginia 22161, 96 pp.

DE LAAT, P. J. M., 1980, *Model for Unsaturated Flow Above a Shallow Water-Table, Applied to a Regional Sub-Surface Flow Problem*, Agric. Research Rept. 895, Pudoc Centre for Agric. Publ. and Doc., 126 pp., UNIPUB, New York.

DeMEIER, W. V., REISENAUER, A. E., and KIPP, K. L., 1974, *Variable Thickness Transient Ground-Water Flow Model User's Manual*, BNWL-1704, Battelle Pacific Northwest Lab., Richland, Washington 99352. (finite difference).

DeVRIES, R. N., and KENT, D. C., 1973, *Sensitivity of Groundwater Flow Models to Vertical Variability of Aquifer Constants*, Water Resources Bull., v. 9, no. 5, pp. 998-1005.

DOMENICO, P. A., and PALCIAUSKAS, V. V., 1982, *Alternative Boundaries in Solid Waste Management*, Ground Water, v. 20, no. 3, pp. 303-311 (plume management model).

DUNLAP, W.J., and others, 1984, *Transport and Fate of Organic Pollutants in the Subsurface: Current Perspectives*, Robert S. Kerr Environmental Research Lab, Ada, OK, PB84-190552, NTIS, Springfield, Virginia 22161, 38 pp.

EDELMAN, J. H., 1947, *Over de Berekening van Grondwaterstroomingen*, Doctorate Thesis, Delft Tech. Univ. Netherlands, 77 pp. (numerical analysis).

EHLIG, CHRISTINE, and HALEPASKA, J. C., 1976, *A Numerical Study of Confined-Unconfined Aquifers Including Effects of Delayed Yield and Leakage*, Water Resources Res., v. 12, no. 6, pp. 1175-1183.

EMSHOFF, J. R., and SISSON, R. L., 1970, *Design and Use of Computer Simulation Models*, Macmillan Ltd., London.

ENFIELD, C. G., and others, 1982, *Approximating Pollutant Transport to Ground Water*, Ground Water, v. 20, no. 6, pp. 711-722.

EVANS, D. H., HARLEY, B. M., BRAS, R., 1972, *Application of Linear Routing Systems to Regional Ground-Water Problems*, Massachusetts Institute of Technology, Ralph M. Parson Lab. Rept. no. 155, 197 pp.

EVENSON, D. E., and JOHNSON, A. E., 1975, *Parameter Estimation Program*, Texas Water Development Board Contract with Water Resources Engineers, Inc., Walnut Creek, California. (inverse problem).

EVENSON, D. E., and others, 1974, *Ground-Water Quality Models: What They Can and Cannot Do*, Ground Water, v. 12, no. 2, pp. 97-101.

FANG, C. S., WANG, S. N., and HARRISON, W., 1972, *Groundwater Flow in a Sandy Tidal Beach 2, Two-Dimensional Finite Element Analysis*, Water Resources Res., v. 8, no. 1, pp. 121-128.

FARLEKAS, G. M., 1979, *Geohydrology and Digital Simulation Model of the Farrington Aquifer in the Northern Coastal Plain of New Jersey*, U. S. Geol. Survey Water Resources Inv. 79-106. 430 Federal Bldg., P.O. Box 1238, Trenton, New Jersey 08607.

FAUST, C. R., and MERCER, J. W., 1980, *Ground-Water Modeling: Numerical Models*, Ground Water, v. 18, no. 4, Jul.-Aug., pp. 395-409.

FAUST, C. R., and MERCER, J. W., 1980, *Ground-Water Modeling: Recent Developments*, Ground Water, v. 18, no. 6, Nov.-Dec., pp. 569-577.

FAYERS, F. J. and SHELDON, J. W., 1962, *The Use of a High-Speed Digital Computer in the Study of the Hydrodynamics of Geologic Basins*, J. Geophys. Res., v. 67, pp. 2421-2431 (finite difference).

FLECK, W. B., and MCDONALD, M. G., 1978, *Three-Dimensional Finite-Difference Model of Ground-Water System Underlying the Muskegon County Wastewater Disposal System, Michigan*, U. S. Geol. Survey, J. Res., v. 6, no. 3, pp. 307-318.

FOGG, G. E., and others, 1979, *Aquifer Modeling by Numerical Methods Applied to an Arizona Ground-Water Basin*, PB-298 NTIS, Springfield, Virginia 22161.

FOGG, G.E., and SENGER, R.K., 1985, *Automatic Generation of Flow Nets with Conventional Groundwater Modeling Algorithms*, Ground Water, v. 23, no. 3, pp. 336-344.

Models

FREEZE, R. A., 1971, *Three-Dimensional, Transient, Saturated-Unsaturated Flow in a Ground Water Basin*, Water Resources Res., v. 7, no. 2, pp. 347-366.

FREEZE, R. A., 1972, *A Physics-Based Approach to Hydrologic Response Modeling: Phase I: Model Development*, Completion Rept., Contract No. 41-31-001-3694, Office of Water Resources Research, Washinton, D. C. (finite difference).

FREEZE, R. A., 1972, *Regionalization of Hydrogeologic Parameters for Use in Mathematical Models of Ground-Water Flow*, paper presented at 24th Internat. Geol. Congress, Montreal, Quebec.

FREEZE, R. A., 1975, *A Stochastic-Conceptual Analysis of One-Dimensional Groundwater Flow in Nonuniform Homogeneous Media*, Water Resources Res., v. 11, no. 5, pp. 725-741.

FREEZE, R. A., and HARLAN, R. L., 1969, *Blueprint for a Physically-Based, Digitally-Simulated Hydrologic Response Model*, J. Hydrology, v. 9, pp. 237-258.

FRIND, E. O., and MATANGA, G. B., 1985, *The Dual Formulation of Flow for Contaminant Transport Modeling, 1, Review of Theory and Accuracy Aspects*, Water Resources Res., vol. 21, no. 2, Feb., Paper 4W1314, pp. 159-169; 2, *The Borden Aquifer*, Paper 4W1315, pp. 170-182.

FRIND, E. O., and PINDER, G. F., 1973, *Galerkin Solution of the Inverse Problem for Aquifer Transmissivity*, Water Resources Res., v. 9, no. 5, pp. 1397-1410.

FRIND, E. O. and VERGE, M. J., 1978, *Three-Dimensional Modeling of Ground-Water Flow Systems*, Water Resources Res., v. 14, no. 5, Oct.

GAMBOLATI, G. and FREEZE, R. A., 1973, 1974, *Mathematical Simulation of the Subsidence of Venice*, Water Resources Res., v. 9, pp. 721-733, 1973, v. 10, pp. 563-577, 1974.

GATES, J. S., and KISIEL, C. C., 1974, *Worth of Additional Data to a Digital Computer Model of a Groundwater Basin*, Water Resources Res. v. 10, no. 5, pp. 1031-1038.

GELHAR, L. W., and WILSON, J. L., 1974, *Ground-Water Quality Modeling*, Ground Water, v. 12, no. 6, pp. 399-408.

GELHAR, L. W., and others, 1974, *Stochastic Modeling of Ground-Water Systems*, Massachusetts Institute of Technology, Ralph M. Parsons Lab. Rept. No. 189, 313 pp. (stochastic technique).

GILHAM, R. W., and FARVOLDEN, R. N., 1974, *Sensitivity Analysis of Input Parameters in Numerical Modeling of Steady-State Regional Groundwater Flow*, Water Resources Res., v. 10, no. 3, pp. 529-538.

GORELICK, S. M., 1982, *A Model for Managing Sources of Groundwater Pollution*, No. 2W0775, Am. Geophys. Union, 2000 Florida Ave., NW, Washington, DC 20009, 9 pp.

GRAY, W. G., and PINDER, G. F., 1976, *An Analysis of the Numerical Solution of the Transport Equation*, Water Resources Res., v. 12, p. 547.

GREEN, D. W., DABIN, H., and KHARE, J. D., 1972, *Numerical Modeling of Unsaturated Ground-Water Flow Including Effects of Evapotranspiration*, Completion Rept., Contract No. 14-31-0001-3084, Office of Water Resources Research, Washington, D. C. (finite difference).

GUPTA, S. K., and TANJI, K. K., 1976, *A Three-Dimensional Galerkin Finite Element Solution of Flow Through Multiaquifers in Sutter Basin, California*, Water Resources Res., v. 12, no. 2, pp. 155-162.

GUREGHIAN, A. B., 1975, *A Study by the Finite-Element Method of the Influence of Fractures in Confined Aquifers*, Society of Petroleum Engineers J., v. 15, pp. 181-191 (finite element).

GUREGHIAN, A. B., WARD, D. S., and CLEARY, R. W., 1981, *A Finite Element Model for the Migration of Leachate from a Sanitary Landfill in Long Island, New York*, Part 1, Water Resources Bull. v. 16, no. 5, pp. 900-906, Part 2, *Application*, Water Resources Bull, v. 17, no. 1, pp. 62-66.

GUYMON, G. L., and others, 1970, *A General Numerical Solution of the Two-Dimensional Diffusion-Convection Equation by the Finite Element Method*, Water Resources Res., v. 6, pp. 1611-1617.

HARRISON, W., FANG, C. S., and WANG S. N., 1971, *Groundwater Flow in a Sandy Tidal Beach*, 1. *One-Dimensional Finite Element Analysis*, Water Resources Res. v. 7, no. 5, pp. 1313-1322.

HEFEZ, E., and others, 1975, *Forecasting Water Levels in Aquifers by Numerical and Semihybrid Methods*, Water Resources Res. v. 11, no. 6, pp. 988-992.

HUANG, C., and EVANS, D. D., 1985, *3-Dimensional Computer Model to Simulate Fluid Flow and Contaminant Transport through a Rock Fracture System*, Arizona Univ., Tucson, Dept. of Hydrology and Water Resources, NUREG/CR-4042/WEP, NTIS, Springfield, Virginia 22161, 116 pp.

HUNT, BRUCE, 1984, *Mathematical Analysis of Groundwater Resources*, Butterworth Publishers, 80 Montvale Ave., Stoneham, Massachusetts 02180, 288 pp.

HUNTOON, P. W., 1974, *Finite Difference Methods as Applied to the Solution of Ground-Water Flow Problems*, Wyoming Water Resources Research Institute Univ. Wyoming, Laramie (finite difference).

HUYAKORN, P. S., and PINDER, G. F., 1983, *Computational Methods in Subsurface Flow*, Academic Press, Inc., 111 Fifth Ave., New York, New York 10003.

INTERCOMP RESOURCE DEVELOPMENT AND ENGINEERING INC., 1976, *A Model for Calculating Effects of Liquid Waste Disposal in Deep Saline Aquifers, Part I - Development, Part II - Documentation*, U. S. Geol. Survey Water-Res. Invest 76-61, 263 pp.

INTERNATIONAL ASSOCIATION FOR HYDRAULIC RESEARCH, (ed.), 1972, *Fundamentals of Transport Phenomena in Porous Media*, Am. Elsevier Publ. Co., Inc., New York, 400 pp.

JAVENDEL, K., and WITHERSPOON, P. A., 1967, *Use of Thermal Model to Investigate the Theory of Transient Flow to a Partially Penetrating Well*, Water Resources Res., v. 3, pp. 591-597.

JOHNSTON, R. H., and LEAHY, P.P., 1977, *Combined Use of Digital Aquifer Models and Field Base-Flow Data to Identify Recharge-Leakage Areas of Artesian Aquifers*, U. S. Geol. Survey, J. Res., v. 5, no. 4, pp. 491-496.

KARPLUS, W. J., 1967, *Hybrid Computer Simulation of Ground-Water Basins*, Am. Water Resources Assoc. National Symp. on Ground-Water Hydrology, San Francisco, California, pp. 289-299 (resistor-digital).

KASHEF, A. I., and others, 1952, *Numerical Solutions of Steady-State and Transient Flow Problems, Artesian and Water-Table Wells*, Purdue Univ. Eng. Exp. Sta. Bull. 117, Lafayette, Indiana, 116 pp.

KENT, D. C., PETTYJOHN, W. A., and PRICKETT, T. A., 1985, *Analytical Methods for the Prediction of Leachate Plume Migration*, Ground Water Monitoring Rev., v. 5, no. 2, pp. 46-59.

KLEINECKE, DAVID, 1971, *Use of Linear Programing for Estimating Geohydrologic Parameters of Groundwater Basins*, Water Resources Res., v. 7, no. 2, pp. 367-374.

KLEINECKE, DAVID, 1971, *Mathematical Modeling of Fresh-Water Aquifers Having Salt-Water Bottoms*, PB-204 545, NTIS, Springfield, Virginia 22161.

KLEMT, W. B., and others, 1979, *Ground-Water Resources and Model Applications for the Edwards (Balcones Fault Zone) Aquifer in the San Antonio Region, Texas*, Texas Dept. of Water Resources, Rep. 239.

KNOWLES, T. R., CLABORN, B. J., and WELLS, D. M., 1972, *A Computerized Procedure to Determine Aquifer Characteristics*, Water Resources Center Pub. WRC-72-5, Texas Tech. Univ., Lubbock, Texas (inverse problem).

KONIKOW, L. F., 1975, *Preliminary Digital Model of Ground-Water Flow in the Madison Group, Powder River Basin and Adjacent Areas, Wyoming, Montana, South Dakota, North Dakota, and Nebraska*, U. S. Geol. Survey Water-Resources Inv. 63-75.

KONIKOW, L. F., 1977, *Modeling Chloride Movement in the Alluvial Aquifer at the Rocky Mountain Arsenal, Colorado*, U. S. Geol. Survey Water-Supply Paper 2044, 43 pp.

KONIKOW, L. F., and BREDEHOEFT, J. D., 1973, *Simulation of Hydrologic and Chemical-Quality Variations in an Irrigated Stream-Aquifer System — A Preliminary Report*, Colorado Water Resources Circ. 17, 43 pp.

KONIKOW, L. F., and BREDEHOEFT, J. D., 1974, *Modeling Flow and Chemical Quality Changes in an Irrigated Stream-Aquifer System*, Water Resources Res., v. 10, no. 3, pp. 546-562.

KONIKOW, L. F., and BREDEHOEFT, J. D., 1978, *Computer Model of Two-Dimensional Solute Transport and Dispersion in Ground Water*, U. S. Geol. Survey Techniques of Water Resources Inv., Book 7, Chapter C2.

KONIKOW, L. F., and GROVE, D. B., 1977, *Derivation of Equations Describing Solute Transport in Ground Water*, U. S. Geol. Survey Water-Resources Inv. 77-19, 34 pp.

KRISHNAMURTHI, N., SUNADA, D. K., and LONGENBAUGH, R. A., 1977, *Mathematical Modeling of Natural Groundwater Recharge*, Water Resources Res., v. 13, no. 4, pp. 720-724.

LAND, L. F., 1977, *Utilizing a Digital Model to Determine the Hydraulic Properties of a Layered Aquifer*, Ground Water, v. 15, no. 2, Mar.-Apr., pp. 153-159.

LEWIS, R. W. and SCHREFLER, B., 1978, *A Fully Coupled Consolidation Model of the Subsidence of Venice*, Water Resources Res., v. 14, no. 2, Apr.

LI, R. M., and others, 1984, *Modeling Physics and Chemistry of Contaminant Transport in Three-Dimensional Unsaturated Groundwater Flow*, Simons, Li and Associates, Inc., Fort Collins, CO, 122 pp.; PB85-160653/WEP, NTIS, Springfield, Virginia 22161.

LIDDAMENT, M. W., and others, 1978, *The Use of a Ground-Water Model in the Design, Performance Assessment and Operation of a River Regulation Scheme*, Proc. Internat. Symp. on Logistics and Benefits of Using Mathematical Models in Hydrologic and Water Resource Systems, Pisa, Italy.

LIN, C. L., 1972, *Digital Simulation of the Boussinesq Equation for a Water Table Aquifer*, Water Resources Res., v. 8, no. 3, pp. 691-698.

LIN, C. L., 1973, *Digital Simulation of an Outwash Aquifer*, Ground Water, v. 11, no. 2, pp. 38-43.

LOVELL, R. E., 1971, *Collective Adjustment of the Paramenters of the Mathematical Model of a Large Aquifer*, Hydrology and Water Resources Rep. No. 4, Univ. of Arizona, Tucson, Arizona.

LUTHIN, J. N., and GASKELL, R. A., 1950, *Numerical Solution for Tile Drainage of Layered Soils*, Trans. Am. Geophys. Union, v. 31, pp. 595-602.

LUTHIN, J. N. and ORHUN, A., 1975, *Coupled Saturated-Unsaturated Transient Flow in Porous Media: Experimental and Numeric Model*, Water Resources Res., v. 11, no. 6, pp. 973-978.

LUZIER, J. E., 1980, *Digital Simulation and Projection of Head Changes in the Potomac-Raritan-Magothy Aquifer System, Coastal Plain, New Jersey*, U. S. Geol. Survey Water Resources Inv. 80-11, Trenton, New Jersey 08607.

MADDOCK, T., III, 1972, *Algebraic Technological Function from a Simulation Model*, Water Resources Res., v. 8, no. 1, pp. 129-134.

MADDOCK, T., III, 1973, *Management Model as a Tool for Studying the Worth of Data*, Water Resources Res., v. 9, no. 2, pp. 270-280.

MARINO, M. A., 1975, *Digital Simulation Model of Aquifer Response to Stream Stage Fluctuation*, J. Hydrology, v. 25, no. 1/2, pp. 51-58.

McELWEE, C. D., and YUKLER, M. A., 1978, *Sensitivity of Groundwater Models with Respect to Variations in Transmissivitiy and Storage*, Water Resources Res., v. 14, no. 3, pp. 451-459.

McLEOD, R. S., 1974, *A Digital-Computer Model for Estimating Drawdowns in the Sandstone Aquifer in Dane County Wisconsin*, U. S. Geol. Survey Open-File Rept.

MERCADO, A., 1976, *Nitrate and Chloride Pollution of Aquifers: A Regional Study with the Aid of a Single Cell Model*, Water Resources Res., v. 14, pp. 409.

MERCER, J. W., and FAUST, C. R., 1980, *Ground-Water Modeling: Mathematical Models*, Ground Water, v. 18, no. 3, May-Jun., pp. 212-227.

MERCER, J. W., SILKA, L. R., and FAUST, C. R., 1983, *Modeling Ground-Water Flow at Love Canal, New York*, J. Environmental Engineering, Am. Soc. Civil Engrs., v. 109, no. 4.

METRY, A. A., 1981, *Predictive Tools for Contaminant Transport in Groundwater*, in *"Permeability and Groundwater Contaminant Transport."* Special Tech. Publ. 746, Am. Soc. for Testing and Materials.

MIDO, K. W., 1980, *Use of Microcomputers in Ground-Water Basin Modeling*, Ground Water, v. 18, no. 3, May-Jun., pp. 230-235.

MIDO, K. W., 1981, *An Economic Approach to Determining the Extent of Ground-Water Contamination and Formulating a Contaminant Removal Plan*, Ground Water, v. 19, no. 1, Jan.-Feb., pp. 41-47, (flow system kinematics model).

MILLER, C. T., and WEBER, W. J., Jr., 1984, *Modeling Organic Contaminant Partitioning in Ground-Water Systems*, Ground Water, v. 22, no. 5, pp. 584-592.

MOLZ, F.J., 1974, *Practical Simulation Models of the Subsurface Hydrologic System with Example Applications*, Water Resources Res. Inst. Bull. 19, Auburn Univ., Alabama (finite difference).

Models

MOREL-SEYTOUX, H. J., and DALY, C. J., 1975, *A Discrete Kernal Generator for Stream-Aquifer Studies*, Water Resources Res., v. 11, pp. 253-260, (finite difference).

MORRIS, W. J, and others, 1972, *Combined Surface Water-Groundwater Analysis of Hydrological Systems with the Aid of the Hybrid Computer*, Water Resources Bull., v. 8, pp. 63-76.

MOTZ, L. H., 1981, *Well-Field Drawdowns Using Coupled Aquifer Model*, Ground Water, v. 19, no. 2, Mar.-Apr., pp. 172-179.

MOULDER, E. A., and JENKINS, C. T., 1969, *Analog-Digital Models of Stream-Aquifer Systems*, Ground Water, v. 7, no. 5, pp. 19-25.

MURRAY, W. A., and JOHNSON, R. L., 1977, *Modeling of Unconfined Ground-Water Systems*, Ground Water, v. 15, no. 4, pp. 306-312.

MURTY, V. V. N. , 1975, *A Fine Element Model for Miscible Displacement in Ground-Water Aquifers*, Ph.D. Dissertation, Univ. of California, Davis, California.

MURTY, V. V. N., and SCOTT, V. H., 1977, *Determination of Transport Model Parameters in Groundwater Aquifers*, Water Resources Res., v. 13, no. 6, pp. 941-947.

NARASIMHAN, T. N., NEUMAN, S. P., and WITHERSPOON, P. A., 1978, *Finite Element Method for Subsurface Hydrology Using a Mixed Explicit-Implicit Scheme*, Water Resources Res., v. 14, no. 5, Oct.

NARASIMHAN, T. N., and WITHERSPOON, P. A., 1976, *An Integrated Finite Difference Method for Analyzing Fluid Flow in Porous Media*, Water Resources Res., v. 12, no. 1, pp. 57-64.

NARASIMHAN, T. N., and WITHERSPOON, P. A., 1977, *Numerical Model for Saturated-Unsaturated Flow in Deformable Porous Media*, 1. *Theory*, Water Resources Res., v. 13, no. 3, pp. 657-664.

NARASIMHAN, T. N., and WITHERSPOON, P. A., 1978, *Numerical Model for Saturated-Unsaturated Flow in Deformable Porous Media*, 3. *Applications*, Water Resources Res., v. 14, no. 6, Dec.

NEUMAN, S. P., 1973, *Calibration of Distributed Parameter Groundwater Flow Models Viewed as a Multiple-Objective Decision Process Under Uncertainty*, Water Resources Res., v. 9, no. 4, pp. 1006-1021.

NEUMAN, S. P., and WITHERSPOON, P. A., 1970, *Variational Principles for Confined and Unconfined Flow of Ground Water*, Water Resources Res., v. 6, pp. 1376-1382 (finite element).

NEUMAN, S. P., and WITHERSPOON, P. A., 1971, *Analysis of Nonsteady Flow with a Free Surface Using the Finite Element Method*, Water Resources Res., v. 7, no. 3, pp. 611-623.

NUTBROWN, D. A., 1975, *Identification of Parameters in a Linear Equation of Groundwater Flow*, Water Resources Res., v. 11, no. 4, pp. 581-588.

NUTBROWN, D. A., 1975, *Normal Mode Analysis of the Linear Equation of Groundwater Flow*, Water Resources Res., v. 11, no. 6, pp. 979-987.

OLSTHOORN, T. N., 1985, *Computer Notes-The Power of the Electronic Worksheet: Modeling Without Special Program*, Ground Water, v. 23, no. 3, pp. 381-390.

PAPADOPULOS, S. S., and LARSON, S. P., 1978, *Aquifer Storage of Heated Water, Part 2 — Numerical Simulation of Field Results*, Ground Water, v. 16, no. 4, pp. 242-248.

PERRIER, E. R., and GIBSON, A. C., 1980, *Hydrologic Simulation on Solid Waste Disposal Sites*, U. S. Army Engineer Waterways Experiment Station, Vicksburg, Mississippi, EPA-1AG-D7-01097.

PICKENS, J. F., and LENNOX, W. C., 1976, *Numerical Simulation of Waste Movement in Steady Groundwater Flow Systems*, Water Resources Res., v. 12, no. 2, pp. 171-180.

PINDER, G. F., 1970, *A Digital Model for Aquifer Evaluation*, Techniques of Water-Resources Inv. U. S. Geol. Survey, Book 7, Chapter C1, 18 pp. (finite difference).

PINDER, G. F., 1971, *An Iterative Digital Model for Aquifer Evaluation*, U. S. Geol. Survey.

PINDER, G. F., 1973, *A Galerkin-Finite Element Simulation of Ground Water Contamination on Long Island, New York*, Water Resources Res., v. 9, no. 6, pp. 1657-1669.

PINDER, G. F., and BREDEHOEFT, J. D., 1968, *Application of the Digital Computer for Aquifer Evaluation*, Water Resources Res., v. 4, no. 5, pp. 1069-1093.

PINDER, G. F., and BREDEHOEFT, J. D., 1971, *Ground Water Chemistry and the Transport Equation*, Internat. Symp. on Mathematical Models in Hydrology, Internat., Assoc. Sci. Hydrology, Warsaw, July 1971.

Models

PINDER, G. F., and FRIND, E. O., 1972, *Application of Galerkin's Procedure to Aquifer Analysis*, Water Resources Res., v. 8, no. 1, pp. 108-120, Feb.

PINDER, G. F., FRIND, E. O., PAPADOPULOS, S. S., 1973, *Functional Coefficients in the Analysis of Groundwater Flow*, Water Resources Res., v.9, no. 1, pp. 222-226.

PINDER, G. F., and GRAY, W. G., 1976, *Is There a Difference in the Finite Element Method?*, Water Resources Res., v. 12, pp. 105-107.

PINDER, G. F., and GRAY, W. G., 1977, *Finite Element Simulation in Surface and Subsurface Hydrology*, Academic Press, New York, 310 pp.

PINDER, G. F., GRAY, W. G., and WOBBER, F. J., 1984, *Princeton University Workshop on Subsurface Transport of Energy-Related Organic Chemicals*, Princeton Univ., DE85007875/WEP, NTIS, Springfield, Virginia 22161, 29 pp.

PINDER, G. F., and SAUER, S. P., 1971, *Numerical Simulation of Flood Wave Modification Due to Bank Storage Effects, Water Resources Res., v. 7, no. 1, pp. 63-70.*

PRICKETT, T. A., and LONNQUIST, C. G., 1971, *Selected Digital Computer Techniques for Ground Water Resource Evaluation*, Illinois State Water Survey, Bull. 55, Urbana, 62 pp.

PRICKETT, T. A., and LONNQUIST, C. G., 1973, *Aquifer Simulation Model for Use of Disk Supported Small Computer Systems*, Illinois State Water Survey Circ. 114.

PRICKETT, T. A., and LONNQUIST, C. G., 1976, *Metodos de Ordenador para Evaluacion de Recursos Hidraulicos Subterraneous*, Bull. no. 41, Ministerio de Obras Publicas, Direccion General de Obras Hidraulicos, Madrid, Spain (finite difference).

PRICKETT, T. A., NAYMIK, T. G., and LONNQUIST, C. G., 1981, *A 'Random-Walk' Solute Transport Model for Selected Groundwater Quality Evaluations*, Bull. 65, Illinois State Water Survey, 605 E., Springfield Ave., Champaign, Illinois 61820.

RALSTON, A., and WILF, H. S., 1961, *Mathematical Methods for Digital Computers*, Wiley & Sons Inc., Chapter 3.

REED, J. E., and others, 1976, *Simulation Procedure for Modeling Transient Water-Table and Artesian Stress and Response*, U. S. Geol. Survey, Little Rock, Arkansas, Open-file Rept. 76-792.

REEVES, M., and DYGYID, J. O., 1975, *Water Movement through Saturated-Unsaturated Porous Media a Finite-Element Galerkin Model*, NTIS, Springfield, Virginia 22161 (finite element).

REMSON, I., APPEL, C. A., and WEBSTER, W. A., 1965, *Groundwater Models Solved by Digital Computer*, J. Hydraulics Div., Proc. Am. Soc. Civil Engrs., HY3, pp. 133-147.

REMSON, I., HORNBERGER, G. M., and MOLZ, D. J., 1971, *Numerical Methods in Subsurface Hydrology*, John Wiley & Sons, Inc., 389 pp.

RITCHIE, J. D., and COLLINS, M.A., 1985, *An Overview of Computer Technology Used in Ground Water Field Studies: Part I: Basics*, Ground Water Monitoring Rev., v. 4, no. 1, pp. 33-38.

ROBERTSON, J. B., 1974, *Digital Modeling of Radioactive and Chemical Waste Transport in the Snake River Plain Aquifer at the National Reactor Testing Station, Idaho*, U. S. Geol. Survey Open-file Rept., 41 pp.

ROBSON, S. G., 1975, *Feasibility of Digital Water-Quality Modeling Illustrated by Application at Barstow, California,*, U. S. Geol. Survey Water-Resour. Inv. 46-73, 72 pp.

ROBSON, S. G., 1978, *Application of Digital Profile Modeling Techniques to Ground-Water Solute Transport at Barstow, California*, U. S. Geol. Survey Water-Supply Paper W 2050, 28 pp.

ROVEY, C. E. K., 1975, *Numerical Model of Flow in a Stream-Aquifer System*, Colorado State Univ. Hydrology Paper No. 74, 73 pp. (finite difference).

RUBIN, JACOB, and JAMES, R. V., 1973, *Dispersion-Affected Transport of Reacting Solutes in Saturated Porous Media: Galerkin Method Applied to Equilibrium-Controlled Exchange in Unidirectional Steady Water Flow*, Water Resources Res., v. 9, no. 5, pp. 1332-1356.

SALEEM, Z. A., 1973, *Method for Numerical Simulation of Flow in Multiaquifer Systems*, Water Resources Res., v. 9, no. 5, pp. 1465-1470.

SAMMEL, E. A., 1963, *Evaluation of Numerical-Analysis Methods for Determining Variations in Transmissivity*, Internat. Assoc. Sci. Hydrology Pub. no. 64, Subterranean Waters, pp. 239-251.

SCHWARTZ, F. W., and DOMENICO, P. A., 1973, *Simulation of Hydrochemical Patterns in Regional Groundwater Flow*, Water Resources Res., v. 9, p. 707.

SHAMIR, U. Y., and HARLEMAN, D. R. F., 1967, *Numerical Solutions for Dispersion in Porous Mediums*, Water Resources Res., v. 3, pp. 557-581.

SHAW, F. S., and SOUTHWELL, R. V., 1941, *Relaxation Methods Applied to Engineering Problems, VII, Problems Relating the Percolation of Fluids Through Porous Materials*, Proc. Royal Soc., Ser. A, v. 178, pp. 1-17.

SHERWANI, J. K., 1973, *Computer Simulation of Ground-Water Aquifers of the Coastal Plain of North Calolina*, Dept. of Environmental Sciences and Engineering, Univ. of North Carolina at Chapel Hill (finite difference).

STALLMAN, R. W., 1956, *Numerical Analysis of Regional Water Levels to Define Aquifer Hydrology*, Trans. Am. Geophys. Union, v. 37 (finite difference).

STALLMAN, R. W., 1956, *Use of Numerical Methods for Analyzing Data on Ground Water Levels*, Symposia Darcy, Internat. Assoc. Sci. Hydrology Pub. 41, pp. 227-231.

STANISLAV, J. F., 1983, *Mathematical Modeling of Transport Phenomena Processes*, Ann Arbor Science Publishers, The Butterworth Group, 10 Tower Office Park, Woburn, Massachusetts 01801, 254 pp.

STEPHENSON, D. and DEJESUS, A. S. M., 1985, *Estimation of Dispersion in an Unsaturated Aquifer*, J. Hydrology, v. 81, no. 1/2, Oct. 30.

STRAUB, W. A., and LYNCH, D. R., 1982, *Models of Landfill Leaching: Moisture Flow and Inorganic Strength*, J. Environmental Engineering Div., Am. Soc. Civil Engrs., v. 108, no. EE2, Apr.

TAYLOR, G. S., and LUTHIN, J. N., 1969, *Computer Methods for Transient Analysis of Water-Table Aquifers*, Water Resources Res., v. 5, no. 1, p. 144.

TAYLOR, O. J., and LUCKEY, R. R., 1972, *A New Technique for Estimating Recharge Using a Digital Model*, Ground Water, v. 10, no. 6, pp. 22-26.

THOMAS, R. G., 1973, *Groundwater Basin Studies Through the Electronic Circuit Analysis Program of the Digital Computer*, Water Resources Res., v. 9, no. 6, pp. 1685-1688.

THRAILKILL, JOHN, 1972, *Digital Computer Modeling of Limestone Groundwater Systems*, Kentucky Water Resources Res. Inst., Rept. no. 50, PB 212 054, 71 pp.

TODSEN, M., 1971, *On the Solution of Transient Free-Surface Flow Problems in Porous Media by Finite Difference Methods*, J. Hydrology, v. 12, pp. 177-210.

TRESCOTT, P. C., 1973, *Iterative Digital Model for Aquifer Evaluation*, U. S. Geol. Survey Open-file Rept., 63 pp.

TRESCOTT, P. C., 1975, *Documentation of Finite-Difference Model for Simulation of Three-Dimensional Ground-Water Flow*, U. S. Geol. Survey, Open-file Rept. 75-438.

TRESCOTT, P. C., and LARSON, S. P., 1977, *Comparison of Interactive Methods of Solving Two-Dimensional Groundwater Flow Equations*, Water-Resources Res., v. 13, no. 1, pp. 125-136.

TRESCOTT, P. C., and LARSON, S. P., 1977, *Solution of Three-Dimensional Groundwater Flow Equations Using the Strongly Implicit Procedure*, J. Hydrology, v. 35, no. 1/2, pp. 49-60.

TRESCOTT, P. C., PINDER, G. F., and LARSON, S. P., 1976, *Finite-Difference Model for Aquifer Simulation in Two-Dimensions with Results of Numerical Experiments*, U. S. Geol. Survey Tech. Water Resources Inv. 7-C1, 116 pp.

TSANG, C. F., 1984, *Studies in Thermo-Hydro-Chemical Transport in Porous and Fractured Media*, California Univ., Berkeley, Lawrence Berkeley Lab, DE85012555/WEP, NTIS, Springfield, Virginia 22161, 12 pp.

TYSON, H. N., Jr., and WEBER, E. M., 1963, *Use of Electronic Computers in the Simulation of the Dynamic Behaviour of Groundwater Basins*, presented before the Am. Soc. Civil Engrs., Water Resources Conf., Milwaukee, Wisconsin, May.

TYSON, H. N., Jr., and WEBER, E. M., 1964, *Ground-Water Management for the Nation's Future — Computer Simulation of Ground-Water Basins*, J. Hydraulics Div., Proc. Am. Soc. Civil Engrs., Proc. Am. Soc. Civil Engrs., v. 90, HY4, pp. 59-77. (finite difference).

UNESCO, 1982, *Ground-Water Models, Vol. I, Concepts, Problems and Methods of Analysis with Examples of Their Application*, Studies and Reports in Hydrology 34, UNIPUB, New York, New York 10036, 235 pp.

Models

U. S. ARMY CORPS OF ENGINEERS, 1970, *Finite Element Solution of Steady-State Potential Flow Problems*, Hydrologic Engineering Publ. Center Rep. 723-440, Sacramento District, Davis, California.

VANDENBERG, A., 1974, *Program FLONET, Flownet Contouring for Well Fields near a Recharge Boundary*, Inland Waters Directorate, Water Resources Branch, Ottawa, Canada (mathematical-finite difference).

VAN GENUCHTEN, M. T., 1978, *Simulation Modes and Their Application to Landfill Disposal Siting: A Review of Current Technology*, Proc. Fourth Annual Hazardous Waste Management Symp., San Antonio, Texas.

VEERUIJT, A., 1972, *Solution of Transient Ground-Water Flow Problems by the Finite Element Method*, Water Resources Res., v. 8, pp. 725-727.

VEMURI, V., and KARPLUS, W. J., 1969, *Identification of Nonlinear Paramenters of Ground Water Basins by Hybrid Computation*, Water Resources Res., v. 5, pp. 172-185.

VERGE, MURRAY-JAMES, 1972, *Design of Analog Model for Aquifer Response Studies, Using a Digital Model*, Ground Water, v. 10, no. 5, pp. 33-37, Sept.-Oct.

VOLKER, R. E., 1969, *Nonlinear Flow in Porous Media by Finite Elements*, J. Hydraulics Div., Amer. Soc. Civil Engrs., v. 95, no. HY6, pp. 2093-2114.

WAGNER, J., WATTS, A., and KENT, D. C., 1985, *Plume 3D: Three-Dimensional Plumes in Uniform Ground Water Flow*, Oklahoma State Univ., PB85-214443/WEP, NTIS, Springfield, Virginia 22161, 82 pp.

WALTON, W. C., 1979, *Progress in Analytical Ground-Water Modeling*, J. Hydrology, v. 43, no. 1/4, Oct.

WALTON, W. C., and NEILL, J. C., 1960, *Analyzing Groundwater Problems with Mathematical Models and a Digital Computer*, Internat. Assoc. Sci. Hydrology Pub. 52.

WALTON, W. C., and WALKER, W. H., 1961, *Evaluation of Wells and Aquifers by Analytical Methods*, J. Geophys. Res. v. 66, no. 10.

WANG, H. F., and ANDERSON, M. P., 1982, *Introduction to Ground-Water Modeling: Finite Difference and Finite Element Methods*, W. H. Freeman and Co., San Francisco, California 94104.

WEEKS, J. B., and others, 1974, *Simulated Effects of Oil-Shale Development on the Hydrology of Piceance Basin, Colorado*, U. S. Geol. Survey Prof. Paper 908.

WERNER, P. W., and NOREN, D., 1951, *Progressive Waves in Non-Artesian Aquifers*, Trans. Am. Geophys. Union, v. 32, no. 2.

WIGGERT, D. C., 1974, *Two-Dimensional Finite Element Modeling of Transient Flow in Regional Aquifer Systems*, Michigan State Univ. Institute of Water Research, Tech. Rept. no. 41, 51 pp.

WILSON, J. L., and GELHAR, L. W., 1974, *Dispersive Mixing in a Partially Saturated Porous Medium*, Rept. No. 191 Ralph M. Parsons Lab. for Water Resources and Hydrodynamics, Massachusetts. Institute of Technology (finite difference).

WINTER, T. C., 1976, *Numerical Simulation Analysis of the Interaction of Lakes and Ground Water*, U. S. Geol. Surv. Prof. Paper 1001, 45 pp.

WINTER, T. C., 1978, *Numerical Simulation of Steady State Three-Dimensional Groundwater Flow near Lakes*, Water Resources Res., v. 14, no. 2, pp. 245-254.

WITHERSPOON, P. A., JAVANDEL, I., and NEUMAN, S. P., 1968, *Use of the Finite Element Method in Solving Transient Flow Problems in Aquifer Systems*, in *The Use of Analogue and Digital Computers in Hydrology.*, Vol. II, IASH—UNESCO Publ. no. 81, pp. 687-698.

YANG, S., 1949, *Seepage Toward a Well Analyzed by the Relaxation Method*, Ph.D. Thesis, Harvard Univ., Cambridge, Massachusetts.

YOUNG, R. A., and BREDEHOEFT, J. D., 1972, *Digital Computer Simulation for Solving Management Problems of Conjunctive Ground Water and Surface Water Systems*, Water Resources Res., v. 8, no. 3, Jun.

ZIENKIEWICZ, P., MAYER, P., and CHEUNG, Y. K., 1966, *Solution of Anisotropic Seepage by Finite Elements*, J. Engineering Mechanics Div., Proc. Am. Soc. Civil Engrs., v. 92, EM1, pp. 111-120.

LAND SUBSIDENCE/SINK HOLES/ COMPACTION/UPLIFT

BIBLIOGRAPHY

ANONYMOUS, 1984, *Land Subsidence Due to Ground-Water Pumping, 1977 - Sept. 1984*, (Citations from the Selected Water Resources Abstracts Data Base), PB84-874908, NTIS, Springfield, Virginia 22161.

GENERAL WORKS

BOUWER, HERMAN, 1977, *Land Subsidence and Cracking Due to Ground-Water Depletion*, Ground Water, v. 15, no. 5, pp. 358-364.

BRAH, W. J., and JONES, L. L., 1978, *Institutional Arrangements for Effective Ground-Water Management to Halt Land Subsidence*, Texas Water Resources Institute Rept. TR-95. College Station 77853.

BULL, W. B., 1961, *Causes and Mechanics of Near-Surface Subsidence in Western Fresno County, California*, U. S. Geol. Survey Prof. Paper 424-B, pp. 187-189.

BULL, W. B., 1975, *Land Subsidence Due to Ground-Water Withdrawal in the Los Banos-Kettleman City Area, California*, Part 2, *Subsidence and Compaction of Deposits*, U. S. Geol. Survey Prof. Paper 437-E.

BULL, W. B., and MILLER, R. E., 1975, *Land Subsidence Due to Ground-Water Withdrawal in the Los Banos-Kettleman City Area, California*, Part 1, *Changes in the Hydrologic Environment Conducive to Subsidence*, U. S. Geol. Survey Prof. Paper 437-E.

BULL, W. B., and POLAND, J. F., 1972, *Land Subsidence Due to Ground-Water Withdrawal in the Los Banos-Kettleman City Area, California*, Part 3, *Interrelations of Water-Level Change, Change in Aquifer-System Thickness, and Subsidence*, U. S. Geol. Survey Open-file Rept., 198 pp.

CORAPCIOGLU, M. Y., and BRUTSAERT, WILFRIED, 1977, *Viscoelastic Aquifer Model Applied to Subsidence Due to Pumping*, Water Resources Res., v. 13, no. 3, pp. 597-604.

CORDOVA, R. M., and MOWER, R. W., 1976, *Fracturing and Subsidence of the Land Surface Caused by the Withdrawal of Ground Water in the Milford Area, Utah*, U. S. Geol. Survey, J. Res., v. 4, no. 5, pp. 505-510.

DAVIS, G. H., and others, 1963, *Land Subsidence Related to Decline of Artesian Pressure in the Ocala Limestone at Savannah, Georgia*, Eng. Geol. Case Histories, Geol. Soc. Am., v. 4.

DOMENICO, P. A., and CLARK, G., 1964, *Electric Analogs in Time-Settlement Problems*, Proc. Am. Soc. Civil Engrs. 90 (SM3).

DOMENICO, P. A., and MIFFLIN, M. D., 1965, *Water from Low-Permeability Sediments and Land Subsidence*, Res. Assoc. Desert Res. Inst. Univ. Nevada Pub., v. 1, no. 4.

EGE, J. R., 1984, *Formation of Solution-Subsidence Sinkholes Above Salt Beds*, U. S. Geol. Survey, Circular 0897, 11 pp.

FADER, S. W., 1976, *Land Subsidence Caused by Dissolution of Salt near Four Oil and Gas Wells in Central Kansas*, U. S. Geol. Survey Water-Resour. Inv. 75-27, 33 pp.

FONTES, J. C., and BARTOLAMI, G., 1973, *Subsidence of the Venice Area During the Past 40,000 Years*, Nature, v. 244, no. 5414, Aug. 10.

FOOSE, R. M., 1967, *Sinkhole Formation by Groundwater Withdrawal: Far West Rand, South Africa*, Science, v. 157, pp. 1045-1048.

FOWLER, L. C., 1981, *Economic Consequences of Land Subsidence*, J. Irrig. and Dranage Div., Am. Soc. Civil Engrs., v. 107, no. IR2, pp. 151-160.

FOX, D. J., 1965, *Man-Water Relationships in Metropolitan Mexico*, Geogr. Review, v. 55, pp. 523-545.

GABRYSCH, R. K., 1980, *Approximate Land-Surface Subsidence in the Houston-Galveston Region, Texas 1906-78, 1943-78, and 1973-78*, U. S. Geol. Survey Open-file Services Section Rept. 80-338.

GABRYSCH, R. K., and BONNET, C. W., 1974, *Land-Surface Subsidence in the Area of Burnett, Scott, and Crystal Bays near Baytown, Texas*, U. S. Geol. Survey Water Resources Inv. 21-74, 48 pp.

GABRYSCH, R. K., and BONNET, C. W., 1974, *Land-Surface Subsidence in the Houston-Galveston Region, Texas*, U. S. Geol. Survey Open-file Rept., 37 pp.

GABRYSCH, R. K., and BONNET, C. W., 1977, *Land-Surface Subsidence in the Area of Moses Lake near Texas City, Texas*, U. S. Geol. Survey Water-Resources Inv. 7632, 90 pp.

GABRYSCH, R. K., and BONNET, C. W., 1977, *Land-Surface Subsidence in the Houston-Galveston Region, Texas*, Texas Water Development Board Rept. 188.

GABRYSCH, R. K., and BONNET, C. W., 1977, *Land-Surface Subsidence at Seabrook, Texas*, U. S. Geol. Survey Water-Resources Inv. 76-31, 108 pp.

GAMBOLATI, GIUSEPPE, 1972, *A Three-Dimensional Model to Compute Land Subsidence*, Bull. Internat. Assoc. Sci. Hydrology, v. 17, no. 2, pp. 219-226.

GAMBOLATI, GIUSEPPE, and FREEZE, R. A., 1973, *Mathematical Simulation of the Subsidence of Venice, 1. Theory*, Water Resources Res., v. 9, no. 3, pp. 721-733, Jun.

GAMBOLATI, GIUSEPPE, GATTO, PAOLO, and FREEZE, R. A., 1974, *Mathematical Simulation of the Subsidence of Venice, 2. Results*, Water Resources Res., v. 10, no. 3, pp. 563-577.

GAMBOLATI, GIUSEPPE and VOLPI, GIAMPIERO, 1979, *Ground-Water Contour Mapping in Venice by Stochastic Interpolators, 1. Theory*, Water Resources Res., v. 15, no. 2, Apr.

GIBBS, H. J., 1960, *A Laboratory Testing Study of Land Subsidence*, Proc. Pan Am. Conf. Soil Mech. Foundation Eng. 1st, Mexico City, 1959, v. 1.

GILLULY, JAMES, and GRANT, U. S., 1949, *Subsidence in the Long Beach Harbor Area, California*, Bull. Geol. Soc. Am., v. 60, pp. 461-530.

GREEN, J. H., 1962, *Compaction of the Aquifer System and Land Subsidence in the Santa Clara Valley, California*, U. S. Geol. Survey Prof. Paper 450-D, art. 172, pp. D175-D178.

GREEN, J. H., 1964, *The Effect of Artesian-Pressure Decline on Confined Aquifer Systems and its Relation to Land Subsidence*, U. S. Geol. Survey Water-Supply Paper 1779-T, 11 pp.

HELM, D. C., 1975, *One-Dimensional Simulation of Aquifer System Compaction near Pixley, California, 1. Constant Parameters*, Water Resources Res., v. 11 no. 3, pp. 465-478.

HELM, D. C., 1976, *One-Dimensional Simulation of Aquifer System Compaction near Pixley California, 2. Stress-Dependent Paramenters*, Water Resources Res., v. 12, no. 3, pp. 375-391.

HOLZER, T. J., 1979, *Elastic Expansion of the Lithosphere Caused by Ground-Water Depletion*, J. Geophys. Res., v. 84, pp. 4689-4698.

HOLZER, T. L., 1984, *Ground Failure Induced by Ground-Water Withdrawal from Unconsolidated Sediment*, Reviews in Engineering Geology, v. 6, pp. 67-105.

HOLZER, T. L., (ed.), 1984, *Man-Induced Land Subsidence*, Geol. Soc. of America, P.O. Box 9140, Boulder, Colorado 80301, 230 pp.

INTERNATIONAL ASSOCIATION OF SCIENTIFIC HYDROLOGY/ UNESCO, 1969, *Land Subsidence*, Proc. Tokyo Symp. Sept. 1969, Composite: English/French, v. 1, 324 pp., v. 2, 337 pp. (UNESCO Studies and Reports in Hydrology, 8).

JOHNSON, A. I., and others, 1968, *Physical and Hydrologic Properties of Water-Bearing Deposits in Subsiding Areas in Central California*, U. S. Geol. Survey Water-Supply Paper 497-A.

JOHNSON, A. I. (ed.), 1981, *Legal, Socioeconomic, and Environmental Significance of Land Subsidence*, J. Irrigation and Drainage Div. Am. Soc. Civil Engrs., v. 107, no. IR2, Jun., pp. 113-185.

JORGENSON, D. G., 1981, *Geohydrologic Models of the Houston District*, Ground Water, v. 19, no. 4, pp. 418-428.

KOPPER, WILLIAM and FINLAYSON, DONALD, 1981, *Legal Aspects of Subsidence due to Well Pumping*, J. Irrig. and Drainage Div., Am. Soc. Civil Engrs., v. 107, no. IR2, pp. 137-150.

KREITLER, C. W., 1977, *Fault Control of Subsidence*, Houston, Texas, Ground Water, v. 15, no. 3, pp. 203-214.

KREITLER, C. W., 1978, *Faulting and Land Subsidence from Ground-Water and Hydrocarbon Production, Houston-Galveston, Texas*, University of Texas, Bureau of Economic Geology, Research Note 8, Box X, Austin, Texas 78712.

LOEHNBERG, ALFRED, 1958, *Aspects of the Sinking of Mexico City and Proposed Countermeasures*, J. Am. Water Works Assoc., v. 50, no. 3, pp. 432-440, Mar.

Subsidence

LOFGREN, B. E., 1961, *Measurement of Compaction of Aquifer Systems in Areas of Land Subsidence*, U. S. Geol. Survey Prof. Paper 424-B, art. 24.

LOFGREN, B. E., 1968, *Analysis of Stresses Causing Land Subsidence*, U. S. Geol. Survey Prof. Paper 600-B, pp. 219-225.

LOFGREN, B. E., 1969, *Land Subsidence Due to the Application of Water*, in Varnes, D. J., and Kiersch, G. (eds.), *Reviews in Engineering Geology*, v. 2, Geol. Soc. Am. Boulder, Colorado, pp. 271-303.

LOFGREN, B. E., 1975, *Land Subsidence Due to Ground-Water Withdrawal, Arvin-Manicopa Area, California*, U. S. Geol. Survey Prof. Paper 437D, D1-D55.

LOFGREN, B. E., 1976, *Land Subsidence and Aquifer-System Compaction in the San Jacinto Valley, Riverside County, California — A Progress Report*, U. S. Geol. Survey, J. Res., v. 4, no. 1, pp. 9-18.

LOFGREN, B. E., and KLAUSING, R. L., 1969, *Land Subsidence Due to Ground-Water Withdrawal, Tulare Wasco Area, California*, U. S. Geol. Survey Prof. Paper 437-B.

LOHMAN, S. W., 1961, *Compression of an Elastic Artesian Aquifer*, U. S. Geol. Survey Prof. Paper 424-B.

MACK, L. E., 1971, *Ground Water Management in Development of a National Policy on Water*, U. S. National Water Comm., Arlington, Virginia, 177 pp.

McCAULEY, CHARLES, and GUM, RUSSELL, 1975, *Land Subsidence: An Economic Analysis*, Water Resources Bull., v. 11, no. 1, Feb., pp. 148-154.

MEADE, R. H., 1968, *Compaction of Sediments Underlying Areas of Land Subsidence in Central California*, U. S. Geol. Survey Water-Supply Paper 497-D.

MILLER, R. E., 1961, *Compaction of an Aquifer System Computed from Consolidated Tests and Decline in Artesian Head*, U. S. Geol. Survey Prof. Paper 424-B, B54-B58.

MILLER, R. E., GREEN, J. H., and DAVIS, G. H., 1971,. *Geology of the Compacting Deposits in the Los Banos-Kettleman City Subsidence Area, California*, U. S. Geol. Survey Prof. Paper 497-B, 46 pp.

NEIGHBORS, R. J., 1981, *Subsidence in Harris and Galveston Counties, Texas*, J. Irrig. and Drainage Div. Am. Soc. Civil Engrs., v. 107, no. IR2, Jun.

NEWTON, J. G., 1981, *Induced Sinkholes: An Engineering Problem*, J. Irrig. and Drainage Div., Am. Soc. Civil Engrs., v. 107, no. IR2, pp. 175-186.

NEWTON, J. G., 1984, *Sinkholes Resulting from Ground-Water Withdrawals in Carbonate Terranes; an Overview*, Reviews in Engineering Geology, v. 6, pp. 195-202.

PETERSON, L. L., and others, 1980, *Analysis of Aquifer-System Compaction in the Orange County Ground-Water Basin*, California Dept. Water Res., Sacramento, California 95801.

PIERCE, R. L., 1970, *Reducing Land Subsidence in the Wilmington Oil Field by Use of Saline Waters*, J. Water Resources Res., v. 6, no. 5, pp. 1505-1514, Oct.

POLAND, J. F., 1960, *Land Subsidence in the San Joaquin Valley, California, and its Effect on Estimates of Ground-Water Resources*, Internat. Assoc. Sci. Hydrology Pub. 52, pp. 324-335.

POLAND, J. F., 1961, *The Coefficient of Storage in a Region of Major Subsidence Caused by Compaction of an Aquifer System*, U. S. Geol. Survey Prof. Paper 424-B.

POLAND, J. F., 1972, *Land Subsidence in the Western States Due to Ground-Water Overdraft*, Water Resources Bull, v. 8, no. 1, Feb.

POLAND, J. F., 1972, *Subsidence and its Control, in Underground Waste Management and Environmental Implications*, Houston, Texas, Am. Assoc. Petrol. Geol. Mem. 18, pp. 50-71.

POLAND, J. F., 1981, *Subsidence in United States due to Ground-Water Withdrawal*, J. Irrig. and Drainage Div., Am. Soc. Civil Engrs., v. 107, no. IR2, pp. 115-136.

POLAND, J. F., 1984, *Guidebook to Studies of Land Subsidence Due to Groundwater Withdrawal*, UNESCO, 7 Place de Fontenoy, 75700 Paris, France, 327 pp.

POLAND, J. F., and DAVIS, G. H, 1956, *Subsidence of the Land Surface in the Tulare-Wasco (Delano) and Los Banos-Kettleman City Area, San Joaquin Valley, California*, Trans. Am. Geophys. Union, v. 37, pp. 287-295.

POLAND, J. F., and DAVIS, G. H., 1969, *Land Subsidence Due to Withdrawal of Fluids,* in Varnes, D. H., and Kiersch, G., (eds.) *Reviews in Engineering Geology,* v. 2, Geol. Soc. of America, pp. 187-269.

POLAND, J. F., and GREEN, J. H., 1962, *Subsidence in the Santa Clara Valley, California, A Progress Report,* U. S. Geol. Survey Water-Supply Paper 1619-C.

POLAND, J. F., and others, 1972, *Glossary of Selected Terms Useful in Studies of the Mechanics of Aquifer Systems and Land Subsidence Due to Fluid Withdrawal,* U. S. Geol. Survey Water-Supply Paper 2025.

POLAND, J. F., and others, 1973, *Land Subsidence in the San Joaquin Valley as of 1972,* U. S. Geol. Survey Prof. Paper 437-H, 141 pp.

PROKOPOVICH, N. P., and MAGLEBY, D. C., 1968, *Land Subsidence in Pleasant Valley Area, Fresno County, California,* J. Am. Water Works Assoc., v. 60, p. 413, Apr.

ROLL, J. R., 1967, *Effect of Subsidence on Well Fields,* J. Am. Water Works Assoc., v. 59, p. 80, Jan.

SHELTON, M. J., and JAMES, L. B., 1959, *Engineer-Geologist Team Investigates Subsidence,* J. Pipeline Div., Am. Soc. Civil Engrs., v. 85, no. PL2, 18 pp.

SHIBASAKI, T., KAMATA, A., and WADA, M., 1971, *Application of the Water Balance Simulation for Predicting Land Subsidence — A Digital Computer Approach,* Memoirs of Internat. Assoc. Hydrogeologists, Tokyo Congress, pp. 197-202.

SHIBASAKI, T., and SHINDO, T., 1969, *The Hydrologic Balance in the Land Subsidence Phenomena,* IASH/UNESCO Tokyo Symp. on Land Subsidence, pp. 201-215.

SINCLAIR, W. C., 1982, *Sinkhole Development Resulting from Ground-Water Withdrawal in the Tampa Area, Florida,* PB-82 185141, NTIS, Springfield, Virginia 22161.

SMITH, C. G., and KAZMANN, R. G., 1978, *Subsidence in the Capital Area Ground-Water Conservation District — An Update,* Capital Area Ground-Water Conservation Commission Bull. no. 2, P. O. Box 64526, Baton Rouge, Louisiana 70806.

TOLMAN, C. F., and POLAND, J. F., 1940, *Ground-Water Salt-Water Infiltration and Ground-Surface Recession in Santa Clara Valley, Santa Clara County*, California Trans. Am. Geophys. Union, 21st Ann. Meeting pt. 1.

WARREN, J. P., and others, 1974, *Costs of Land Subsidence Due to Ground-Water Withdrawal*, Texas A & M University Water Resources Institute Tech. Rep. no. 57, College Station, Texas 77843.

WIER, W. W., 1950, *Subsidence of Peat Lands of the Sacramento-San Joaquin Delta, California*, Hilgardia, v. 20, pp. 37-56.

WILSON, G., and GRACE, H., 1942, *The Settlement of London due to Under-Drainage of the London Clay*, J. Inst. Civil Engrs., v. 19, no. 2.

WINSLOW, A. G., and DOYEL, W. W., 1954, *Land-Surface Subsidence and its Relation to the Withdrawal of Ground Water in the Houston-Galveston Region, Texas*, Econ. Geol. v. 49, pp. 413-422.

YAMAMOTO, S., and others, 1969, *Simulation of Groundwater Balance as a Basis of Considering Land Subsidence in the Koto Delta, Tokyo*, IASH/UNESCO Tokyo Symp. on Land Subsidence, pp. 215-224.

ZEEVAERT, LEONARDO, 1963, *Foundation Problems Related to Ground Surface Subsidence in Mexico City*, in *Field Testing of Soils*, Am. Soc. for Testing and Materials Special Pub. no. 322, pp. 57-66.

GROUNDWATER LAW/REGULATIONS/ LEGAL ISSUES

(See also Groundwater Protection and Management Section)

BIBLIOGRAPHY

ANONYMOUS, 1984, *Ground-Water Law, 1977 - Nov. 1984*, (Citations from the Selected Water Resources Abstracts Data Base), PB85-850824/WNR, NTIS, Springfield, Virginia 22161.

JACOBSTEIN, J. M. and MERSKY, R. M., 1966, *Water Law Bibliography 1847-1965*, Jefferson Law Book Co., Silver Spring, Maryland.

NATIONAL WATER WELL ASSOCIATION, 1970, *Bibliography of State Well Drilling Laws*, Natl. Water Well Assoc., Dublin, Ohio.

GENERAL WORKS

AIKEN, J. D., 1984, *Evaluation of Legal and Institutional Arrangements Associated with Groundwater Allocation in the Missouri River Basin States*, Nebraska Water Res. Center, Lincoln, Nebraska, 88 pp.

ANONYMOUS, 1967, *Water Law Atlas*, State Bur. of Mines and Mineral Resources, New Mexico Inst. of Mining and Tecnology, Socorro, New Mexico, Circ. 95.

ANONYMOUS, 1972, *The Consulting Engineer as an Expert Witness*, Consulting Engineers Assoc. of California, 1308 Bayshore Highway, Burlinghame, California 94010, 50 pp.

ANONYMOUS, 1981, *Environmental Quality Issues: The Relation of Federal Laws to State Programs*, Resource and Environmental Quality Div., Chamber of Commerce of the United States, 1615 H St. N.W., Washington, D. C. 20062.

ANONYMOUS, 1982, *Ground Water Law in Puerto Rico,* Water Well J., v. 36, no. 6, pp. 64-66.

ANONYMOUS, 1982, *Superfund Handbook,* Pollution Engineering Book Shelf, P.O. Box 1096, Rockville, Maryland 20850, 120 pp. (text of law and analysis).

BARTELT, R. E., 1979, *State Ground-Water Protection Programs — A National Summary,* Ground Water, v. 17, no. 1, Jan.-Feb., p. 89-94.

BLACK, A. P., 1947, *Basic Concepts in Ground Water Law,* J. Am. Water Works Assoc., v. 39, pp. 989-1002.

BLISS, J. H., 1951, *Administration of the Ground-Water Law of New Mexico,* J. Am. Water Works Assoc., v. 43.

BROOKS, R. O., 1979, *Ground water Law in Vermont: Planning for Uncertainty, Pluralism and Conflict,* Vermont Water Resources Center, Burlington, PB80-220601, NTIS, Springfield, Virginia 22161, 203 pp.

BUREAU OF NATIONAL AFFAIRS, *Environment Reporter,* 1231 25th St., NW, Washington, DC 20037.

CLARK, R. E., 1982, *Overview of Groundwater Law and Institutions in United States Border States,* Nat. Resources J., v. 22, pp. 1007-1015.

CLARK, S.D., 1979, *Ground-Water Law and Administration in Australia,* Australian Water Resources Council Tech. Paper No. 44. Australian Government Publishing Service, P.O. Box 84, Canberra, ACT 2600.

CONKLING, H., 1937, *Administrative Control of Underground Water: Physical and Legal Aspects,* Trans. Am. Soc. Civil Engrs., v. 102 pp. 753-837.

CONNALL, D. D., 1982, *A History of the Arizona Groundwater Management Act,* Arizona State Law J., pp. 313-344.

CORKER, C. E., 1972, *Ground-Water Law Management and Administration,* N.W.C.-L72-026 Legal Study No. 6, National Water Comm., Washington, D. C., 360 pp.

CORKER, C. E. and CROSBY, J. W., 1972, *Ground Water Law, Management and Administration,* PB 205 527, NTIS, Springfield, Virginia 22161.

COX, W. E., and WALKER, W. R., 1973, *Ground-Water Implications of Recent Federal Law,* Ground Water, v. 11, no. 5, pp. 15-18.

CRITCHLOW, H. T., 1948, *Policies and Problems in Controlling Ground-Water Resources*, J. Am. Water Assoc., v. 40, pp. 775-783.

DEUTSCH, MORRIS, 1963, *Ground-Water Contamination and Legal Controls in Michigan*, U. S. Geol. Survey Water-Supply Paper 1691, 79 pp.

DEWSNUP, R. L., JENSEN, D. W., and SWENSON, R. W., 1973, *A Summary - Digest of State Water Laws*, National Water Comm. Arlington, Virginia, 826 pp.

EDISON ELECTRIC INSTITUTE, 1984, *Trends In U.S. Groundwater Law, Policy and Administration*, EEI, 1111 Nineteenth Street, N.W., Washington, DC 20036, 140 pp.

ELY, NORTHCUTT, 1963, *Legal Problems in Development of Water Resources*, 57th Meeting of Princeton Univ. Conf., pp. 89-94 May.

FETTER, C. W., Jr., 1981, *Interstate Conflict Over Ground Water: Wisconsin-Illinois*, Ground Water, v. 19, no. 2, Mar.-Apr., pp. 201-213.

FOOD AND AGRICULTURE ORGANIZATION OF THE UNITED NATIONS, 1964, *Groundwater Legislation in Europe*, FAO Legislative Series no. 5, 175 pp.

GLEASON, V. E., 1978, *The Legalization of Ground Water Storage*, Bull. Am. Water Resources Assoc., v. 14, no. 3, pp. 532-541. (aquifer storage rights in California).

GOLDSHORE, LOUIS, 1984, *New Jersey Water Supply Handbook*, County and Municipal Government Study Comm., 115 West State St., Trenton, New Jersey 08625, (water law and regulation).

GRANT, D. L., 1981, *Reasonable Groundwater Pumping Levels Under the Appropriation Doctrine: Underlying Economic Goals*, Natural Resources J., v. 21, pp. 1-36.

GRANT, D. L., 1983, *Reasonable Groundwater Pumping Levels Under the Appropriation Rule: Underlying Social Goals*, Natural Resources J., v. 23, pp. 53-75.

HANKS, E. H. and HANKS, J. L., 1970, *The Law of Water in New Jersey, Groundwater*, Rutgers Law Review, v. 24, no. 4, pp. 621-671.

HARDING, S. T., 1936, *Water Rights for Irrigation*, Stanford Univ. Press, Stanford, California, 176 pp.

HARDING, S. T., 1953, *United States Water Law*, Trans. Am. Soc. Civil Engrs., v. CT, pp. 343-356.

HARDING, S. T., 1955, *Statutory Control of Ground Water in the Western United States*, Trans. Am. Soc. Civil Engrs., v. 120, pp. 490-498.

HARRIS, LINDA, 1984, *New Mexico Water Rights*, Water Resources Res. Inst., Box 3167, NMSU, Las Cruces, New Mexico 88003, 60 pp.

HAYTON, R. D., 1982, *The Ground Water Legal Regime as Instrument of Policy Objectives and Management Requirements*, Natural Resources J., v. 22, pp. 119-137.

HENDERSON, T. R., TRAUBERMAN, T., and GALLAGHER, T., 1984, *Ground Water Strategies for State Action*, Environmental Law Inst., Washington D. C., 353 pp.

HOBERG, A. C., 1982, *The Nature and Extent of Groundwater Management and Groundwater Problems: A Survey*, Agric. Law J., v. 4, pp. 404-442.

HOPPING, W. L., and PRESTON, W. D., 1984, *The Water Quality Assurance Act of 1983 - Florida's "Great Leap Forward" into Groundwater Protection and Hazardous Waste Management*, Florida State Univ. Law Review, v. 11, pp. 599-641.

HUGHES, W. F., 1945, *Proposed Ground-Water Conservation Measures in Texas*, Texas J. Sci., v. 2, pp. 35-45.

HUTCHINS, W. A., 1942, *Selected Problems in the Law of Water Rights in the West*, U. S. Dept. Agric. Misc. Pub. 418, 513 pp.

HUTCHINS, W. A., 1946, *The Hawaiian System of Water Rights*, Bd. of Water Supply of Honolulu.

HUTCHINS, W. A., 1955, *Trends in the Statutory Law of Ground Water in the Western States*, Texas Law Review, v. 34, pp. 157-191.

HUTCHINS, W. A., 1955, *The New Mexico Law of Water Rights*, State Engineer of New Mexico (in cooperation with U. S. Dept. of Agr.).

HUTCHINS, W. A., 1956, *Irrigation Water Rights in California*, Agric. Exp. Sta. Circ. 452, Univ. California, Berkeley, 56 pp.

HUTCHINS, W. A., 1956, *The California Law of Water Rights*, State Calif., Sacramento, 571 pp.

HUTCHINS, W. A., 1971, *Water Rights Laws in Nineteen Western States*, U. S. Dept. Agric., Washington, D. C., 650 pp.

KELLY, W. E., 1980, *Ground and Surface Water Interaction — Legal Aspects*, J. Water Resources Planning and Management Div., Am. Soc. Civil Engrs. v. 106, no. WR1, Mar.

KIERSCH, G. A., 1969, *The Geologist and Legal Cases — Responsibilty, Preparation and the Expert Witness*, in *Legal Aspects of Geology in Engineering Practice*, Geol. Soc. Am. Eng. Case Histories, no. 7, pp. 1-6.

LEHR, J. H. and others, 1976, *A Manual of Laws, Regulations and Institutions for Control of Ground Water Pollution*, U. S. Environmental Protection Agency, PB 257-808, NTIS.

LOTTERMAN, E. D., and WAELTI, J. J., 1983, *Efficiency and Equity Implications of Alternative Well Interference Policies in Semi-Arid Regions*, Natural Resources J., v. 23, pp. 323-334.

McCRAY, KEVIN, 1982, *The Federal Response to Ground Water Protection*, Water Well J., v. 36, no. 6, pp. 42-44.

McCRAY, KEVIN, 1982, *Ground Water and the States*, Water Well J., v. 36, no. 6, pp. 47-50.

McGUINNESS, C. L., 1945, *Legal Control of Use of Ground Water*, Water Works Eng., v. 98, pp. 475, 508, 510, 512.

McGUINNESS, C. L., 1951, *Water Law with Special Reference to Ground Water*, U. S. Geol. Survey Circ. 117, 30 pp.

MILLER, JAMES, 1980, *The Legal Implications of Ground-Water Heat Pump Use*, Water Well J., 34, no. 7, Jul., pp. 66-73.

NATIONAL CONFERENCE OF STATE LEGISLATURES, 1983, *Hazardous Waste Management: A Survey of State Legislation 1982*, NCSL, Suite 1500, PR2, 1125 Seventeenth St., Denver, Colorado 80202, (laws passed from 1976-82).

NATIONAL RESOURCES PLANNING BOARD, 1943, *State Water Law in the Development of the West*, Water Resources Committee, Subcommittee on State Water Law, Washington, D. C., 138 pp.

NATIONAL WATER COMMISSION, 1972, *Ground-Water Law, Management and Administration*, PB-205 527, NTIS Springfield, Virginia 22161.

NATIONAL WATER WELL ASSOCIATION, 1970, *Model Law for Proper Regulation of the Well Drilling Business*, Natl. Water Well Assoc., Dublin, Ohio.

O'BYRNE, J. C., 1956, *Symposium — Water Use and Control*, Iowa Law Review, v. 41, no. 2, Iowa City, Iowa.

PAGAN, A. R., 1978, *Groundwater Law — The Riparian Problem*, J. Am. Water Works Assoc., v. 70, no. 3, pp. 153-155.

PASSAIC VALLEY GROUND WATER PROTECTION COMMITTEE, 1983, *A Model Municipal Ordinance for Control of Toxic and Hazardous Materials*, Passaic Valley Ground Water Protec. Comm., 246 Madisonville Rd., Basking Ridge, New Jersey 07920.

PIPER, A. M., 1959, *Requirements of a Model Water Law*, J. Am. Water Works Assoc., v. 51.

PIPER, A. M., 1960, *Interpretation and Current Status of Ground-Water Rights*, U. S. Geol. Survey Circ. 432.

PIPER, A. M. and THOMAS, H. E., 1958, *Hydrology and Water Law — What is Their Future Common Ground?*, Water Resources and the Law, Univ. Michigan Law School.

PRACTICING LAW INSTITUTE, 1982, *Hazardous Waste Litigation 1982*, P.L.I., 810 Seventh Ave., New York, New York 10019, 708 pp. (apportionment negotiation)

RADIAN CORPORATION, 1983, *Individual Civil and Criminal Liability for Environmental Damages*, Environmental Risk Control Services, The Hartford Steam Boiler Inspection and Insurance Co., Hartford, Connecticut 06102, (compilation of statutory and common law bases).

ROBERTS, R. S., and BUTLER, L. M., 1984, *Information for State Groundwater Quality Policymaking*, Natural Resources J., v. 24, pp. 1015-1041.

SAVAGE, R. J., 1986, *Groundwater Protection: Working Without A Statute*, J. Water Pollution. Control Fed., v. 58, no. 5, pp. 340-342.

PRESIDENT'S WATER RESOURCES POLICY COMMISSION, 1950, *Water Resources Law*, v. 3, Washington, D. C., 777 pp.

SELIG, E. I., 1981, *Rights and Liabilities of Water Suppliers Arising from Ground-Water Pollution*, J. Am. Water Works Assoc., v. 73, no. 4, Apr.

STATE OF CALIFORNIA, 1951, *Water Code*, Sacramento, 756 pp.

TECLAFF, L. A., and UTTON, A. E., 1981, *International Groundwater Law*, Oceana Publ., London, 490 pp.

Law / Regulations

TEMPLER, O. W., 1973, *Water Law and the Hydrologic Cycle: A Texas Example*, Water Resources Bull., v. 9, no. 2, pp. 273-283.

THOMAS, H. E., 1955, *Water Rights in Areas of Ground-Water Mining*, U. S. Geol. Survey Circ. 347, 16 pp.

THOMAS, H.E., 1958, *Hydrology vs. Water Allocation in the Eastern United States*, in *The Law of Water Allocation in the Eastern United States*, The Ronald Press Co., New York.

THOMAS, H. E., 1961, *Ground Water and the Law*, U. S. Geol. Survey Circ. 446, 6 pp.

THOMAS, H. E., 1970, *Water Laws and Concepts*, U. S. Geol. Survey Circ. 629.

THOMPSON, D. G. and FIEDLER, A. G., 1938, *Some Problems Relating to Legal Control of Use of Ground Waters*, J. Am. Water Works Assoc., v. 30. pp. 1049-1091.

TOLMAN, C. F. and STIPP, A. C., 1941, *Analysis of Legal Concepts of Subflow and Percolating Waters*, Trans. Am. Soc. Civil Engrs., v. 106, pp. 882-933.

TRAVIS, C., and ETNIER, E. L., (eds.), 1984, *Groundwater Pollution: Environmental and Legal Problems*, Papers from AAS Selec. Symp. 95 Westview Press, 5500 Central Ave., Boulder, Colorado 80301, 149 pp.

TRELEASE, F. J., 1974, *Water Law — Cases and Material on Water Law — Resource Use and Environomental Protection*, West Publishing Co., 50 W. Kellogg Blvd., St. Paul, Minnesota 55102.

TRELEASE, F. J., 1979-80, *Legal Solutions to Groundwater Problems - A General Overview*, Pacific Law J., v. 11, pp. 863-875.

UTTON, A. E., 1982, *The Development of International Groundwater Law*, Natural Resources J., v. 22, pp. 95-118.

VAN DER LEEDEN, FRITS, 1973, *Groundwater Pollution Features of Federal and State Statutes and Regulations*, Office of Research and Development, U. S. Environmental Protection Agency, EPA-600/4-73-0012 a, 88 pp.

WALKER, W. R., and COX, W. E., 1977, *Deep Well Injection of Industrial Wastes — Government Controls and Legal Constraints*, Virginia Water Resources Center, Blacksburg, Virginia 24061.

WARNER, D. L., 1970, *Regulatory Aspects of Liquid Waste Injection into Saline Aquifers*, J. Water Resources Res., v. 6, no. 5, pp. 1458-1462, Oct.

WEINBERG, D. B., GOLDMAN, G. S., and BRIGGUM, S. M., 1984, *Hazardous Waste Regulation Handbook: A Practical Guide to RCRA and Superfund*, Executive Enterprises Publ. Co., Inc., 33 West 60th St., New York, New York 10023. 379 pp.

WICKERSHAM, GINIA, 1981, *A Preliminary Survey of State Ground-Water Laws*, Ground Water, v. 19, no. 3, pp. 321-327.

GROUNDWATER MANAGEMENT, PROTECTION, AND CONSERVATION/ CONJUNCTIVE USE/ECONOMICS/ ZONING

BIBLIOGRAPHY

BATTELLE MEMORIAL INSTITUTE, 1966, *Bibliography on Socio-Economic Aspects of Water Resources*, Contract no. 14-01-0001-822, U. S. Office of Water Resources Res., 453 pp.

MAKNOON, REZA, and BURGES, S. J., 1978, *Conjunctive Use of Ground and Surface Water, (includes literature review)*, J. Am. Water Works Assoc., v. 70, no. 8, Aug., pp. 419-425.

TURNQUIST, SUSAN, 1985, *Community Guide to Groundwater Protection and Management, An Annotated Bibliography*, Northeast Regional Center for Rural Development, 293 Roberts Hall, Cornell Univ., Ithaca, New York 14853, 65 pp.

WOLFF, NED, 1985, *Groundwater Protection-A Selected Bibliography*, U. S. Dept. of Justice, Washington, D.C., PB86-130341, NTIS, Springfield, Virginia 22161, 42 pp.

GENERAL WORKS

AGUADO, EDUARDO and others, 1974, *Optimal Pumping for Aquifer Dewatering*, J. Hydraulics Div., Am. Soc. Civil Engrs. v. 100, no. HY7, Jul.

ALLISON, S. V., 1967, *Cost, Precision, and Value Relationships of Data Collection and Design Activities in Water Development Planning*, Calif. Water Resources Center Contrib. 120, Berkeley, California, 142 pp.

AMBROGGI, R. P., 1977, Underground Reservoirs to Control the Water Cycle, Sci. American, v. 236, no. 5, pp. 21-27.

AMERICAN PETROLEUM INSTITUTE, 1984, *Recommended Practice for Underground Petroleum Product Storage Systems at Marketing and Distribution Facilities*, RP 1635, API, 1220 L St. N.W., Washington, DC 20005.

AMERICAN SOCIETY OF CIVIL ENGINEERS, 1972, *Ground Water Management*, Manual Engrng. Practice 40, New York, 216 pp.

ANDERSON, M. G., and BURT, T. P., 1985, *Hydrological Forecasting*, John Wiley & Sons, New York.

ANONYMOUS, 1956, *Water Use in the United States, 1900-1975*, Supplement to Willing Water 38, Am. Water Works Assoc., 8 pp.

ANONYMOUS, 1983, *Siting Manual for Storing Hazardous Substances: A Practical Guide for Local Officials*, N.Y. State Dept. Env. Conversation, Bur. of Water Resources, 50 Wolf Rd. Albany, New York 12233, 98 pp.

ANONYMOUS, 1984, *A Citizens Handbook on Groundwater Protection*, Natural Resources Defense Council (NRDC), 122 E. 42nd St., New York, New York 10168.

ANONYMOUS, 1984, *Groundwater: A Community Action Guide*, Concern, Inc., 1794 Columbia Rd. N.W., Washington, DC 20009, 22 pp.

ANONYMOUS, 1985, *Proceedings of the NWWA Eastern Regional Conference on Ground Water Management, Orlando, FL., Oct. 1983*, NWWA, Dublin, Ohio 43017.

ANONYMOUS, 1985, *Proceedings of the NWWA Western Regional Conference on Ground Water Management, San Diego, CA., Nov. 1983*, NWWA, Dublin, Ohio 43017.

ARON, GERT., 1968, *Optimization of Conjunctively Managed Surface and Groundwater Resources*, Water Resources Center Contrib. no. 130, Univ. Calif., Berkeley, California.

BANKS, H. O., 1953, *Utilization of Underground Storage Reservoirs*, Trans. Am. Soc. Civil Engrs., v. 118, pp. 220-234.

BANKS, H. O., 1960, *Priorities for Water Use*, Proc. Natl. Conf. on Water Pollution, pp. 153-156, 179, 181, U. S. Public Health Service, Washington, D. C.

BANKS, H. O., 1981, *Management of Interstate Aquifer Systems*, J. Water Resources Planning and Mangement Div., Am. Soc. Civil Engrs., v. 107. no. WR2, Oct.

BARKSDALE, H. C. and REMSON, IRWIN, 1954, *The Effect of Land Management Practices on Ground Water*, Internat. Assoc. Sci. Hydrology, General Assembly Rome, Pub. 37, v. 2, pp. 520-525.

BEAR, JACOB and LEVIN, O., 1967, *The Optimal Yield of an Aquifer, Artificial Recharge and Mangement of Aquifers*, Proc. Symp. of Haifa, Internat. Assoc. Sci. Hydrology, Pub. 72, pp. 402-412.

BEAVER, D. W., 1974, *Cost of Importing Deep Sandstone Water to Eliminate Ground-Water Deficits in Northeastern Illinois*, Illinois State Water Survey Circ. 120.

BEAVER, J. A. and FRANKEL, M. L., 1969, *Significance of Ground-Water Management Strategy — A Systems Approach*, Ground Water, v. 7, no. 3, pp. 22-26.

BIRD, J. C., 1985, *Groundwater Protection: Emerging Issues and Policy Challenges*, Env. and Energy Study Institute, Washington, D.C., 42 pp.

BISHOP, A. B., and others, 1973, *Social, Economic, Environmental, and Technical Factors Influencing Water Reuse*, Utah Water Research Lab, PRJER025-1.

BISHOP, A. B., and others, 1974, *Evaluating Water Reuse Alternatives in Water Resources Planning*, Utah Water Research Lab, PRWG123-1, 1974.

BITTINGER, M. W., 1964, *The Problem of Integrating Ground-Water and Surface Water Use*, Ground Water, v. 2, no. 3, Jul., pp. 33-38.

BITTINGER, M. W., 1972, *Survey of Interstate and International Aquifer Problems*, Ground Water, v. 10, no. 2, pp. 44-54.

BITTINGER, M. W. & ASSOCIATES, 1970, *Management and Administration of Ground Water in Interstate and International Aquifers, Phase 1*, P. O. Box 1592, Fort Collins, Colorado 80521.

BOSTER, M. A. and MARTIN, W. E., 1977, *Economic Analysis of the Conjunctive Use of Surface Water and Ground Water of Differing Prices and Qualities: A Coming Problem for Arizona Agriculture*, Agric. Experiment Station Tech. Bull. 235. Tucson, Arizona 85721.

BOUWER, HERMAN, 1976, *Zoning Aquifers for Tertiary Treatment of Wastewater*, Ground Water, v. 14, no. 6, pp. 386-392.

BREDEHOEFT, J. D. and YOUNG, R. A., 1970, *The Temporal Allocation of Ground Water — A Simulation Approach*, Water Resources Res., v. 6, no. 1, p. 3.

BROWN, GARDNER, Jr., and DEACON, ROBERT, 1972, *Economic Optimization of a Single-Cell Aquifer*, Water Resources Res., v. 8, no. 3, pp. 557-564.

BROWN, R. F., and others, 1978, *Artificial Ground-Water Recharge as a Water-Management: Technique on the Southern High Plains of Texas and New Mexico*, Texas Dept. Water Resources Rept. 220, 32 pp.

BRUTSAERT, W. F., and others, 1975, *C. E. Jacob's Study on the Prospective and Hypothetical Future of the Mining of the Ground Water Deposited Under the Southern High Plains of Texas and New Mexico*, Ground Water, v. 13, no. 6, pp. 492-505.

BRUTSAERT, W. F., and GEBHARD, T. G., Jr., 1975, *Conjunctive Availability of Surface and Ground Water in the Albuquerque Area, New Mexico: a Modeling Approach*, Ground Water, v. 13, no. 4, pp. 345-353.

BURAS, N., 1963, *Conjunctive Operation of Dams and Aquifers*, J. Hydraul. Div., Proc. Am. Assoc. Civil Engrs., v. 89, no. HY6, pp. 111-131, Nov.

BURAS, N., 1966, *Dynamic Programming in Water Resources Development*, in *Advances in Hydroscience*, Ven Te Chow, (ed.), v. 3, Academic Press, New York, pp. 367-412.

BURAS, N. and BEAR, Jacob, 1964, *Optimal Utilization of a Coastal Aquifer*, Proc. 6th Internat. Congr. of Agric. Engrs., Lausanne, Switzerland, Sept.

BURBY, R. J., KAISER, E. J., MILLER, T. L., and MOREAU, D. H., 1984, *Drinking Water Supplies—Protection Through Watershed Management*, Butterworth Publishers, 80 Monvale Ave., Stoneham, Massachusetts 02180, 273 pp.

BURDICK, C. B., 1942, *Ground Water — A Vital National Resource, Midwest Problems*, J. Am. Water Works Assoc., v. 37.

BURKE, K. J., and others, 1984, *Interstate Allocation And Management of Nontributary Groundwater*, Western Governors Assoc., Denver, CO, 168 pp.

BURT, O. R., 1964, *Optimal Resource Use over Time with an Application to Groundwater*, Management Sci., v. 11, pp. 80-93.

BURT, O. R., 1966, *Economic Control of Groundwater Reserves*, J. Farm. Econ., 48, pp. 632-647.

BURT, O. R., 1967, *Groundwater Management under Quadratic Criterion Functions*, Water Resources Res., v. 3, no. 3, pp. 673-682.

BURT, O. R., 1967, *Temporal Allocation of Groundwater*, Water Resources Res., v. 3, no. 1, pp. 45-56.

BURT, O. R., 1970, *Groundwater Storage Control under Institutional Restrictions*, Water Resources Res., v. 6, no. 6, pp. 1540-1548.

BURT, O. R., and CUMMINGS, R. G., 1977, *Natural Resources Management, the Steady-State, and Approximately Optimal Decision Rules*, Land Economics, v. 53, no. 1, pp. 1-22.

CALIFORNIA DEPARTMENT OF WATER RESOURCES, 1968, *Planned Utilization of Ground Water Basins: Coastal Plain of Los Angeles County*, Bull. 104, Sacramento, 25 pp., plus apps.

CARTWRIGHT, K. R., GILKESON, H., and JOHNSON, T. M., 1981, *Geological Considerations in Hazardous-Waste Disposal*, J. Hydrology, v. 54, no. 1/3, Dec.

CASEY, H. E., 1972, *Salinity Problems in Arid Lands Irrigation, Office of Arid Lands Studies*, Univ. Arizona, Tucson, 300 pp.

CEDERSTROM, D. J., 1970, *Cost Analysis of Ground-Water Supplies in the North-Atlantic Region*, U. S. Geol. Survey Water-Supply Paper 2034, 47 pp., also 1974 Water Well J., v. 28, no. 8, Aug.

CHANOUX, J. K., 1982, *Groundwater Protection on the Local Level: Integrating the Fragments of Regulation*, J. New England Water Works Assoc., v. 96, no. 4.

CHARBENEAU, R. J., (ed.), 1983, *Regional and State Water Resources Planning and Management*, Proc. San Antonio, Texas, Symposium, 366 pp.

CHUN, R. Y. D. and others, 1964, *Optimum Conjuctive Operation of Ground Water Basins*, J. Hydraul. Div., Proc. Am. Soc. Civil Engrs. v. 90, no. HY4, pp. 79-95.

CHUN, R. Y. D. and others, 1966, *Planned Utilization of Ground Water Basins, Coastal Plain of Los Angeles County, Operation and Economics*, Bull. 104, Appendix C, Dept. of Water Resources, Sacramento, California.

CHUN, R. Y. D., WEBER, E. M. and MIDO, K. W., 1967, *Planned Utilization of Groundwater Basins: Studies Conducted in Southern California*, Publ. no. 72 Symp. Haifa, Internat. Assoc. of Sci. Hydrology, Mar.

CLENDENEN, F. B., 1955, *Economic Utilization of Ground Water and Surface Water Storage Reservoirs*, Paper presented before meeting of Am. Soc. Civil Engrs., San Diego.

COCHRAN, G. F. and BUTCHER, W. S., 1970, *Dynamic Programming for Optimum Conjunctive Use*, Water Resources Res., v. 6, no.3.

COHEN, P. and others, 1968, *An Atlas of Long Island's Water Resources*, New York Water Resources Bull. 62, New York Dept. of Environmental Conservation, Albany, New York.

COMPTROLLER GENERAL OF THE UNITED STATES, 1980, *Ground-Water Overdrafting Must be Controlled*, Report to Congress, Supt. of Documents, U. S. Govt. Printing Office Washington, D. C. 20402.

CONKLING, H., 1946, *Utilization of Ground-Water Storage in Stream - System Developments*, Trans. Am. Soc. Civil Engrs., v. 111, pp. 275-354.

CONOVER, C. S., 1961, *Ground-Water Resources — Development and Management*, U. S. Geol. Survey Circ. 442, 7 pp.

CONSERVATION FOUNDATION, 1985, *Groundwater-Saving the Unseen Resource*, CF, 1225 23rd St N.W., Washington, DC 20037.

CONSERVATION FOUNDATION, CHEMICAL MANUFACTURERS ASSOC. AND NATIONAL AUDUBON SOCIETY, 1984, *Siting Hazardous Waste Management Facilities*, NAS, 115 Indian Mound Trail, Tavernier, Florida 33070.

CRAWFORD, N. H. and LINSLEY, R. K., 1966, *Digital Simulation in Hydrology: Stanford Watershed Model IV*, Tech. Rept. no. 39, Dept. Civil Eng., Stanford Univ., California.

CUMMINGS, R. G., 1971, *Optimum Exploitation of Ground-Water Reserves with Saltwater Intrusion*, Water Resources Res., v. 7, no. 6 pp. 1415-1424.

CUMMINGS, R. G., and McFARLAND, J. W., 1974, *Groundwater Management and Salinity Control*, Water Resources Res., v. 10, no. 5, pp. 909-916.

DAMM, M. V., 1980, *Hydrologic Effects From Changes in Ground Water Pumpage*, Water Resources Bull. v. 16, no. 5, pp. 907-913. (computer simulation of stream-aquifer system to study effect of allocation scheme).

De RIDDER, N. A., and EREZ, A., 1977, *Optimum Use of Water Resources*, Pub. No. 21, Internat. Institute for Land Reclamation and Improvement, Wageningen, The Netherlands, 250 pp. (planning methodology for joint use of surface and ground water for irrigation.)

DEVINE, M. D. and others, 1984, *Ground-Water Management in the Southeastern United States*, PB84-152768, NTIS, Springfield, Virginia 22161, 533 pp.

DOMENICO, P. A., 1967, *Economic Aspects of Conjunctive Use of Water*, Smith Valley, Nevada, USA, Internat. Assoc. Sci. Hydrology Publ. 72, pp. 474-482.

DOMENICO, P. A., ANDERSON, D. V., and CASE, C. M., 1968, *Optimal Ground-Water Mining*, Water Resources Res., v. 4, no. 2, Apr, pp. 247-255.

DORFMAN, R. and others, 1965, *Waterlogging and Salinity in the Indus Plain: Some Basic Considerations*, Harvard Univ. Water Resources Reprint no. 77.

DOWNING, R. A., and others, 1974, *Regional Development of Groundwater Resources in Combination with Surface Water*, J. Hydrology, v. 22, pp. 155-177.

DRACUP, J. A., 1966, *The Optimum Use of a Ground-Water and Surface Water System: A Parametric Linear Programming Approach*, Water Resources Center Contrib. no. 107, Univ. California, Berkeley, Jul.

DREIZIN, Y. C., and HAIMES, Y. Y., 1977, *A Hierarchy of Response Functions for Groundwater Management*, Water Resources Res., v. 13, no. 1, pp. 78-86.

EHRHARDT, R. E. and LEMONT, S., 1979, *Institutional Arrangements for Intrastate Ground-Water Management*, OWRT No. 14-34-0001-8410, JBF Scientific Corp., Arlington, Virginia.

ESHETT, A. and BITTINGER, M. W., 1965, *Stream-Aquifer System Analysis*, J. Hydraul. Div., Proc. Am. Soc. Civil Engrs., v. 91, no. HY6, Nov, pp. 153-164.

EVERETT, L. G., and SCHMIDT, K. D., 1976, *Monitoring Ground Water Quality: Methods and Costs*, U. S. Environmental Protection Agency, Environmental Monitoring Series, EPA 600/4-76-023.

FALK, L. H., 1970, *Economic Aspects of Ground-Water Basin Control*, Louisiana Water Resources, Res. Inst. Bull. GT-3, Louisiana State Univ., Baton Rouge, Louisiana.

FERLAND, R.K., 1983, *The Protection of Groundwater Quality in the Western States-Regulatory Alternatives and the Mining Industry*, Rocky Mountain Law Inst., v. 29, pp. 899-976.

FIERING, M., 1967, *A Groundwater-Precipitation Model for Streamflow Synthesis*, Proc. IBM Sci. Computing Symp. on Water and Air Resource Management, Yorktown Heights, New York, Oct. 23-25, pp. 203-213.

FLORES, W., E. Z., GUTJAHR, A. L. and GELHAR, L. W., 1978, *A Stochastic Model of the Operation of a Stream-Aquifer System*, Water Resources Res., v. 14, no. 1, pp. 30-44.

FOOSE, R. M., 1951, *Ground-Water Conservation and Development*, Monthly Bull., Penn. Dept. Internal Affairs, Harrisburg, v. 19, no. 2, pp. 17-28.

FORSTE, R. H., 1973, *Verification of Groundwater Capital Costs*, Water Resources Res. Center, University New Hampshire.

FOWLER, L. C., 1964, *Ground Water Management for the Nation's Future — Ground Water Basin Operation*, J. Hydraul. Div., Proc. Am. Soc. Civil Engrs, v. 90, no. HY4, July, pp. 51-57.

FRANKE, O. L. and McCLYMONDS, N. E., 1972, *Summary of the Hydrologic Situation on Long Island, New York, as a Guide to Water-Management Alternatives*, U. S. Geol. Survey Prof. Paper 627-F, 59 pp.

FRENZEL, S. A., 1984, *Methods for Estimating Groundwater Pumpage for Irrigation*, U. S. Geol. Survey Water-Resources Inv. Report 83-4277, 11 pp.

FRENZEL, S. A., 1985, *Comparison of Methods for Estimating Ground-Water Pumpage for Irrigation*, Ground Water, v. 23, no. 2, pp. 220-226.

GARCIA-BENGOCHEA, J. I., *Protecting Water Supply Aquifers in Areas Using Deepwell Wastewater Disposal*, J. Am. Water Works Assoc., v. 75, pp. 288-291.

GERAGHTY & MILLER, INC., 1979, *Site Location and Water-Quality Protective Requirements for Hazardous Treatment, Storage, and Disposal Facilities*, Rept. prepared for USEPA Office of Solid Waste, Contract No. 68-01-4636., 194 pp.

GIBB, J. P., 1971, *Cost of Domestic Wells and Water Treatment in Illinois*, Illinois State Water Survey Circ. 104, 23 pp.; also Ground Water, v. 9, no. 5, Sept.-Oct, pp. 40-48.

Management / Protection

GIBB, J. P. and SANDERSON, E. W., 1969, *Cost of Municipal and Industrial Wells in Illinois, 1964-1966,* Illinois State Water Survey Circ. 98.

GIESE, R. G., 1982, *A State Ground-Water Management Program,* Ground Water Monitoring Review, v. 2, no. 1, pp. 26-30.

GILLIES, N.P., 1974, *Ground-Water Protection in Pennsylvania, A Model State Program,* Geraghty & Miller Inc., Spec. Rep., Plainview, New York 11803.

GISSER, MICHA and MERCADO, ABRAHAM, 1972, *Integration of the Agricultural Demand Function for Water and the Hydrologic Model of the Pecos Basin,* Water Resources Res., v. 8, no. 6, pp. 1373-1384.

GISSER, MICHA, and SANCHEZ, D. A., 1980, *Competition Versus Optimal Control in Groundwater Pumping,* Water Resources Res., v. 16, no. 4, pp. 638-642.

GREYDANUS, H. W., 1978, *Management Aspects of Cyclic Storage of Water in Aquifer Systems,* Water Resour. Bull., v. 14, no. 2, pp. 477-480.

HAIMES, Y. Y., and DREIZIN, Y. C., 1977, *Management of Groundwater and Surface Water via Decomposition,* Water Resources Res., v. 13, no. 1, pp. 69-77.

HALL, W. A. and DRACUP, J. A., 1967, *The Optimum Management of Ground-Water Resources,* Proc. Internat. Conf. on Water for Peace, Washington, D. C., May 23-31.

HALL, C.W., PIWONI, M.D., and PETTYJOHN, W.A., 1983, *Research for Groundwater Quality Management,* PB83-152256, NTIS, Springfield, Virginia 22161, 58 pp.

HAM, H. H., 1971, *High Capacity Wells for Conjunctive Use of Water,* v. 9, no. 5, pp. 4-11.

HAMDAN, A. S., and MEREDITH, D. D., 1975, *Screening Model for Conjunctive-Use Water Systems,* J. Hydraul. Div., Am. Soc. Civ. Eng., v. 101 (HY10), pp. 1343-1355.

HANSEN, H. J., 1970, *Zoning Plan for Managing a Maryland Coastal Aquifer,* J. Am. Water Works Assoc., v. 62, no. 5, May, pp. 286-292.

HARTMAN, L. M., 1965, *Economics and Ground-Water Development,* Ground Water, v. 3, no. 2, pp. 4-8.

HEIDARI, MANOUTCHEHR, 1982, *Application of Linear Systems Theory and Linear Programming to Ground-Water Management in Kansas*, Water Resources Bull., v. 18, no 6.

HELWEG, O. J., 1976, *Salinity Management Strategy for Stream-Aquifer Systems*, Colorado State Univ. Hydrology Paper No. 84, Fort Collins, Colorado 80521.

HELWEG, O. J., 1977, *A Nonstructural Approach to Control Salt Accumulation in Ground Water*, Ground Water, v. 15, pp. 51-57.

HELWEG, O. J., 1978, *Regional Ground-Water Management*, Ground Water, v. 16, no. 5, pp. 318-321.

HELWEG, O. J., 1982, *Economics of Improving Well and Pump Efficiency*, Ground Water, v. 20, no. 5, pp. 556-562.

HELWEG, O. J., and LABADIE, J. W., 1978, *Accelerated Salt Transport Method for Managing Groundwater Quality*, Water Resources Bull., v. 12, no. 4, pp. 681-693.

HOWARD, J. F., 1974, *The Cost of Ground Water vs. Surface Water*, Water Well J., v. 28, no. 8, Aug.

HOWELLS, D. H., 1980, *Southeast Conference on Ground-Water Management Held at Birmingham, Alabama January 30-31, 1980*, North Carolina Water Resources Res. Inst., Raleigh, PB81-119489, NTIS, Springfield, Virginia 22161, (State ground-water programs in Alabama, Florida, Georgia, North Carolina, South Carolina, Tennessee, and Virginia.)

HOWITT, R. E., 1979, *Is Overdraft Always Bad?*, Proc. 12th Biennial Conference on Groundwater, California Water Resources Center, Rep. No. 45, Davis, California.

HUFFMIRE, M. M., FRANKEL, L., and REITMAN, F.C., 1984, *Regulation of Land Use Practices for Areas Surrounding Aquifers; Economic and Legal Implications*, PB84-159763, NTIS, Springfield, Virginia 22161, 358 pp.

HUGHES, G. M. and CARTWRIGHT, KEROS, 1972, *Scientific and Administrative Criteria for Shallow Waste Disposal*, Civil Engineering, v. 42, no. 3, pp.70-73.

ILLINOIS STATE WATER PLAN TASK FORCE, 1984, *Strategy for the Protection of Underground Water in Illinois*, Environmental Protection Agency Spec. Rept no. 8, Springfield, Illinois 62706.

JACQUETTE, D. J., 1981, *Efficient Water Use in California: Conjunctive Management of Ground and Surface Reservoirs*, J. Hydrology, v. 51, no. 1/4, May.

JONES, E. E., Jr., and MURRAY, C. M., 1977, *Improving the Sanitary Protection of Ground Water in Severely Folded, Fractured, and Creviced Limestone*, Ground Water, v. 15, no. 1, pp. 66-74.

JONES, O. R., and SCHNEIDER, A. D., 1972, *Ground-Water Management on the Texas High Plains*, Water Resources Bull., v. 8, no. 3, June, pp. 416-522.

KASHEF, A. I., 1971, *On The Management of Ground Water in Coastal Aquifers*, Ground Water v. 9, no. 2, pp. 12-20.

KAZMANN, R. G., 1958, *Problems Encountered in the Utilization of Ground-Water Reservoirs*, Trans. Am. Geophys. Union, v. 39, pp. 94-99.

KAZMANN, R. G., KIMBLER, O. K., and WHITEHEAD, W. R., 1974, *Management of Waste Fluids in Salaquifers*, J. Irrigation and Drainage Div. Am. Soc. Civil Engrs., 100, pp. 413-424.

KELSO, M. M., 1961, *The Stock Resource Value of Water*, J. Farm Econ., v. 43, no. 5., Dec, pp. 112-1129.

KENT, D. C., NANEY, J. W., and WITZ, F. E., 1982, *Prediction of Economic Potential for Irrigation Using a Ground-Water Model*, Ground Water, v. 20, no. 5, pp. 577-585.

KISIEL, C. C. and DUCKSTEIN, L., 1968, *Water Resources Development and Management in the Tucson Basin*, paper presented at 1968 Spring Joint ORSA/TIMS meeting, May 1-3, San Francisco, California.

KNAPP, K. C., and FEINERMAN, ELI, *The Optimal Steady-State in Groundwater Management*, 1985 Water Resources Bull., v. 21, no. 6, pp. 967-975.

KUDELIN, B. I., 1957, *The Principles of Regional Estimation of Underground Water Natural Resources and the Water Balance Problem*, Internat. Assoc. Sci. Hydrology, General Assembly Toronto, Pub. 44, v. 2, pp. 150-167.

LEACH, S. D., KLEIN, H., and HAMPTON, E. R., 1972, *Hydrologic Effects of Water Control and Management of Southeastern Florida*, Florida Bureau of Geol. Rept. of Inv. 60, 115 pp.

LEGRAND, H. E., 1964, *Management Aspects of Groundwater Contamination*, J. Water Pollution Control Fed., v. 36, no. 9, pp. 1133-1145.

LEOPOLD, L. B., 1958, *Water and the Conservation Movement*, U. S. Geol. Survey Circ. 402.

LOHMAN, S. W., 1953, *Sand Hills Area, Nebraska*, in *Sub-Surface Facilities of Water Management and Patterns of Supply — Type Area Studies*, U. S. Congress, House of Rep., Interior and Insular Affairs Comm., pp. 79-91.

LUCE, C. F., 1973, *Water Policies for the Future — Final Report to the President and to the Congress of the U. S.*, U. S. National Water Commission, Arlington, Virginia, Water Information Center, Plainview, New York, 579 pp.

LUZIER, J. E., and BURT, R. J., 1974, *Hydrology of Basalt Aquifers and Depletion of Ground Water in East-Central Washington*, Washington Dept. Ecology Water-Supply Bull. 33, 53 pp.

MACK, L. E., 1971, *Ground Water Management in Development of a National Policy on Water*, Rept. NWC-EES-71-004 Natl. Water Comm., Washington, D. C., 179 pp.

MACK, L. E., 1972, *Ground Water Management*, PB 201 536, NTIS, Springfield, Virginia 2211.

MacKICHAN, K. A., 1957, *Estimated Use of Water in the United States, 1955*, J. Am. Water Works Assoc., v. 49, pp. 369-391.

MacNISH, R. D., MYERS, D. A., and BARKER, R. A., 1973, *Appraisal of Ground-Water Availability and Management Projections, Walla Walla River Basin, Washington and Oregon*, Washington Dept. Ecology Water-Supply Bull. 37, 25 pp.

MADDAUS, W. O., and AARONSON, M. A., 1972, *A Regional Groundwater Resource Management Model*, Water Resources Res., v. 8, no. 1, pp. 231-237.

MADDOCK, THOMAS, III, 1974, *The Operation of a Stream-Aquifer System under Stochastic Demands*, Water Resources Research, v. 10, no. 1, pp. 1-10.

MADDOCK, THOMAS, III, 1976, *Cost Functions for Additional Groundwater Development*, Water Resources Bull., v. 12, no. 3, pp. 539-545.

MADDOCK, THOMAS, III, and HAIMES, Y. Y., 1975, *A Tax System for Groundwater Management*, Water Resources Res., v. 11, no. 1, pp. 7-14.

MALMBERG, G. T., 1975, *Reclamation by Tubewell Drainage in Rechna Doab and Adjacent Areas, Punjab Region, Pakistan*, U. S. Geol. Survey Water-Supply Paper 1608-0, 72 pp. J. Am. Water Works Assoc., v. 70, no. 8, Aug., pp. 419-425.

MANN, J. F., Jr., 1960, *Safe Yield Changes in Groundwater Basins*, Internat. Geol. Congress, XXI Session, Pt. XX, pp. 17-23.

MANN, J. F., Jr., 1963, *Factors Affecting the Safe Yield of Ground-Water Basins*, Trans. Am. Soc. Civil Engrs., v. 128, Pt. III, pp. 180-190.

MANN, J. F., Jr., 1968, *Concepts in Ground Water Management*, J. Am. Water Works Assoc., v. 60, Dec, p. 1336.

MANN, J. F., Jr., 1969, *Ground-Water Management in the Raymond Basin, California*, in *Legal Aspects of Geology in Engineering Practice*, Geol. Soc. Am. Eng. Geology Case Histories, no. 7, pp. 61-74.

MANNING, J. C., 1967, *Resume of Ground Water Hydrology in the Southern San Joaquin Valley*, J. Am. Water Works Assoc., v. 59, Dec., pp. 1513-1526.

MARTIN, W. E., and ARCHER, T., 1971, *Cost of Pumping Irrigation Water in Arizona, 1891 to 1967*, Water Resources Res., v. 7, no. 1, pp. 23-31.

MAUGHAN, W. D., 1980, *Reflections on Ground-Water Management*, Irrigation J., v. 30, no. 1, Jan.-Feb.

MAXEY, G. B. and DOMENICO, P. A., 1967, *Optimum Development of Water Resources in Desert Basins*, Am. Water Resources Assoc., Proc. Ser. 4, Symp. on Groundwater Hydrology, San Francisco, pp. 84-90.

McCLESKEY, G. W., 1972, *Problems and Benefits in Ground-Water Management*, Ground Water, v. 10, no. 2, pp. 2-5.

McGAUHEY, P. H., 1968, *Engineering Management of Water Quality*, McGraw-Hill, New York, 295 pp.

McGUINNESS, C. L., 1946, *Recharge and Depletion of Ground-Water Supplies*, Proc. Am. Soc. Civil Engrs., v. 72, pp. 963-984.

McKELVEY, V. E., 1972, *Underground Space — An Unappraised Resource*, Underground Waste Management and Environmental Implications Mem. 18, Am. Assoc. Petrol. Geol., Tulsa, Oklahoma, pp. 1-5.

McMILLAN, W. D., 1966, *Theoretical Analysis of Groundwater Basin Operations*, Water Resources Center Contrib. no. 114, Univ. California, Berkeley, Nov.

MICHIGAN DEPT. OF NATURAL RESOURCES, 1982, *Groundwater Management Strategy for Michigan: Development of Administrative Rules for Groundwater Quality*, Water Quality Div., Lansing, Michigan; PB82-208216, NTIS, Springfield, Virginia, 22161, 50 pp.

MILLER, D. W., 1958, *The Ground-Water Phase of Well Field Management*, TAPPI, v. 41, no. 10, pp. 174A-178A.

MILLER, D. W., and SCALF, M. R., 1974, *New Priorities for Ground-Water Quality Protection*, Ground Water, v. 12, no. 6, pp. 335-347.

MOENCH, A. F., and SCHICHT, R. J., 1971, *Projected Groundwater Deficiencies in Northeastern Illinois, 1980-2020*, Ill. Water Survey Circ. 101.

MOENCH, A. F., and VISOCKY, A. P., 1971, *A Preliminary "Least Cost" Study of Future Groundwater Development in Northeastern Illinois*, Illinois State Water Survey Circ. 102.

MOENCH, A. F., SCHICHT, R. J., and VISOCKY, A. P., 1972, *A Systems Study of Future Ground-Water Development Costs in the Chicago Region*, Water Resources Bull., v. 8, no. 2, p. 328.

MOORE, C. V., and SNYDER, J. H., 1969, *Some Legal and Economic Implications of Sea Water Intrusion: A Case Study in Ground Water Management*, Natural Resources J., v. 9, pp. 401-419.

MOOSEBURNER, G. J., and WOOD, E. F., 1980, *Management Model for Controlling Nitrate Contamination in the New Jersey Pine Barrens Aquifer*, Water Resources Bull., v. 16, no. 6, p. 971.

MOREL-SEYTOUX, H. J., 1975, *A Simple Case of Conjunctive Surface-Ground-Water Management*, Ground Water, v. 13, no. 6, pp. 506-515.

MUNDORFF, M. J., and others, 1976, *Hydrologic Evaluation of Salinity Control and Reclamation Projects in the Indus Plain, Pakistan — A Summary*, U. S. Geol. Survey Water-Supply Paper 1608-Q, 59 pp.

MURRAY, C. R., 1968, *Estimated Use of Water in the United States, 1965*, U. S. Geol. Survey Circ. 556, 53 pp.

MURRAY, C. R., and REEVES, E. B., 1977, *Estimated Use of Water in the United States in 1975*, U. S. Geol. Survey Circ. 765, 39 pp.

NACE, R. L., 1960, *Water Management, Agriculture, and Ground-Water Supplies*, U. S. Geol. Survey Circ. 415, 12 pp.

NACE, R. L., 1965, *Global Thirst and the International Hydrological Decade*, J. Am. Water Works Assoc., v. 57, no. 7.

NATIONAL ACADEMY OF SCIENCE, 1986, *Ground Water Quality Protection-State and Local Strategies*, National Academy Press, 2101 Constitution Ave, NW, Washington, DC 20418, 320 pp.

NATIONAL WATER WELL ASSOCIATION, 1983, *State, County, Regional and Municipal Jurisdiction of Ground Water Protection*, Proc. 6th Nat. Ground Water Quality Symp., Atlanta, Georgia, Sept., 1982; NWWA, Dublin, Ohio 43017, 312 pp.

NATIONAL WATER WELL ASSOCIATION,1985, *The Impact of Mining on Ground Water', in Proc. Western Regional Ground Water Conf., January, 1985, Reno, Nevada, NWWA, Dublin, Ohio 43017, pp.80-129.*

NELSON, A. G. and BRUSH, C. D., 1967, *Cost of Pumping Irrigation Water in Cetral Arizona*, Ariz. Agr. Exp. Sta., Tech. Bull. 182, p. 48.

NIESWAND, G. H., and GRANSTROM, M. L., 1971, *A Chance-Constrained Approach to the Conjunctive Use of Surface Waters and Groundwaters*, Water Resources Res., v. 7, no. 6, pp. 1425-1426.

NIGHTINGALE, H. I., and BIANCHI, W. C., 1977, *Ground-Water Chemical Quality Management by Artificial Recharge*, Ground Water, v. 15, pp. 15-22.

NOVITZKI, R. P., 1973, *Improvement of Trout Streams in Wisconsin by Augmenting Low Flows with Ground Water*, U. S. Geol. Survey Water-Supply Paper 2017, 52 pp.

NUTBROWN, D. A., 1976, *Optimal Pumping Regimes in an Unconfined Coastal Aquifer*, J. Hydrology, v. 31, no. 3/4, Dec.

ORLOB, G. T., and DENDY, B. B., 1973, *Systems Approach to Water Quality Management*, J. Hydraulics Div., Amer. Soc. Civil Engrs., v. 99, no. HY4, pp. 573-587.

OWEN, L. W., 1968, *Ground Water Management and Reclaimed Water*, J. Am. Water Works Assoc., v. 60, no. 2, Feb, pp. 135-144.

PARKER, G. G., 1973, *Highlights of Water Management in the Southwest Florida Water Management District*, Ground Water, v. 11, no. 3, pp. 16-25.

PARKER, G. G., 1975, *Water and Water Problems in the Southwest Florida Water Management District and Some Possible Solutions*, Water Resources Bull., v. 11, no. 1, Feb.

PARKER, G. G. and others, 1964, *Water Resources of the Delaware River Basin*, U. S. Geol. Survey Prof. Paper 381, 200 pp.

PERALTA, R. C., 1985, *Conjunctive Use/Sustained Groundwater Yield Design*, Proc. Symp. Computer Applications in Water Resources, ASCE, Buffalo, NY, pp. 1391-1400.

PERALTA, R. C., and PERALTA, A. W., 1984, *Arkansas Groundwater Management Via Target Levels*, Trans. of ASAE, v. 27, no. 6, pp. 1696-1703.

PETERS, H. J., 1972, *Groundwater Management*, Water Resources Bull., v. 8, no. 1, pp. 188-197.

PETERSON, D. F., 1968, *Ground Water in Economic Development*, Ground Water, v. 6, no. 3, pp. 33-41.

POJASEK, R. B., (ed.), 1977, *Drinking Water Quality Enhancement Through Source Protection*, Ann Arbor Science, Ann Arbor, 614 pp.

PORTER, N. W., 1941, *Concerning Conservation of Underground Water with Suggestions for Control*, Trans. Am. Soc. Heat. Vent. Engrs., v. 47, pp. 309-322.

RALSTON, D. R., 1973, *Administration of Ground Water as Both a Renewable and Nonrenewable Resource*, Water Resources Bull., v. 9, no. 5, pp. 908-917.

RAUCHER, R. L., 1983, *A Conceptual Framework for Measuring the Benefits of Groundwater Protection*, Water Resources Res., v. 19, pp. 320-326.

RAYMOND, L. S., Jr., 1981, *Groundwater Use Management in the Northeastern States: Legal and Institutional Issues*, PB84-132463, NTIS, Springfield, Virginia 22161.

RAYNER, F. A., 1972, *Ground-Water Basin Management on the High Plains of Texas*, Ground Water, v. 10, no. 5, Sept.-Oct, pp. 12-17.

RENSHAW, E. F., 1963, *The Management of Ground Water Reservoirs*, J. Farm Econ., v. 45, no. 2, pp. 285-295.

REVELLE, R., and LAKSHMINARAYANA, V., 1975, *The Ganges Water Machine*, Science, v. 188, pp. 611-616.

ROCKAWAY, J. D. and JOHNSON, R. B., 1967, *Statistical Analysis of Ground Water Use and Replenishment*, Tech. Rept. no. 2, Water Resources Res. Center, Purdue Univ., Lafayette, Indiana, Sept.

SAHUQUILLO, ANDRES, 1971, *Conjunctive Use of the Tajo-Segura Aqueduct Surface System and Aquifers of the La Mancha Area*, Internat. Symp. on Mathematical Models in Hydrology, Warsaw, Internat. Assoc. Sci. Hydrology, c/o AGU, Washington, D. C. 20006.

SASMAN, R. T., and SCHICHT, R. J., 1978, *To Mine or Not to Mine Groundwater*, J. Am. Water Works Assoc., v. 70, pp. 156-161.

SCHICHT, R. J. and MOENCH, A. F., 1971, *Future Demands on Ground Water in Northeastern Illinois*, Ground Water, v. 9, no. 2, pp. 21-28.

SCHMIDT, K. D., 1975, *Salt Balance in Groundwater of the Tulare Lake Basin, California*, in *Hydrology and Water Resources in Arizona and Southwest*, v. 5, Proc. Am. Water Resources Assoc., Arizona Section, Tempe, pp. 177-184.

SELLERS, J. H., 1973, *Tax Implications of Ground-Water Depletion*, Ground Water, v. 11, no. 6, pp. 27-35.

SHELTON, M. L., 1982, *Ground-Water Management in Basalts*, Ground Water, v. 20, no. 1, pp. 86-93.

SHIFRIN, N. S., and NOLAN, M. N., 1981, *Ground-Water Protection by Recharge Zone Management: Institutional Arrangements*, PB82-197948, NTIS, Springfield, Virginia 22161, 66 pp.

SHUVAL, H. I., (ed.), 1981, *Water Quality Mangement Under Conditions of Scarcity*, Academic Press, Inc. New York, New York, 352 pp. (water management in Israel)

SLOGGETT, GORDON, 1981, *Prospects for Groundwater Irrigation: Declining Levels and Rising Energy Costs*, U. S. Dept. of Agric., Econ. Res. Ser., AER-478, 43 pp.

SLOGGETT, GORDON, and MAPP, H. P., Jr., 1984, *An Analysis of Rising Irrigation Costs in the Great Plains*, Water Resources Bull., v. 20, no. 2, pp. 229-233.

SMITH, Z. A., 1985, *Interest Group Interaction and Groundwater Policy Formation in the Southwest*, Univ. Press of America, Inc., 4720 Boston Way, Lanham, Maryland 20706, 228 pp.

SPIEGEL, ZANE, 1972, *Regional Zoning — An Aid in Maintaining a Proper Water Budget*, Ground Water, v. 10, no. 6, pp. 48-49.

STULTS, H. M., 1966, *Predicting Farmer Response to a Falling Water Table: an Arizona Case Study*, Conf. Proc., Comm. Econ. Water Resources of the Western Agric. Econ. Research Council, Rept. no. 15, Las Vegas, Nevada, Dec.

SUMMERS, K. V., RUPP, G. L., and GHERINI, S. A., 1981, *Management Options for Ground Water in Urbanizing Areas*, PB82-198052, NTIS, Springfield, Virginia 22161, 125 pp.

SUPALLA, R. J., and others, 1982, *Economic Evaluations of Ground-Water Policy Alternatives in the Northern Great Plains*, PB83-125260, NTIS, Springfield, Virginia 22161.

SUTER, M. and others, 1959, *Preliminary Report on Ground Water Resources of the Chicago Region*, Ill. State Water Survey and Geol. Survey Cooperative Ground-Water Rept. 1.

TAYLOR, O. J. and LUCKEY, R. R., 1974, *Water-Management Studies of a Stream-Aquifer System, Arkansas River Valley*, Colorado, Ground Water, v. 12, no. 1, pp. 22-38.

TEMPLER, O. W., 1980, *Conjunctive Management of Water Resources in the Context of Texas Water Law*, Water Resources Bull., v. 16, no. 2, Apr.

THEIS, C. V., 1940, *The Source of Water Derived from Wells, Essential Factors Controlling the Response of an Aquifer to Development*, Civil Engineering, May.

THOMAS, H. E., 1972, *Water-Management Problems Related to Ground-Water Rights in the Southwest*, Water Resources Bull., v. 8, no. 1, Feb.

THOMAS, H. E. and SCHNEIDER, W. J., 1970, *Water as an Urban Resource and Nuisance*, U. S. Geol. Survey Circ. 601-D.

TODD, D. K., 1971, *Systems Analysis for Ground-Water Management, Phase II*, PB-201969, NTIS, Springfield, Virginia.

TRIPP, J. B. and JAFFE, A. B., 1979, *Preventing Ground-Water Pollution: Towards a Coordinated Strategy to Protect Critical Recharge Zones*, Harvard Environ. Law. Rev., v. 3, no. 1.

TYSON, H. N. and WEBER, E. M., 1964, *Ground-Water Management for the Nation's Future — Computer Simulation of Ground-Water Basins*, J. Hydraul. Div., Proc. Am. Soc. Civil Engrs., v. 90, no. HY4, Jul., pp. 59-77.

UPSON, J. E., 1967, *Plans of the U. S. Geological Survey Water Resources Division for Research Investigations and Data Collection in Groundwater*, Ground Water, v. 5, no. 2.

U. S. AD HOC PANEL ON HYDROLOGY, 1962, *Scientific Hydrology*, Federal Council for Science and Technology, Washington, D. C., 37 pp.

U. S. COMPTROLLER GENERAL, 1980, *Report to Congress: Ground Water Overdrafting Must be Controlled*, U. S. General Accounting Office, Washington, D. C., 52 pp.

U. S. ENVIRONMENTAL PROTECTION AGENCY, 1980, *Ground Water Protection: A Water Quality Management Report*, USEPA, Washington, D. C., 36 pp.

U. S. ENVIRONMENTAL PROTECTION AGENCY, 1984, *Ground Water Protection Strategy*, USEPA Office of Ground Water Protection, Washington, D. C., 56 pp.

U. S. ENVIRONMENTAL PROTECTION AGENCY, 1984, *Community Relations in Superfund: A Handbook (Interim Version)*, PB84-209378, NTIS, Springfield, Virginia 22161, 132 pp.

U. S. ENVIRONMENTAL PROTECTION AGENCY, 1985, *Selected State and Territory Ground-Water Classification Systems*, USEPA Office of Ground-Water Protection, Washington, D. C.

U. S. ENVIRONMENTAL PROTECTION AGENCY, OFFICE OF GROUND WATER PROTECTION, 1985, *Overview of State Ground Water Program Summaries (Vol.1)*, Stock No. 055-000-00246-1; *State Ground Water Program Summaries (Vol.2)*, Stock No. 055-000-00247-9; Supt. Documents, Washington, DC 20401.

U. S. GENERAL ACCOUNTING OFFICE,1984, *Federal and State Efforts to Protect Ground Water*, Report to the Subcommittee on Commerce, Transportation, and Tourism, Committee on Energy and Commerce, House of Representatives, Govt, Printing Office, 80 pp.

U. S. OFFICE OF TECHNOLOGY ASSESSMENT, 1984, *Protecting the Nation's Groundwater from Contamination*, OTA, Washington, D. C. 20510, 2 vols, 503 pp.

U. S. WATER RESOURCES COUNCIL, 1980, *State of the States: Water Resources Planning and Management*, USWRC, 2120 L St., N.W. Washington, D. C. 20037. (guide to ground-water management and protection in 50 states and 5 territories).

U. S. WATER RESOURCES COUNCIL, 1981, *Essentials of Ground-Water Hydrology Pertinent to Water-Resources Planning*, Bull. 16, No. 052-045-00083-5, Supt. of Documents, Washington, D. C. 20402.

U. S. WATER RESOURCES COUNCIL, 1982, *Ground-Water Management: Discussion of Issues*, (report of the Interstate Ground-Water Management Task Force), PB82-181637, NTIS, Springfield, Virginia 22161.

URBISTONDO, R., and BAYS, L. R., 1985, *Ground-Water Management*, Proc. Symp. IWSA, Berlin, FRG, April 1985, IWSA, London, Pergamon Press, London, 254 pp.

VAN DER HEIJDE, PAUL, and others, 1985, *Groundwater Management: The Use of Numerical Models*, Am. Geophys. Union, 2000 Florida Ave., N.W., Washington, D. C. 20009, 180 pp.

VANLIER, K. E., WOOD, W. W., and BRUNETT, J. O., 1974, *Water-Supply Development and Management Alternatives for Clinton, Eaton, and Ingham Counties, Michigan*: U. S. Geol. Survey Water-Supply Paper 1969, 111 pp.

VILLUMSEN, ARNE, and SONDERSKOV, CARSTON, 1982, *Vulnerability Maps: A Promising Tool in Groundwater Protection*, Aqua, v. 5, pp. 466-468.

UTTON, A. E., and ATKINSON, C. K., 1979, *International Ground-Water Management: The Case of the Mexico-United States Frontier, Final Report*, New Mexico State Univ. Water Resources Res. Inst. Rept. 109., Las Cruces, New Mexico 88001.

WALKER, W. H., 1972, *Comparative Costs of Producing Water*, Illinois State Water Survey, Urbana, Illinois.

WALTON, W. C., 1964, *Potential Yield of Aquifers and Ground Water Pumpage in Northeastern Illinois*, J. Am. Water Works Assoc., Feb.

WARREN, JOHN, and others, 1982, *Economics of Declining Water Supplies in the Ogallala Aquifer*, Ground Water, v. 20, no. 1, pp. 73-79.

WEBER, E. M., and HASSAN, A. A., 1972, *Role of Models in Groundwater Management*, Water Resources Bull., v. 8, no. 1, pp. 198-206.

WEEKS, J. B., 1978, *Plan of Study for the High Plains Regional Aquifer-System Analysis in Parts of Colorado, Kansas, Nebraska, New Mexico, Oklahoma, South Dakota, Texas, and Wyoming*, U. S. Geol. Survey Water-Resources Inv. 78-70, 28 pp.

WELBY, C. W., and WILSON, T. M., 1982, *Use of Geologic and Water Yield Data from Ground Water Based Community Water Systems as a Guide for Ground-Water Planning and Management*, Univ. North Carolina, Water Resources Res. Institute, Raleigh, NC, 111 pp.

WELSCH, W. F., 1960, *Ground Water Recharge and Conservation — Conservation in Nassau County, Long Island, New York*, J. Am. Water Works Assoc., v. 52, Dec, p. 1494.

WESCHLER, L. F., 1968, *Water Resources Management: The Orange County Experience*, California Govt. Ser. no. 14, Inst. Govt. Affairs, Univ. California, Davis, Jan, 67 pp.

WICKERSHAM, GINIA, 1980, *Ground-Water Management in the High Plains*, Ground-Water, v. 18, no. 3, May-Jun., pp. 286-290.

WIENER, A., 1967, *The Role of Advanced Techniques of Ground Water Management in Israel's National Water Supply System*, Bull. Internat. Assoc. Sci. Hydrology, v. 12, no. 2, pp. 32-38.

WILLIAMS, D. E., 1970, *Ground Water Development and Management in the Owens Valley*, Annual Conf. Los Angeles Dept. of Water and Power, Los Angeles, California, 12 pp.

WILLIS, R., 1979, *A Planning Model for the Management of Ground-Water Quality*, Water Resources Res., v. 15, no. 6, Dec.

WILSON, L., and HOLLAND, M. T., 1984, *Aquifer Classification for the UIC Program: Prototype Studies in New Mexico*, Ground Water, v. 22, no. 6, pp. 706-716.

WITHERSPOON, P. A., 1973, *Evaluation of Hydrological Properties of Aquitards and Their Role in Ground-Water Management*, Univ. California, Water Resources Center Rep. W-279, Berkeley, California 94700.

WOOD, F. O., 1976, *Sensible Salting Program*, Salt Institute, 206 North Washington Street, Alexandria, Virginia 22314.

WOODS, PHILIP, C., 1967, *Management of Hydrologic Systems for Water Quality Control*, Water Resources Center, Contrib. no. 121, Univ. California, Berkeley, Jun.

WYATT, A. W., BELL, A., and MORRISON, S., 1976, *Analytical Study of the Ogallala Aquifer in Lamb County, Texas: Projections of Saturation Thickness, Volume of Water Storage, Pumping Lifts, and Well Yields.* Texas Water Devel. Bd., Rep. 204, 63 pp.

YAZDANIAN, A. and PERALTA, R. C., 1986, *Sustained Yield Ground-Water Planning by Goal Programming*, Ground Water, v. 24, no. 2, pp. 157-165.

YODER, DOUGLAS, and others, 1981, *Dade County's Plan for Aquifer Management*, J. Environmental Engineering Div., Am. Soc. Civil Engrs. v. 107, no. EE6, Dec.

YOUNG, C. P., and others, 1977, *Prediction of Future Nitrate Concentrations in Ground Water*, Proc. Third National Ground Water Quality Symp., EPA-600/9-77-014, Robert S. Kerr Environmental Research Laboratory, Ada, Oklahoma, pp. 93-98.

YOUNG, R. A., 1970, *Safe Yield of Aquifers: an Economic Reformulation*, J. Irrigation and Drainage Div., Am. Soc. Civil Engrs., v. 96, no. IR4, pp. 377-385.

YOUNG, R. A., and BREDEHOEFT, J. D., 1972, *Digital Computer Simulation for Solving Management Problems of Conjunctive Groundwater and Surface Water Systems*, Water Resources Res., v. 8, no. 3, pp. 533-556.

YU, WANYOUNG and HAIMES, Y. Y., 1974, *Multilevel Optimization for Conjunctive Use of Groundwater and Surface Water*, Water Resources Res., v. 10, no. 4, pp. 625-636.

EDUCATION/TRAINING

AGNEW, A. F., 1968, *The Geological Profession and Ground Water*, Ground Water, v. 6, no. 1, pp. 5-9.

BEAR, JACOB, 1969, *Hydrologic Education in Israel*, Internat. Seminar for Hydrology Professors, Urbana, Illinois, 13-25 July, Preprint, 23 pp.

DeWIEST, R. J. M., 1964, *Educational Facilities in Ground-Water Hydrology and Geology*, Ground Water, v. 2, pp. 18-24.

EATON, E. D., 1969, *Comments on Some Recent Trends in Hydrology Research*, Internat. Seminar for Hydrology Professors, Urbana, Illinois, 13-25 July, Preprint, 21 pp.

EVANS, D. D., and HARSHBARGER, J. W., 1969, *Curriculum Development in Hydrology*, Internat. Seminar for Hydrology Professors, Urbana, Illinois, 13-25 July, Preprint, 20 pp.

HARSHBARGER, J. W. and FERRIS, J. G., 1963, *Inter-Disciplinary Training Program in Scientific Hydrology*, Ground Water, v. 1, no. 2.

KINDSVATER, C. E. and SNYDER, W. M., 1969, *Hydrologic Training for Water Resources Development*, Internat. Seminar for Hydrology Professors, Urbana, Illinois, 13-25 July, Preprint, 12 pp.

MEYBOOM, P., 1969, *Some Aspects of the Scientific Communication System in Hydrology*, Ground Water, v. 7, no. 4, pp. 28-37.

MOORE, W. L., 1969, *Teaching Aids in Hydrology*, Internat. Seminar for Hydrology Professors, Urbana, Illinois, 13-25 July, Preprint, 18 pp.

NATIONAL WATER WELL ASSOCIATION, 1981, *Water Well Education*, Water Well J., v. 35, no. 4, Apr., pp. 58-60 (schools offering programs in water well technology).

UNIVERSITIES COUNCIL ON WATER RESOURCES, 1967, *Education in Hydrology, United States Universities, early 1966*, Water Resources Center, Univ. California, Los Angeles, 44 pp.

WALTON, W. C., 1964, *Education Facilities in Ground-Water Geology and Hydrology in the United States and Canada — 1963*, Ground Water, v. 2, no. 3.

WATER WITCHING

BAUM, JOSEPH, 1974, *The Beginner's Handbook of Dowsing*, Crown Publishers, 419 Park Ave. South, New York, New York 10016.

CHADWICK, D. G., and JENSEN, L., 1971, *The Detection of Magnetic Fields Caused by Ground Water and the Correlation of Such Fields with Water Dowsing*, Utah State Univ. Water Research Lab. Rept. WG78-1, Logan 84321.

ELLIS, A. J., 1917, *The Divining Rod — A History of Water Witching*, U. S. Geol. Survey Water-Supply Paper 416, 59 pp.

EMMART, B. D., 1952, *All-Purpose Dowsing*, Atlantic Monthly, v. 190, no. 1, pp. 90-92.

MARTIN, MICHAEL, 1983, *A New Controlled Dowsing Experiment*, The Skeptical Inquirer, v. 8, no. 2, pp. 138-140.

RANDI, JAMES, 1979, *A Controlled Test of Dowsing Abilities*, The Skeptical Inquirer, v. 4, no. 1, pp. 16-20.

RIDDICK, T. M., 1952, *Dowsing — An Unorthodox Method of Locating Underground Water Supplies or an Interesting Facet of the Human Mind*, Proc. Am. Philosophical Soc., v. 96, pp. 526-534.

ROBERTS, KENNETH, 1951, *Henry Gross and His Dowsing Rod*, Doubleday & Co., Garden City, New York.

RYDER, L. W., 1949, *The Case for Water Witching*, J. New England Water Works Assoc., v. 63, pp. 232-237.

VOGT, E. Z., and HYMAN, RAY, 1959, *Water Witching U. S. A.*, Univ. of Chicago Press, 259 pp.

ZIEMKE, P. C., 1949, *Water Witching*, Water and Sewage Works, v. 96, p. 136.